THE FASCINATING WORLD OF MATHEMATICS

THE FASCINATING WORLD OF MATHEMATICS

Chander Mohan

MV Learning
A Viva Books imprint

3, Henrietta Street
London WC2E 8LU
UK

4737/23, Ansari Road,
Daryaganj, New Delhi 110 002
India

ISBN: 978-93-87486-37-9

Printed and bound in India.

Dedicated to

The Budding
Ramanujans, Newtons and Einsteins
of the world

Contents

Preface

This book is about mathematics and not a book on mathematics. The book will be of interest to those who have knowledge of school level mathematics and want to be aware of its fascinating and charming aspects which they might not have been exposed to, or might have missed in school days. The book is also intended to motivate in particular the youngsters into feeling attracted, fascinated and excited about mathematics instead of dreading it.

The book is primarily a collection of school level topics of mathematics which lend themselves to relatively simple mathematical analysis. An effort has been made to present certain basic areas of school mathematics in a way that may provide intellectual stimulus and excitement to the reader. The book has been written at a level where it can be appreciated by school students as well as common people having school level background of mathematics. It has been the intention of the author to inform, entertain and make common people comfortable with mathematics instead of getting scared of it. Justifications and explanations have been kept simple so that these can be appreciated by general reader. Author has chosen to be somewhat light hearted in the analysis of certain problems so that these are easily comprehended and appreciated even by non-mathematicians. It seemed to be a good idea to be not serious about everything.

All major aspects of even school mathematics have not been touched upon. Instead an effort has been made to take the reader on a 'walking tour' of certain exciting and fascinating aspects of the popular areas of school level mathematics in order to provide the reader intellectual stimulus and excitement about the subject. It is the wish of the author that more and more persons in the society share such beautiful morsels of mathematics.

The book has been divided into two parts. The first part is titled 'Charm and beauty in mathematics'. In this part an effort has been made to highlight charming, exciting and fascinating aspects of school level mathematics and its applications. It is spread over the first eleven chapters. The second part of the book is titled 'Mathematical diversions and recreations'. It introduces the reader to a variety of games and puzzles based on mathematics which provide not only good pastime and refreshment to the mind but also help in sharpening the mental faculties for critical and fast thinking. This part of the book contains five chapters.

It is hoped that after reading this book common persons, particularly youngsters, will discard their dread of mathematics and the elders may change their perception of the subject and start regarding mathematics as a subject to be appreciated and admired and not detested.

Acknowledgments

The author would like to express his gratitude to all the authors whose work he consulted while collecting the material for this book. He would also like to thank all those who helped him, in one way or the other, in type setting the manuscript, in particular my faculty colleague, Mr. Rajeev Goel and Ms. Jaspal Bhardula of Panchkula. He would also like to thank the management of Ambala college of Engineering and Applied Research, Devsthali, Ambala for providing him the necessary facilities. The moral support provided by my wife Smt. Tripta Kumari and children and grand children is also gratefully acknowledged. Infact it was the probing questions in the related topics put to him by his young grand children Shambhavy, Snigdha, Apporva and Pranvi which prompted him to take up the task of writing this book.

Chander Mohan

(chander_mohan2@rediffmail.com)

Introduction

Mathematics is both fascinating and frustrating. For most of us it has an air of mystery. However all of us realize that mathematics is inextricably woven into our modern society and one can not escape using it. We use mathematics while buying or selling, building houses, bridges and tunnels, in constructing rockets and satellites, in finance and accounts, in fact in every sphere of human activity. Indeed these days it is very difficult to think of situations where mathematics is not required or mathematics is not being used in one way or the other. However, the worrying part is that most of us feel intimidated by it even though we realize the necessity of being proficient in it.

Most of us regard mathematics as a necessary evil. People despise the subject but feel the necessity of undergoing the ordeal of learning it in schools. Author of the book, *'Mathematics Charms'* writes, 'When I meet people socially and they discover that my field of interest is mathematics, I am usually confronted with the statement: 'Oh, you are a mathematician! However I was always terrible in mathematics.' To compound this lack of popularity of the subject, a child who may not be doing well in mathematics is often consoled by the parents that they too were not so good in mathematics during their school days. Such a negative model generally has a deleterious effect on the youngster's motivation towards the subject. Why such a distaste of mathematics? Why being weak in mathematics is regarded as a badge of honour?

Most of us want to acquire some knowledge and familiarity with the basics of mathematics and knowledge of its those aspects which permeate modern society. Most of us have the ability to master its basic concepts. However most of us view the subject as an uncontrollable beast, an inscrutable subject having a language which can be understood only by a few. It will be far better to regard mathematics as a subject that all of us use in varying degrees in our daily lives, and merely acknowledge that some persons are more adept in mathematics than others. Mathematics does not hide its secrets from all but a few. Instead most of us have a phobia about the subject which needs to be dispelled.

Why is mathematics so unpopular and regarded as a necessary evil without which modern education is not possible? Those who appreciate mathematics are fine and enjoy and relish it. However those who somehow develop distaste for it feel great hardship in undergoing the drudgery of learning mathematics in their school days. An eminent mathematician in his book on history of mathematics remarks, 'Mathematicians as a race apart are supposed to possess supernatural powers. While it is very flattering for successful mathematicians, it is very bad for those who for one reason or the other are trying to learn mathematics.'

Since childhood we are told that just as medicine, howsoever bitter and unpalatable it may be, has to be taken when a person is sick, so is the case with the study of mathematics. What are the reasons for such an unpopular view of the subject? Something certainly needs to be done by successful mathematicians to remove mathematics phobia amongst public. People need to be made aware of the beauty and charm of mathematics, so that youngsters feel eager and excited to study the subject rather than abhor it. The present book is an attempt in that direction.

1. What is Mathematics

Mathematics means different things to different persons. While some regard mathematics as an art, others consider it to be a science. Twentieth century eminent British mathematician and philosopher, Bertrand Russell, once remarked, 'Mathematics possesses not only truth but also supreme beauty cold and austere.' Like the beauty of a sculpture, it is pure and capable of stern perfection! Mathematics has been called the queen and at the same time the handmaid of science. These statements demonstrate the wide spectrum which mathematics now occupies. Ranging from the abstract to the most realistic applications, it now straddles a very wide spectrum.

Classical mathematics primarily comprises of fields devoted to numbers and figures. Whereas algebra and arithmetic deal with numbers, geometry and analysis deal with figures. However, in recent years, the scope of the subject has considerably widened. Its domain of enquiry now extends from understanding and analyzing phenomena of physical world to the analysis of abstract mathematical structures.

Domain of mathematics is now quite vast. Depending upon one's perceptions and exposure to the subject, it may be considered as an activity of discovery (an art subject) or an act of invention (a science subject). On the one hand, the subject is now a product of developments which are rooted in problems that are classical and concrete, on the other hand, it is also manifestation of an activity which finds its expression in constructions that are abstract.

Mathematics straddles both familiar and unfamiliar. We use it in our everyday dealings. Professionals such as engineers and economists rely on more complex higher mathematics. Scientists view mathematics as a supreme intellectual endeavor ranking much above other physical sciences as it transcends physical world and provides basis for modern science's attempt to understand and comprehend nature and universe and perhaps even God!

According to Henrich Hertz (1857–1894), 'One can not escape the feeling that these mathematical formulas have an independent existence and an intelligence of their own.

They are wiser than we are, wiser even than their discoverers and that we get more out of them than was originally put into them.'

2. Origin and Growth of Mathematics

Mathematics did not appear out of thin air at a given point of time of human civilization because of some discovery or invention. It gradually developed as the human urge for better understanding of natural phenomena and better inter human relationship grew with the development of human civilization and human society. The subject has evolved in fits and spurts as ancient civilizations in Asia and Africa fumbled and stumbled trying to formulate most fundamental concepts of mathematics so as to systematically deal with their day to day necessities and requirements. Unlike other physical sciences, earlier contributions to mathematics have not been discarded in favour of the later discoveries. Instead, these have usually served as corner stones for checking validity of later constructed more sophisticated concepts.

The growth of the subject has been gradual. Starting with numbers needed for counting it gradually evolved as human society evolved. In fact it took ancients long time before they moved beyond counting on fingers. As human society became complex and its requirements grew, the subject also developed and evolved to fulfil these requirements.

3. Nature of Mathematics

Mathematics is a field of knowledge which has several remarkable features. Perhaps the most remarkable of all its features is the one which enables humans to describe, measure and evaluate different aspects of endeavor that involve and effect humanity in a variety of ways.

Whereas the world described by natural and physical sciencies is real, concrete and perceptible, the world of mathematics is abstract one. It is built upon ideas and concepts and can only be perceived and appreciated with mind's eyes. Mathematics has enabled science to comprehend and visualise the vastness of the cosmos as well as infinitely small domains of fundamental particles. It straddles almost every field of human knowledge. Its domain of enquiry includes both real and physical world as well as abstract structures.

Mathematics is unique in the sense that it is neither supernatural (God sent, such as teachings of religions are claimed to be), nor natural (based on observation of natural phenomena, such as physics, chemistry, biology, etc.). It is also not a chronology of events or record of places such as history and geography. This is perhaps the only subject which has developed out of the necessity of human society to manage its affairs systematically and at the same time comprehend and understand natural phenomena. It has developed and grown over the years as the humanity developed and evolved, and its urge to better utilize natural resources and better appreciate and understand causes behind various observed natural phenomena increased. New disciplines appeared and developed as human capacity for better appreciation and understanding of natural phenomena increased.

Mathematics is more than a successful career in accounts or engineering. One needs to appreciate the way the subject is structured. Mathematicians generally prefer the simple over complex. If a single rule can perform work of two rules then mathematicians prefer to replace those two rules by single one. Mathematicians prefer definitions which are universal and hold good everywhere under all circumstances. Two plus two has always been four everywhere. Definitions of line and circle have been same everywhere over the ages. Definitions of addition, subtraction, multiplication and division have been consistently same over the ages and at all places. Mathematicians become skeptical if a definition has to be modified to fit different situations.

Even though mathematics is rooted in experiences of the real world, it is the most conceptual of all the subjects. Its worth is totally dependent on its internal rigour. Mathematicians use proofs to check the validity of their concepts. Mathematician has to work within accepted structures. If a mathematician is unable to provide quantitative support to his ideas and concepts, he is ignored regardless of whether his ideas get ultimately accepted or not. Pure thought underpins mathematics. This is a bit frustrating for the majority.

Mathematics is based on deductive reasoning. A mathematical proof consists of a series of deductive arguments which consist of premises and conclusions. Statements such as if A then B else C are quite common in mathematics. Deductive reasoning avoids the evidentiary trap inherent in inductive reasoning. Inductive reasoning is an intellectual process whereby the thinker makes various observations in the world around and then forms conclusions based on these observations. A major shortcoming of such inductive reasoning is that in spite of several observations made, one can not be sure whether the conclusions drawn will hold true in all circumstances. Deductive reasoning is particularly well suited for mathematics as it does not look upto the real world for validation. It also enables one to draw general conclusions from specific statements of facts. Mathematical reasoning is less dependent upon how the statements relate to the real world. It depends instead upon internal consistency of the statements.

Proofs are of paramount importance in mathematics. They illustrate how problems be solved in a logical manner. In mathematics, we also have axioms and postulates. Whereas axioms are general assumptions which are valid for all branches of mathematics, postulates are assumptions specific to geometry. Axioms are in a way statement of facts which are self evident and do not require any proof. For instance a quantity equals to itself, whole equals sum of all its parts. The assumption that only a single straight line can pass through two given points is a postulate. However, modern mathematicians do not make much difference between the two. Axioms and postulates serve as basis from where to start deductive reasoning of a proof for proving a stated result using logical reasoning.

French philosopher Denis Diderest wrote of mathematicians, 'They resemble those who gaze out from the tops of mountains whose summits are lost in clouds. Objects on the plains have disappeared from their view. They are left with their own thoughts and are conscious of the heights to which they have risen where it is not possible for every one to follow and breathe in thin air.' Such statements may amuse successful mathematicians. However they scare the common people. No doubt there is lot of abstraction in mathematics but the subject also has a lot of practical relevance and real life applications. It is these aspects of the subject of which the public need be more aware of and not be scared by its abstraction.

Mathematics is essentially an activity of generating problems and then trying to solve them. In some cases we are able to obtain their solutions while in some others we are neither able to prove nor disprove a stated result. Such statements are called 'conjectures'. Conjectures which have remained unproved or disproved are often referred to as 'open problems' in literature. Such open problems serve as challenges for the professional mathematicians (and even amateurs) to try their hand upon.

Mathematicians generally feel that problems formulated or posed by them are solvable and their solution will be found sooner or later. Eminent mathematician Hilbert put it thus in his address to conference of mathematicians in Paris: 'The conviction that every problem posed in mathematics has a solution is a powerful incentive for the researchers. In our hearts, we often feel a call that there is a problem whose solution can be found using reason alone so let us try.' Hilbert also believed that the proof that posed problem has no solution is also an acceptable solution of the problem and sometimes it has lead to new developments in the subject.

Negative solutions can be found in the history of mathematics. For example, the equation $x^2 = 2$ has no solution in the set of rational numbers lead to the discovery of irrational numbers. It was only in nineteenth century that the mathematicians could show that certain geometrical constructions such as squaring a circle or trisecting an angle using compass and ruler only were not possible. Thus, sometimes a problem which looks simple and interesting may turn out to be unsolvable.

According to Paul Thomas Mann, mathematics elevates a person above the lusts of flesh. In this connection he once said, 'I tell them that if they occupy themselves with study of mathematics, they will find in it the best remedy against the lusts of flesh.' Albert Einstein also once remarked, 'How can it be that mathematics being after all a product of human thought independent of experience is so admirably adopted to the objects of reality.'

4. Interaction with other Subjects

Since its origin, mathematics has been intimately linked with humanities urge to better understand and comprehend nature, better utilize natural resources and make inter-personal exchanges and communication convenient, simple and systematic. It is therefore but natural to expect that this subject must be having intimate relationship with other spheres of human knowledge and activity. Origin of basic branches of mathematics such as arithmetic, algebra, geometry, trigonometry and calculus is due to the specific needs of development of such subjects felt by human society to understand and comprehend natural phenomena, and to better utilize natural resources. Not only physical sciences such as physics and chemistry but also natural sciences such as botany and zoology and even subjects such as economics and commerce have provided stimulus for growth of mathematics and origin of new disciplines in it. Physicists like Newton and Einstein needed mathematical tools to appropriately formulate and test their theories. Subjects such as statistics and operations research were developed to more efficiently utilize natural resources. Even now, we hear of developments of new theories such as fuzzy set theory, theory of strings and knots, theory of chaos to comprehend more complex phenomena of nature. The process of interaction is an ongoing process. Moreover, it is not a one way process. Mathematics enriches other subjects and in return gets enriched by them.

5. Dread of Mathematics

Unfortunately most of us are scared of the subject. Mathematics is regarded tricky and even terrifying! Most of us would rather prefer to do without it. Why is there such a widespread distaste of mathematics? Reason for distaste lies not merely in the nature of the subject but also in the way it is presented and taught in our schools. Rather than being made to observe and appreciate its elegance and beauty, students are generally advised to look upon the subject as a necessary evil consisting of a set of rules and operations which are arbitrary and have no relevance to reality. They are advised to learn and remember these rules by heart (without appreciating their significance and relevance) if they are to do well in the subject.

What is about the subject which causes most people to freeze? Most of the students at very early stage form an adverse opinion about the subject in that it is a subject where a set of rules which have no relevance to reality, have to be crammed and mugged up and then applied to situations most of which have no relevance to real life. If we recall our school days, then arithmetic which mostly deals with real life problems did not appear that distasteful as algebra and geometry which did not seem to have much relevance to real life situations.

In fact mathematics is like classical music. Classical music sounds jarring to most of us. However, when we hear bhajans and songs laced with classical music we relish it. Similarly if mathematics is taught and presented to children in schools in a way that they appreciate its necessity and utility in real life, then it would not appear distasteful. With enough of practical illustrations, one is likely to form a more positive view of the subject and even develop a liking for it. On observing its practical relevance and inner beauty, the student may even start asking why he was not made aware of such lovely and exciting aspects of the subject while in school. Reason for the dread of the subject does not lie in the subject but rather in the way it is presented to children in schools.

Taken in right spirit, mathematics as such is not a boring or appalling subject. It has its own charm and beauty and utility. Transition from abstract fundamentals of the subject to its concrete and practical uses is sometimes accidental and unexpected and thus all the more thrilling. My object in writing this book is to highlight the charm and beauty of the subject as well as highlight some of its exciting and fascinating aspects.

PART I

Charm and Beauty in Mathematics

Chapter 1
Counting and Number System

1. Introduction

Mathematics historians surmise that the earliest necessity of counting must have arisen when ancients wanted to keep record of their family members, children, flocks of cattle and sheep. Initially, fingers of hand must have served the purpose. However when families and flocks grew in size, necessity must have been felt of having some means to record numbers greater than ten (ten being total number of fingers on two hands).

Natural numbers were perhaps the earliest mathematical notion developed by humans. Difference between a single cow and a pack of cows, a single stone and a heap of stones, etc. must have lead early humans to systematize the concept of numbers. In fact in Sanskrit (and also in some other Indo-European languages) after one and two, we have many. In Sanskrit grammar we have एकवचन (single), द्विवचन (two) and बहुवचन (many). One, two, many and no distinction beyond two. We have for instance नर, नारौ, नरा: for one person, two persons and many persons. It was only at some later stage that necessity of distinct numbers beyond two was felt to indicate that there is some difference between different numbers and plurality. Later we find use of expressions such as पाँच पांडव (five Pandavas), five fingers, five mangoes etc. We notice that in these 'five' is common. This is the count of elements of these sets. If we match elements of one set with the elements of the other set, we observe one to one correspondence. So, we say these sets are equinumerous. It was this naturality between different equivalent sets which led to the notion of natural numbers which we now write as 1, 2, 3, 4, 5, etc.

2. Roman Numerals

Natural numbers have not always been written as 1, 2, 3, 4, 5,.... the way we write these now. In fact in the western world another system of writing numbers as I, II, III, IV, V, ... was in vogue around 2000 years back. These were called Roman Numerals. In this system special letters of alphabet were used to denote some specific numbers and then these were suitably manipulated to generate rest of the numbers. In this system one was denoted by capital alphabet letter I. Writing it twice II denoted two and writing it thrice III denoted three. Alphabet capital V was symbol for five. Writing I to its left was used to denote four, writing I once to its right (VI) denoted six, writing I two times to the right of V (i.e. VII) denoted seven and writing three times I to the right of V (i.e. VIII) denoted eight. Again alphabet letter capital X denoted ten, IX denoted nine, XI eleven, XII twelve, XIII thirteen, XIV fourteen, XV fifteen and so on. In fact specific letters of alphabet were chosen to represent numbers, one (I), five (V), ten (X), fifty (L), hundred (C) and thousand (M), etc. These were then manipulated by above types of rules to generate remaining numbers. The convention is that if a letter denoting a smaller number is written to the left of a number denoting larger number, then it means subtracting the smaller number from the larger. However, if a smaller number is written to the right of a larger number it means addition. Whereas a smaller number can be written to the left of a larger number only once, it can be written to its right upto three times. Thus whereas IV denotes four $(5 - 1 = 4)$, VII denotes seven $(5 + 2 = 7$, IX denotes nine $(10 - 1 = 9)$, XIV denotes fourteen $[10 + (5 - 1) = 14]$, XLIV denotes forty four $[(50 - 10) + (5 - 1) = 44]$, CCXLVI denotes two hundred and forty six $[100 + 100 + (50 - 10) + 5 + 1 = 246]$ and so on. However, difficulty is felt in writing very large numbers. For instance 1984 has to be written as MCMLXXXIV $[1000 + (1000 - 100) + 50 + 10 + 10 + 10 + (5 - 1)]$.

3. Arabic Numerals

In the western world numerals written as 1, 2, 3, 4, 5, ... etc. are called 'Arabic numerals' as the western world came to know about these from Arabs. In Arabic world, these are known as 'Hindse' as they learnt about these from India. Compared to Roman numerals these are more versatile. That is why these have now almost taken over Roman numerals even in the west. Roman numerals now have only historical and ritual value even in the western world. For instance in Olympics they still use these to denote the year and number of Olympic games. Similarly many schools still have classrooms numbered I, II, III, IV and so on.

In fact it was the French mathematician Fibonacci (actual name Filuis Bonacci (1180-1250)) who introduced the western world to Arabic numerals. He had traveled extensively through Arab world of his times. He introduced these formally through manuscript 'Liber abaci' in 1202 which was later revised in 1228. It was first published in 1850 as 'Scutti Leonardo Pisano'. The book begins with the sentence, 'These are the nine figures of the Indians: 9, 8, 7, 6, 5, 4, 3, 2, 1. With these nine figures and sign 0, which in Arabic world is called Zephirum (word Zephirum evolved from the Arabic word 'as-sifer' which comes from the Sanskrit word 'shunya' used liberally in India as early as fifth century) it is possible to write any number how so ever large it be. Shunya (called zero) refers to nothingness, or emptiness or void. All of us are now well familiar as to how these ten figures are used to write not only natural numbers but even much more.

3.1 Origin of Arabic Numerals

More than twelve hundred years ago (767 A.D.) Abdullah Al Mansoor, the second Abbasid Khalif, celebrated the founding of his new capital Baghdad by organizing an international scientific conference to which he invited Greeks, Jews and Hindu scholars. The theme of the conference was observational astronomy. At this conference a Hindu astronomer, Kankah by name, presented a paper on Hindu numerals which were thus far unknown outside India. This paper changed the subsequent course of thinking in the rest of the world.

Although we are now taught this system from elementary school stage itself, it is worth noting its origin and reviewing its basics.

3.2 Significance of Place Value in Arabic Numerals

Whereas the western world had been using Roman numerals as late as two thousand years back, Hindus in India had been using decimal system even much earlier than that. It is not known exactly as to when this system came into operation in India.

In this system place value of a digit has great significance. Decimal system of Hindus uses ten digits 0, 1, 2, 3, 4, 5, 6, 7, 8 and 9 to generate every conceivable number big or small. Decimal system might have evolved as human beings have a total of ten fingers on both hands. When digits one to nine are exhausted at 9, next number is created by writing one to the left of zero (10). We continue writing like this for the subsequent numbers by writing 1, 2, 3, ... to the right of one till we reach number 19 (nineteen). To write the next number twenty, we write two to the left of zero (20) and continue like this till we reach 99 (ninety nine). To write the next higher number (hundred) we put two zeros to the right of one (100) and so on.

In this system the value of a digit in a number is decided by its place in the number. Digit 4 written at first place on right of a number implies 4, the same digit when written at second place from right implies 40 and at third place from right 400 and so on.

System is also capable of denoting numbers smaller than one. To denote such numbers, we put a dot '.' (different from number 0) to the right of the digit. Thus .2 is one tenth of 2, .02 is one hundredth of 2 and so on. For example in the number 4312.407, values of different digits appearing in it starting from left are: 4 is 4000, 3 is 300, 1 is 10, 2 is 2, 4 is 4×10^{-1}, next 0 is 0 (0×10^{-2}) and next 7 is 7×10^{-3}.

4. Natural Numbers

Numbers 1, 2, 3, 4, 5, ... are called **natural numbers**. Natural numbers have been grouped into different classes depending upon some commonality in their properties. Name of the class to which a number belongs is generally indicative of the properties of the class to which it belongs. A number which is divisible by 2 is called an **even number**. A number not divisible by 2 is called an **odd number**. A natural number which is not divisible by any other natural number except one and itself is called a **prime number**. A number which can be expressed as product of two or more natural numbers is called a **composite number**. A number which equals the sum of its divisors is called a **perfect number**. A **divisor** or a **factor** of a natural number is a number which divides it exactly yielding a natural number as a quotient and zero as remainder. For instance 5 is an odd number, 8 an even number. Again 11 is prime but 18 is a composite number. Also 6 is a perfect number ($6 = 1 + 2 + 3$) and so on. We also have some more classes of natural numbers which we shall learn later on as we observe interesting properties of different classes of natural numbers in a subsequent chapter.

5. Cardinal and Ordinal Numbers

Development of counting gave birth to the concept of cardinal and ordinal numbers. Whereas **cardinal numbers** are unordered collection of numbers such as ten pieces of marble on floor, fifteen chairs in the room, two stools etc., ordinal numbers are **ordered numbered** such as first, second, fifth, sixteenth, etc.

Natural numbers as we write 1, 2, 3, 4, 5, ... etc. are written in a specific order. Number two is greater than one, three greater than two and so on. Number 42 we know is greater than 36, but less than 47 or 59. Numbers when written in ascending (increasing) or descending (decreasing) order are referred to as ordinal numbers. However if we are not interested in their relative ranking and randomly write a set of natural numbers such as 47, 36, 54, 58, 3, 29, 87, then we call these cardinal numbers. A natural number a greater than natural number b is expressed as $a > b$. However, if a is less than b, we write $a < b$. Thus $8 < 12$ but $13 > 4$.

6. Addition and Subtraction Operations

Once people had mastered the basic idea of counting and numbers, it was only a matter of time before they began to group objects together and carry out operations of addition and subtraction. As the human civilization progressed, it was realized that just having a set of natural numbers did not fulfill the day to day requirements of the society. For example, if a farmer had three cows and purchased two more then what is the total number of cows which he now had. Similarly, if a farmer had 8 sheep and sold five of these, with how many sheep was he left? This lead to formalization of arithmetic operations of addition and subtraction of natural numbers. Symbol '+' is now used for addition and '–' for subtraction. Thus in the first case the farmer wants to know the natural number $3 + 2$ and in the later case he wants to know the natural number $8 - 5$. In the first case it was obvious that after having bought two new cows if he started counting total number of cows that he then had, he would get number 5. So $3 + 2 = 5$. In the second case, if he counted the sheep that he was left with after selling 5 sheep, the number that he would get is 3. Thus $8 - 5 = 3$. Over the years these processes of addition and subtraction gradually evolved and got formalized and made systematic as we know them now. Whereas to add two natural numbers, we have to simply start counting forward from one of these two numbers till we have counted as many natural numbers as is the second natural number, in the case of subtraction we have to start counting backward from first number till we have counted upto the number to be subtracted without bothering whether these refer to cows, sheep, trees, etc. or any other countable set.

Addition and subtraction are now little more than ordinary arithmetic operations for most of us. We accept and use these as absolute truths to be accepted without questioning. However, there are underlying philosophical questions such as: Are operations of addition and subtraction mirror images of each other?

6.1 Negative Numbers

Although addition of any two natural numbers always yields a natural number, difficulty was felt when a larger natural number was subtracted from a smaller one. For instance 5 – 3 = 2 but what about 3 – 5? For getting result of 3 – 5, we have to count 5 backwards from 3. But there are only two natural numbers 1 and 2 before 3. Therefore, it was decided to extend number system backward and symbol '–' was prefixed to a number if it did not belong to the set of natural numbers. Thus when a larger natural number was to be subtracted from a smaller natural number, first smaller natural number was subtracted from the larger and '–' sign prefixed to it. Thus 3 – 5 = –(5 – 3) = –2, 12 – 17 = –5. Symbol '–' indicates that there is so much of deficit when a larger number is subtracted from the smaller one. If a person has 15 rupees and he gets 7 more then he now has 15 + 7 = 22 rupees. However, if a person has 7 rupees and he owes 15 rupees to some other person and wants to pay these, then he has a deficit of 15 – 7 = 8 rupees which is indicated by –8. So symbol '–' indicates deficit. No symbol prefixed to the result of a subtraction operation means there is still surplus.

6.2 Number 'Zero'

A ticklish situation arises when a person has 15 rupees and he owes 15 rupees. How much money is he left with when he makes this payment. This results in a number which is neither a positive natural number nor a negative number. There is balance between what he has and what he owes. A new symbol '0' called 'zero' was invented to indicate this balance between what a person had and what he owed. Thus 15 – 15 = 0. Similarly 27 – 27 is also zero and –3 – (–3) is also zero.

Zero is in fact a special type of number. It does not belong to set of natural or negative numbers. It is regarded as dividing point between the set of positive and negative numbers. India is credited with having invented this special number zero (शून्य). Set of positive natural numbers 1, 2, 3, ... and set of negative numbers –1, –2, –3, ... taken together are referred to as set of integers. 1, 2, 3, 4, ... are positive integers and –1, –2, –3, ... negative integers. Set of integers (positive, negative and zero) is usually denoted by symbol Z (Z is the first letter of German word Zalhen which means numbers). Because zero represents 'nothingness', it is used to indicate place value of a digit. For example 2, 20, 200, .02 represent different values which same digit 2 assumes.

7. Multiplication and Division Operations

Multiplication operation may be regarded as faster addition, same number being added to itself several times. Multiplication of a natural number a with a natural number b means adding number a to itself b times. It is usually denoted by $a \times b$ or $a \cdot b$. Thus 7×4 means $(7 + 7 + 7 + 7) = 28$. One could also say that one is adding 4 seven times. For smaller numbers, it is alright even if we perform actual additions. However, if we have to multiply two large numbers say 148×231, then this addition operation becomes cumbersome. For such cases, more systematic rules have been framed and students are taught these in schools. In fact multiplication tables have been prepared for fast results. The reader is expected to be familiar with these and our object here is not to recount those rules. The operation was later on extended to negative numbers also and we now know how to multiply two integers a and b whether both are positive, both negative or one positive and one negative.

Just as subtraction is reverse of addition operation, similarly division is reverse of multiplication operation. Necessity for designing this operation arose when at the death of a person his estate had to be divided amongst his siblings. Whereas in multiplication operation, we add a number to itself specified number of times, in division operation, we subtract one number (called divisor) from the other number (called dividend) as many times as possible till we are left with a number smaller than the divisor. We call the left over number as remainder and number of times the second number was subtracted from the first quotient. Thus when we divide 47 by 6, we can subtract 6 seven times from 47 before we are left with a number lesser than 6 which is 5. In this case 6 is divisor, 47 dividend, 7 quotient and 5 remainder. If no remainder is left, we say second number exactly divides the first. For example 6 exactly divides 48. In this case quotient is 8 and remainder zero. Like other arithmetic operations, division operation has also been now formalized and we can divide any integer a by another integer b, be these integers positive or negative. We usually write division of a by b as $a \div b$ or a/b.

Operations of addition, subtraction, multiplication and division are the four basic arithmetic operations which serve most of the common needs of the human society. Symbol '+' is used for addition, '–' for subtraction, '×' for multiplication and '÷' for division.

8. Fractions

When we perform operations of addition, subtraction or multiplication on two elements of Z (two integers positive or negative) we get an element in Z. However, it is not always so in case of division operation. No doubt $27 \div 3 = 9$, $-15 \div 3 = -5$, $-12 \div -4 = 3$ but what about 7 ÷ 3 or 5 ÷ 27. Such situation arises when divisor is not an exact multiple of dividend. Whereas in the first case, we get an integer quotient (two) with an integer remainder (one), even that is not possible when 5 is divided by 27. No doubt one could say that in this case quotient is zero and remainder 5 but such a statement does not serve any useful purpose. In order to handle such situations, number system was extended further to include what we now call '**fractions**'.

For example $\frac{2}{3}, \frac{12}{8}, -\frac{3}{4}, \frac{28}{3}, \frac{2}{5}$ are all fractions.

A fraction is said to be proper if p and q have no common factor and $p < q$. For example, 8/12 is not a proper fraction as numerator and denominator have a common factor 4. Its proper fraction form is 2/3.

If $p > q$, we can divide p by q and write it as

$x\frac{y}{q}$ where x is quotient and y remainder when p is divided by q.

Thus $\frac{28}{3} = 9 + \frac{1}{3}$ or $9\frac{1}{3}$.

There is an interesting saying comparing a human with a fraction. 'A man is like a fraction whose numerator is what he actually is and whose denominator is what he thinks of himself. The larger the denominator, the smaller the fraction.'

8.1 Operations on Fractions

Performing operations on fractions is daunting to many of us. Whereas we easily appreciate 1 + 2 = 3 or 5 – 3 = 2, many of us are not comfortable with operations such as

$$\frac{1}{2} + \frac{3}{4} = \frac{5}{4}, \frac{3}{4} - \frac{1}{2} = \frac{1}{4}, \text{ etc.}$$

The simplest way to perform addition and subtraction operations on fractions is to first make the denominators of all fractions involved identical by multiplying both numerator and denominator of each fraction with an appropriate factor and then performing desired arithmetic operations on new numerators. The final result is the simplified form of the resulting fraction obtained after canceling out common factors from numerator and denominator.

For example $\frac{1}{3} + \frac{2}{5} = \frac{1}{3} \times \frac{5}{5} + \frac{2}{5} \times \frac{3}{3} = \frac{5}{15} + \frac{6}{15} = \frac{11}{15}$

$$\frac{1}{3} - \frac{2}{5} = \frac{5}{15} - \frac{6}{15} = -\frac{1}{15}$$

In case of multiplication simply multiply numerator terms and denominator terms.

For example $\frac{4}{3} \times \frac{5}{8} = \frac{20}{24} = \frac{5}{6}$

For division, we interchange the numerator and denominator terms of the dividing fraction and then multiply.

$$\frac{4}{3} \div \frac{5}{8} = \frac{4}{3} \times \frac{8}{5} = \frac{32}{15} = 2\frac{2}{15}$$

In fact ancient Egyptians were quite adept and skillful in performing arithmetic operations on fractions.

9. Exponents

It is possible to apply multiplication operation on the same number itself a number of times. For example $a \times a \times .... \times a$ multiplied n times, is denoted by a^n, where n is called **exponent**. Thus, a^n is a multiplied itself n times. In particular

For $n = 2$ $\quad a^2 = a \times a$ is called **square** of a.

For $n = 3$ $\quad a^3 = a \times a \times a$ is called **cube** of a.

It may be noted that exponent n need not always be a positive integer. It can be a positive as well as a negative integer or even a fraction. Thus we can have a^2, a^3, $a^{3/4}$, a^{-2}, $a^{-1/2}$, etc.

In fact $\quad a^{-2} = (a^2)^{-1} = \frac{1}{a \times a}$

$a^{1/2} = \sqrt{a}$ (called **square root** of a)

$a^{-1/2} = \frac{1}{\sqrt{a}}$ (**reciprocal** of square root of a)

$a^{3/4} = (a^3)^{1/4}$ (or fourth root of a^3)

Exponent n can also be zero. In fact $a^0 = 1$.

Geometrically a^2 is the area of a square whose each side is of length a, a^3 is volume of a cube whose each edge is of length a, and $a^{1/2}$ is length of a side of a square whose area is a units. Also, a^{-1} means reciprocal of a and a^{-2} means reciprocal of area of a square of side a each.

There are now well established rules for performing operations on exponents.

For example $\quad a^2 \times a^3 = a^{2+3} = a^5$, $a^2 \times a^{-1/4} = a^{2-1/4} = a^{7/4}$

$a^{3/4} \div a^{1/2} = a^{3/4} \times a^{-1/2} = a^{3/4 - 1/2} = a^{1/4}$ etc.

However, such simplifications are not possible in case of addition or subtraction operations on numbers in exponent form. For example, we can not simplify $a^{1/2} + a^{3/4}$ or $a^3 - a^{2/3}$.

Extended set Z of integers which also includes fractions is denoted by Q. Operations of addition, subtraction, multiplication and division (except division by zero) can be performed on these numbers without needing any further extension of number system.

10. Order of Operations

Mathematics is rooted in the idea that expressions involving same arithmetic operations in a prescribed sequence must yield same result to all, at all places, and at all times. This invariability is made possible by sticking to a hierarchy of rules relating to the sequence in which arithmetic operations be performed.

Suppose we have the expression $5 \times 4 + 6 - 3$. Different persons may end up with different results if a prescribed hierarchy of operation of rules is not followed. One person may first multiply 5 with 4 then add 6 and mext subtract 3 to obtain 23 as answer. Another could first add 4 and 6, then subtract three and then multiply the result with 5 to obtain 35 as the answer. Another could first add 4 and 6, then multiply with 5 and finally subtract 3 to obtain 47 as his answer.

The hierarchy in which arithmetic operations are performed is as follows. First perform operations within parenthesis if any (starting with the innermost parenthesis if these are nested) and work way outwards. Next perform exponential operation if any. Then carry out multiplication and division and finally addition and subtraction operations in the order they appear from left to right.

Supposing we have the expression

$$2 + \{(8 - 2 \times 3)^2 + 4(5 + 6)\} - 6,$$

then this is equal to $2 + \{(8 - 6)^2 + 4 \times 11\} - 6 = 2 + \{4 + 44\} - 6 = 2 + 48 - 6 = 44$.

11. Number Base

Decimal number system (also sometimes called Deca system) in common use has 10 as the base. When we write 1023041 in this system, it actually means the number

$$1 \times 10^6 + 0 \times 10^5 + 2 \times 10^4 + 3 \times 10^3 + 0 \times 10^2 + 4 \times 10^1 + 1.$$

Its origin and popularity is to some extent based on the fact that humans have ten fingers on the two hands and when people shifted from counting with fingers to writing numbers on paper this seemed obvious choice. However, there is nothing sacronatic about it. A base other than ten such as five and two can be chosen. Base five (fingers of one hand) has often been used in performing arithmetic operations in Vedic Mathematics and base two (also called binary base) is now used in performing arithmetic operations on computers. Bases such as 4 and 8 have also been used.

It may be noted that when a natural number n is chosen as base then only numbers 0, 1, 2, ..., $n - 1$ can be used in writing the number in this base. That is why we have digits 0, 1, 2, 3, 4, 5, 6, 7, 8, 9 appearing in numbers written in decimal form. However, if base is 5, then only digits 0, 1, 2, 3 and 4 can be used and if base is 2 only digits 0 and 1 can be used. Base 2 is used in performing arithmetic operations on computers because 0 can mean switch off (no current flowing) and 1 switch on (current flowing). We can definitely say that number 1476 is not written in base 2 (or even 5) as digits such as 4 and 6 greater than these bases appear in it. However, it can be a number in base 8 or 9 or 10. Similarly 1432 can be a number in a base greater than 4 but not in base 2, 3 or 4. In general a number $a_k a_{k-1} \ldots\ldots\ldots a_2 a_1 a_0$ written in base n is

$$a_k n^k + a_{k-1} n^{k-1} + \ldots\ldots\ldots + a_2 n^2 + a_1 n^1 + a_0 n^0$$

where $a_i (i = 1, 2, ..., k)$ can take values from 0 to $n - 1$.

Thus number 1476 in decimal base when written in base 5 will be 21401 and when written in base 2 will be 10111000100. A simple way to express a number written in decimal system into a system with base n is to keep on dividing the number successively by n till a quotient less than n is achieved. Now starting with the first remainder, write the remainders from right to left ending with the last quotient (if it is non zero).

12. Binary Number System

The binary number system has the advantage of being the simplest using only two digits 0 and 1. Computers make use of it as zero may correspond to off position (no current flowing) and 1 to on position (flow of current). Another advantage which makes it more suited for use in computers is its correspondence with logic. In logic a statement is either true (Yes) or false (No). True could correspond to 1 and false to 0. Hence logic status of any statement can be described by a sequence of binary numbers. Thus a number like 1100011 in binary base will mean that computer can display seven switches (nodes) out of which first two and last two are on and middle three off.

Binary number system has a disadvantage also. It requires a large number of digit places to represent even small numbers. For example 45 needs six digit places and 83 seven. To overcome this disadvantage sometime octal system is used for manipulating large numbers. Octal base 8 is closely related to binary base 2 (8 is 2^3). Moreover 8 being close to base 10 of decimal system, numbers generated with this base are not enormously large and unwieldy even for manual manipulations. For example

45 in octal base is $\quad 45 = 5 \times 8 + 5 \times 8^0 = 55.$

12.1 Binary Arithmetic

With the advent of computers, binary arithmetic has now gained importance. In fact it forms the very basis of how computers work and perform arithmetic operations.

In order to perform an arithmetic operation of addition, subtraction, multiplication or division of two numbers given in decimal form, these have to be first expressed in binary form and then desired operation performed on these. These operations are essentially performed in the same manner as in decimal case except that now we have to keep in mind that a carry over will occur as soon as number exceeds one instead of carryover occurring when number exceeds nine in decimal system.

For illustration let us perform arithmetic operations of addition, subtraction, multiplication and division on numbers 113 and 83 which are in base 10.

First we express these two numbers in binary base. 113 when expressed in binary base is 1110001 and 83 when expressed in binary base is 1010011.

Addition operation (113 + 83 = 196)

Binary Operation

```
 1110001
+1010011
--------
11000100
```

Resulting decimal form

$2^7 + 2^6 + 2^2 = 128 + 64 + 4 = 196$

Substraction operation (113 − 83 = 30)

Binary Operation

```
 1110001
-1010011
--------
 0011110
```

Resulting decimal form

$2^4 + 2^3 + 2^2 + 2^1 = 16 + 8 + 4 + 2 = 30$

Multiplication (113 × 83 = 9379)

Binary Operation

```
             1110001
            ×1010011
            --------
             1110001
            1110001
           0000000
          0000000
         1110001
        0000000
       1110001
      --------------
      10010010100011
```

Resulting decimal form

$2^{13} + 2^{10} + 2^7 + 2^5 + 2^1 + 2^0 = 9379$

Division $\left(113 \div 83 = 1\frac{30}{83}\right.$ i.e., quotient 1, remainder 30)

```
             1
1010011 | 1110001     Quotient 1
          1010011     Remainder 2^4 + 2^3 + 2^2 + 2^1 = 30
          -------
          0011110
```

Quotient 1

Remainder $2^4 + 2^3 + 2^2 + 2^1 = 30$

Similarly if number 113 is to be divided by 15 quotient is 7 and remainder 8.

Now to perform this in binary format, binary format of 113 is 1110001 as earlier. Binary format of 15 is 1111. Hence desired division operation is

```
       111
1111 | 1110001
       1111
       -----
       11010
        1111
       -----
        10111
         1111
       -----
         1000
```

So in decimal form, quotient is $2^2 + 2^1 + 1 = 7$ and remainder is $2^3 = 8$

12.2 Multiplication of two Numbers using Multiplication and Division by 2 only

We are familiar with the method of multiplying two numbers. Suppose we want to multiply 45 with 78.

Then

```
   45
  ×78
 ----
  360
 +315
 ----
 3510
```

However it is possible to achieve this result by performing multiplications and divisions with 2 only. The method is as follows:

Write the two numbers in two separate columns. Now (i) divide the number in the first column by two writing remainder in parenthesis, (ii) multiply the second number in the second column with two, (iii) continue this process of division of entries in the first column and multiplication of entries in the second column iteratively by two till quotient in the first column becomes zero, (iv) write those entries of the second column in a new third column which have one as remainder in the first column and add these. This is the required product.

Applying this method to find the product of 45 with 78 we observe

45	78	
22 (1)	78	78
11 (0)	156	
5 (1)	312	312
2 (1)	624	624
1 (0)	1248	
0 (1)	2496	2496
		3510

Method is essentially writing one number in binary format and then multiplying it with the other number.

13. Decimal Representation of Fractions

The convention that the first written digit on the right represents unit place, next digit to its left tenth place, next left hundredth place and so on used in writing any positive or negative integer (howsoever large or small it be), can not be used in writing fractions, such as

$$\frac{3}{4},\ \frac{-2}{3},\ 2\frac{2}{7} \text{ etc., in the same format.}$$

In order to express a fraction also in the form in which integers are written, decimal convention is used. In this convention we put a dot (.) to the right of the right most unit digit. (This dot is a symbol and not zero).

This dot separates the integer part of a number from its fractional part which is now written to the right of this dot. First digit to the right of dot represents that digit multiplied with 1/10 (i.e., 10^{-1}) next the digit multiplied with 1/100 (i.e., 10^{-2}) and so on. Thus as we move from this dot to left each shift multiplies the value of written digit with 10^0, 10^1, 10^2, 100^3 and so on when we move to the right, the value of digit written gets multiplied with 10^{-1}, 10^{-2}, 10^{-3}, and so on.

Thus for example 243.147 means

$$2 \times 100 + 4 \times 10 + 3 \times 1 + 1 \times \frac{1}{10} + 4 \times \frac{1}{100} + 7 \times \frac{1}{1000}$$

It is possible to convert fractional component of a number into this decimal format. For this we first divide the numerator of the fraction by the denominator. Quotient is integer part. Now multiply remainder with 10 and divide by denominator. New quotient is first integer to right of decimal. Again multiply the new remainder with 10 and divide by the denominator. Quotient is the digit at second place of decimal. The process is continued till remainder becomes zero or we have got the decimal representation to desired number of decimal places.

For example
$$121\frac{3}{4} = 121.75$$
$$3\frac{7}{8} = 3.875$$
$$\frac{2}{3} = 0.6666...$$
$$\frac{16}{7} = 2.285714,\ 285714,\ ...$$

Whereas in the first two cases we have terminal decimal representation in the last two cases it is non-terminating. We can continue as long as we like. However we notice that in these two cases digits start repeating in the decimal part. In the first case (2/3) 6 starts repeating from the very first place itself. In the second case (16/7), repetition is after every sixth place. Group of digits 285174 keeps on repeating iteratively.

For brevity we write these as $0.\overline{6}$, $\overline{2.285714}$. A bar at the top of digits indicates that these digits are recurring repeatedly. In these two cases we have non-terminating recurring decimal representation. Such representation is called **recurring decimal**. We can also have a non-ending no-recurrent representation.

13.1 Representation of Numbers on a Straight Line

It is possible to represent numbers on a straight line called real line. Draw a straight line and choose some arbitrary point on it as origin to represent number zero. Now decide a unit of length and mark points on this line both to the right and left of 0. Then these points can be used to represent integers.

P_3 P_2 P_1
−4 −3 −2 −1 0 1 2 3 4

Points to the right of 0 are positive integers and points to left of 0 are *−ve* integers. A point 4 units to right of 0 is number 4 and a point 2 units to left of 0 is number −2. 0 represents zero. Fractions can be represented by points between integers.

For example $2\frac{3}{4} = 2.75$ is a point P_1 between 2 and 3 such that distance from 2 to P_1 is $\frac{3}{4}$ of distance between 2 and 3. $\frac{1}{2}$ is represented by P_2 which is mid point of 0 and 1, and $-3\frac{2}{3}$ is represented by P_3 which is a point between −3 and −4 such that distance between −3 and P_3 is $\frac{2}{3}$ of distance between −3 and −4.

14. Rational and Irrational Numbers

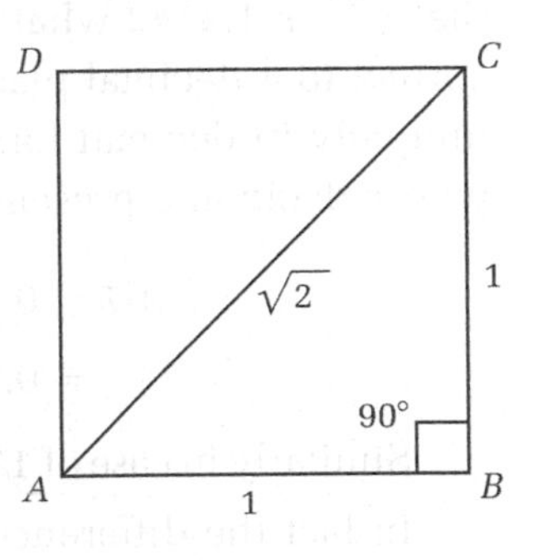

Integers and fractions taken together are called rational numbers. As seen above, these are expressible in the form p/q where p is a positive or negative integer and q a positive integer. These can be represented geometrically by points on a straight line called real line. A question naturally arises: are all points on the real line rational numbers or are there some points on real line which are not expressible in the form p/q. The answer is yes. Those numbers which can not be expressed in form p/q are called **'irrational numbers'**. Consider for example a square *ABCD* whose each side is of unit length. Join diagonal *AC*. Then since *ABC* is a right angled triangle

$$AC^2 = AB^2 + BC^2$$

i.e. $$AC = \sqrt{AB^2 + BC^2} = \sqrt{1+1} = \sqrt{2}$$

Thus length *AC* represents number $\sqrt{2}$. Which is a real length. So $\sqrt{2}$ can be represented on the real line by marking a point distant *AC* to the right of *O*. This lies between 1 and 2. Similarly $\sqrt{3}$, $\sqrt{5}$, $\sqrt{7}$ etc. are all real numbers which can be marked on a real line. It can be shown that such numbers can not be expressed in the form p/q. For instance in the case of $\sqrt{2}$, let if possible

$$\sqrt{2} = p/q$$

where p/q is a proper fraction (i.e., $q > 0$) and p and q have no common factor.

Then squaring $\quad 2 = p^2/q^2$

or $\quad p^2 = 2q^2$

Hence p^2 is even and therefore p is also even. Let $p = 2r$. Substituting

$$4r^2 = 2q^2$$

or $\quad q^2 = 2r^2$

Hence q^2 and therefore q is also even.

So we have shown that p and q are both even. Hence they have a common factor 2. This contradicts our assumption that p and q have no common factor. Thus $\sqrt{2}$ is not expressible in form p/q and is thus not a rational number. Numbers such as $\sqrt{2}$, $\sqrt{3}$, $\sqrt{5}$. Which are real but not expressible in form of p/q are called **irrational numbers.**

In fact $\sqrt{2}$ when expressed in decimal form is

$$\sqrt{2} = 1.41421356237095......$$

An unending non recurring sequence of digits after decimal place. When we say that $\sqrt{2}$ is 1.4142 what we actually mean is that 1.4142 is approximate value of $\sqrt{2}$ correct to 4 decimal places and not its exact value. Its actual value is not expressible uniquely in decimal form upto a finite number of decimal places. This is unlike 1/7 whose decimal representation though non terminating has a recurring pattern

$$1/7 = 0.142857, 142857, 142857, ...$$
$$= 0.\overline{142857}$$

Similarly in case of 1/3 also we have a non-terminating but recurring pattern $\frac{1}{3} = 0.\overline{3}$

In fact the difference between rational and irrational numbers is that whereas in the case of a rational number when expressed in decimal form we have a terminating or recurring decimal representation, in the case of an irrational number its decimal representation is both non-recurring and non-terminating. It may however be noted that sometimes in the case of even rational numbers recurring pattern may not be visible even for quite a large number of decimal places. For example in case of 1/109 which is a rational number, its decimal representation is

$$\frac{1}{109} = 0.0091743119266055045 87 ......, 009174311 ......$$

The pattern repeats after 108 decimal places!

14.1 Algebraic and Transcendental Numbers

Greeks made distinction between real numbers which are constructable with ruler and compass and those which can not be. For example integers, fractions and irrational numbers such as $\sqrt{2}$ can be marked on a real line using a ruler. Greeks called such constructable numbers algebraic as these are solutions of algebraic equations. For example ¾ is solution of $4x - 3 = 0$, $\sqrt{2}$ is solution of $x^2 - 2 = 0$. However converse is not true. In other words there are algebraic equations whose solutions are not constructable with ruler.

For example cube root of two i.e. $2^{1/3}$ is a solution of algebraic equation $x^3 - 2 = 0$. However it is not constructable using a ruler even though it is a real number. All those numbers which are solutions of algebraic equations are called **algebraic numbers**.

Numbers which are not algebraic but are still real are called **transcendental numbers** (transcendental means heavenly). Euler conjectured in 1748 that number e (base of natural lgorithms) is one such number. It was proved so in 1873 by Charles Hermite. π, the ratio of circumference of a circle and its diameter is another such number. In fact not only e but e^x where x is rational number, is also transcendental. By the end of nineteenth century, not many transcendental numbers besides e and π were known. Hilbert had conjectured that a^b will be transcendental if a is algebraic (different from one) and b irrational. For example $2^{\sqrt{2}}$.

In 1929 Gelfand showed that e^{π} is also transcendental. In 1930 Siegel proved that $2^{\sqrt{2}}$ is indeed transcendental. In 1966 Alen Baker showed that any finite product of a transcendental number of type e^{π} and $e^{\sqrt{2}}$ is also transcendental (Baker received Field's medal for this in 1970).

Despite all these advances, transcendental numbers still remain a mystery. For instance we still do not know that even though e and π are transcendental, what about their sum $e + \pi$ or their product $e\pi$. Are these also transcendental?

14.2 Real Numbers

All those numbers which can be represented on a real line are called **real numbers.** These include rational numbers (integers and fractions both positive and negative) as well as positive and negative irrational and transcendental numbers.

15. Concept of Infinity

When an integer p is divided by some other integer q we obtain a fraction p/q such as 2/3, 7/4, –3/4 etc. However, difficulty arises when q is zero. When an integer p is divided by another integer q, it means we are dividing p into parts of length q each. However what does an integer p divided by zero imply? It essentially means we are dividing p into parts of length zero (no length) each. Obviously these number will be very very large and in fact uncountable whatever integer is p. This has been given a special name **'infinity'** from the word infinite. Whereas any positive number divided by zero is plus infinity (denoted by $+\infty$ or simply ∞) a negative number divided by zero yields minus infinity ($-\infty$). Again it must be kept in mind that like zero, behaviour of infinity is also different from other real numbers. Whereas plus infinity (∞) is a number much much larger than the largest number we can think of (or one expects to get in his computations), minus infinity ($-\infty$) is a number much lesser than a negative number we can think of.

15.1 Nature of Infinity

As mentioned, infinity is not a number like other real numbers. This fact becomes clear from the following. When we want to find the exact number of objects which a set has, we enumerate them. In other words we set one to one correspondence between these objects and positive integers 1, 2, 3, 4, ... etc. In case the parity of last number is with N we say the number of objects in the set is N or we say that the set has **cardinality** N.

Cantor introduced the notion of cardinal numbers associated with the sets. When cardinal numbers of two sets are same we say that two sets have same cardinality. This means that elements of one set can be put in one to one correspondence with elements of the other set.

If we consider set $I = \{1, 2, 3, ...\}$, the set of all positive integers, then this set has infinite elements and so its cardinality is infinite. Cantor called it 'Aleph Null' and denoted it by N_0.

Next consider set of all even integers $E = \{2, 4, 6, 8, ...\}$. If we now set one to one correspondence of elements of I and E we notice

E	2	4	6	8	...	$2n$	...
I	1	2	3	4	...	n	...

Which means that cardinalities of both sets E and I are identical even though I has many more numbers than E. Not only this, set of all odd integers $\{1, 3, 5, 7, ...\}$ can also be put in one to one correspondence with elements of I. Infinity of this set is also same as that of I. Not only this even the set Q^+ of all positive rational numbers can also be shown to have cardinality of I even though I is a proper subset of Q^+.

It may thus appear as if all infinities will have same cardinalities. However it is not so. It may look surprising but mathematicians have shown that the set of all real numbers even between 0 and 1 has a cardinality bigger than that of the set I. In fact a hierarchy of infinities has now been established each succeeding infinity being bigger than the preceding one. But all this is for mathematicians. For a layman infinity is just infinity. However this should set at rest the commonly held notion of infinity being some unique number like other real numbers.

16. Arithmetic Operations with Zero and Infinity

In view of their special nature, rules of arithmetic operations applicable to real numbers have been suitably modified in the case of zero and infinity so as to keep consistency with similar operations as applied to other real numbers.

If zero is added or subtracted from a real number it remains unchanged.

$$a + 0 = a \text{ and } a - 0 = a$$

also, $$0 + a = a \text{ and } 0 - a = -a$$

In the case of infinity

$$a + \infty = \infty \text{ and } a - \infty = -\infty$$

and, $$\infty + a = \infty \text{ and } \infty - a = \infty$$

In case of multiplication any real number a multiplied with zero is zero and any real number multiplied with infinity is infinity.

$$a \times 0 = 0 \times a = 0 \text{ but } a \times \infty = \infty \times a = \infty$$

if a is positive and $-\infty$ is a is negative.

However $$a \times \infty = \infty \times a = \infty \quad \text{if } a \text{ is positive}$$

but $$a \times (-\infty) = -\infty \times a = -\infty \quad \text{is } a \text{ is positive.}$$

also $$0^n = 0, \text{ and } \infty^n \text{ is } \infty \text{ where } n \text{ is } a \text{ +ve integer.}$$

In case n is a negative integer say $-m$ (m +ve) then

$$\infty^n = \infty^{-m} = \frac{1}{\infty^m} = \frac{1}{\infty} = 0$$

and
$$\infty^{-n} = \infty^m = \frac{1}{0^m} = \frac{1}{0} = \infty$$

A ticklish situation arises when an arithmetic operation is to be performed between 0 and ∞.

$$0 + \infty = \infty,\ 0 - \infty = -\infty,\ 0 \div \infty = 0 \times \frac{1}{\infty} = 0 \times 0 = 0.$$

However $0 \times \infty$ or $0/0$ or ∞/∞ are undefined. Their exact value is not uniquely defined. However in specific cases, their limiting values can be obtained and these are not unique. They change from situation to situation.

For example $\frac{x^2 - 1}{x - 1}$ is $0/0$ for $x = 0$.

However if we factorise numerator and simplify by canceling out $x - 1$ from numerator and denominator before putting $x = 1$ we have

$$\frac{x^2 - 1}{x - 1} = \frac{(x-1)(x+1)}{x-1} = x + 1 = 1 \text{ for } x = 0.$$

Again $\frac{x^2 - 4}{x - 2} = x + 2 = 2$ for $x = 0$.

One may feel a bit uncomfortable with these results. However, they ensure consistency. As mentioned earlier, these are the limiting values of these fractions as x approaches zero and not their actual values. These are used when necessity of using these along with other real numbers arises. It must be kept in mind that number system and rules of arithmetic operations on numbers are man made which have gradually evolved over the years to take care of various requirements of human society.

17. Imaginary and Complex Numbers

We have thus far considered integers, fractions, rational, irrational and transcendental numbers. All of these are real numbers (called **set of real numbers**). These can be represented on real line. A question now arises are there more types of numbers which can not be represented on real line. The answer is yes. We have now what are call imaginary and complex numbers.

When we take square root of a positive real number we get a real number which can be positive as well as negative. For example square root of 4 is +2 or –2 because square of +2 as well as –2 is 4. Similarly square root of 2 is $\pm\sqrt{2}$ which are again real numbers. However what about square root of a negative number.

For example what is $\sqrt{-2}$? It will be a number whose square is -2. But we do not have any real number whose square is -2. For that matter square root of any negative real number is not a real number. This gives rise to another class of numbers which are not real, since such numbers can not be represented on real line. These have been called **imaginary numbers.** It may be of interest to note that

$$\sqrt{-2} = \sqrt{-1 \times 2} = \sqrt{-1} \times \sqrt{2} \text{ whereas } \sqrt{2} \text{ is real but } \sqrt{-1} \text{ is not real.}$$

Similarly $\quad \sqrt{-3} = \sqrt{-1} \times \sqrt{3}$ whereas $\sqrt{3}$ is real but $\sqrt{-1}$ is not real.

Thus $\sqrt{-1}$ is crucial in case of an imaginary number. Any real number multiplied with $\sqrt{-1}$ yields an imaginary number. For convenience $\sqrt{-1}$ is written as i the first letter of word imaginary. (However electrical engineers prefer to write it j as they generally reserve symbol i for current).

17.1 Complex Numbers

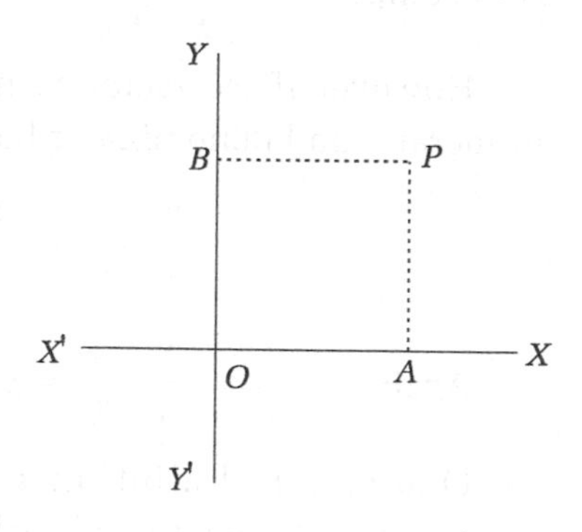

Since imaginary numbers could not be represented on real line $X'OX$, mathematicians started representing them on a line YOY' drawn through origin (number zero of real line) perpendicular to real line. They call $X'OX$, real axis and YOY' imaginary axis. Taking a cue from coordinate geometry, any point A $(x, 0)$ on real axis represents real number x. A point B $(0, y)$ on imaginary axis represents imaginary number y. Origin O where the real axis XOX' and imaginary axis YOY' intersect is zero of both real and imaginary number systems. Consider a point P in XOY plane where lines PA and PB drawn from P perpendicular to the real and imaginary axes respectively make intercepts OA of length x on real axis and intercept OB of length y on imaginary axis. Such a point P represents a number which has two components, one real x and other imaginary y. This is symbolically represented as $z = x + iy$.

z is called a complex number. In case imaginary component y is zero, z is a real number x and in case real component x of z is zero, it is imaginary number y.

Thus complex numbers constitute a more bigger set of numbers. This set has sets of real and imaginary numbers as its subsets. Both imaginary and complex numbers find a lot of practical applications in solution of real life problems.

For example we know that whereas metric (measure of distance) in real space is

$\sqrt{(dx)^2 + (dy)^2 + (dz)^2}$, dx, dy, dz being small distances along X, Y and Z axis,

it is $\quad \sqrt{(dx)^2 + (dy)^2 + (dz)^2 + (icdt)^2}$

$$= \sqrt{(dx)^2 + (dy)^2 + (dz)^2 - c^2 (dt)^2}$$

in space time continuum where x, y, z are space dimensions, t dimension of time and c velocity of light. This metric is vital to theory of relativity.

17.2 Significance of *i*

Although i $(\sqrt{-1})$ is imaginary and is at the root of all imaginary and complex numbers, it has no less practical significance than say 2, 8,107, or $\sqrt{2}$. It has at least as much significance as transcendental numbers e and π. This follows from a well known result in trigonometry according to which

$$e^{ix} = \cos x + i \sin x$$

So that when $x = \pi$, $e^{i\pi} = \cos \pi + i \sin \pi = -1 + i \times 0 = -1$ which is real.

In fact efforts have been made by mathematicians to calculate values of i^i, i.e., $(\sqrt{-1})^{\sqrt{-1}}$ Surprinsingly, i^i is real and not imaginary! Scudder Uhler (1872–1956) published value of i^i to more than fifty decimal places. Its value upto nine decimal places is 0.207879576.

17.3 Operations on Complex Numbers

Like addition, subtraction, multiplication and division of real numbers, we can also perform these operations on complex numbers. Let $z_1 = a + ib$, $z_2 = c + id$ be two complex numbers where a, b, c, d are all real numbers.

Then
$$z_1 + z_2 = (a + ib) + (c + id) = (a + c) + i(b + d)$$
$$z_1 - z_2 = (a + ib) - (c + id) = (a - c) + i(b - d)$$
$$z_1 \times z_2 = (a + ib) \times (c + id) = ac + iad + ibc + i^2bd$$
$$= (ac - bd) + i(ad + bc) \qquad (\text{as } i^2 = -1)$$
$$z_1 \div z_2 = \frac{a + ib}{c + id} = \frac{(a + ib)(c - id)}{(c + id)(c - id)} = \frac{(ac + bd) + i(bc - ad)}{c^2 - (id)^2}$$
$$= \frac{ac + bd}{c^2 + d^2} + i\frac{bc - ad}{c^2 + d^2}$$

Thus each of these four arithmetic operations when performed on complex numbers yields a complex number. In fact arithmetic operations on real and imaginary numbers are particular cases of these operations. If we put imaginary components b and d of z_1 and z_2 zero in the above we get operations on real numbers and when we put real components a and c of z_1 and z_2 zero we get operations on imaginary numbers.

In fact complex numbers are the most general form of number system which have real numbers and imaginary numbers as their subsets.

18. Number Field

Mathematicians define **number field** as a set of numbers which contains zero and one (unit) and which is closed with respect to arithmetic operations of addition, subtraction, multiplication and division. In other words if we choose any two numbers from a set and perform on them any of four basic arithmetic operations, then we should get a number belonging to the same set.

For example the set Q of positive and negative integers (including zero) is not a field because when we divide one integer by another integer we may not get an integer (for example 2/3 is not an integer). However the set of rational numbers is a field. It is denoted by F_1. Set of irrational numbers by themselves do not form a field. However set of real numbers is a field. Similarly set of imaginary numbers by themselves do not form a field, but set of complex numbers is a field.

Mathematicians have shown that given a field of numbers it is possible to construct another field using it. If k_0 is a positive rational number and $\sqrt{k_0}$ is not in field F_1 then it has been shown that we can get another field F_2 by joining $\sqrt{k_0}$ to F_1. Similarly if k_1 is in F_2 but $\sqrt{k_1}$ is not in F_2, then joining $\sqrt{k_1}$ to F_2 we get a new field F_3 and so on. Thus it is possible to generate new fields from existing fields.

19. Some Interesting Paradoxes

We conclude this chapter with some interesting paradoxes on number system. A **paradox** is a kind of fallacy which generally results from an unintentional violation (or oversight) of some basic and fundamental rule or law of mathematics. Apparently they poke fun at mathematics pointing out some inherent contradictions in the reasoning process. However a closer scrutiny generally reveals that some fundamental violation has been committed in the use of some basic rule or law. Paradoxes are entertaining as they yield absurd results and then make one look for the error committed. In some cases the error committed is obvious. In other cases error committed is subtle and even though every step seems to be logically correct yet the obtained result is absurd and not acceptable. Paradoxes are a fun to observe and generally have an important message embedded in them.

We present here some simple paradoxes which arise from improper use of basic arithmetic operations. It will be fun trying to identify the violation.

1. We know that 2 rupees = 200 paise (i)
   Dividing both sides by 4
   ½ rupee = 50 paise (ii)
   Multiplying (i) and (ii)
   $2 \times \frac{1}{2}$ rupee $= 200 \times 50$ paise or 1 rupee $= 1000$ paise!

2. Again 2 feet = 24 inches (i)
   Dividing both sides by 4
   ½ feet = 6 inches (ii)
   Multiplying (i) and (ii)
   1 foot = 24 × 6 = 144 inches!

3. $1 \times 0 = 2 \times 0$
   Dividing both sides by 0
   $1 = 2$?

4. $16 - 36 = -20 = 25 - 45$

$$16 - 36 + \frac{81}{4} = 25 - 45 + \frac{81}{4}$$

or $$4^2 - 2 \times 4 \times \frac{9}{2} + \left(\frac{9}{2}\right)^2 = (5)^2 - 2 \times 5 \times \frac{9}{2} + \left(\frac{9}{2}\right)^2$$

or $$\left(4 - \frac{9}{2}\right)^2 = \left(5 - \frac{9}{2}\right)^2$$

i.e. $$4 - \frac{9}{2} = 5 - \frac{9}{2}$$

or $4 = 5$?

5. $i = \sqrt{-1} = \sqrt{\frac{1}{-1}}$

$$\frac{1}{i} = \frac{1}{\sqrt{-1}} = \frac{\sqrt{1}}{\sqrt{-1}} = \sqrt{\frac{1}{-1}} = \sqrt{-1} = i \quad \text{(i)}$$

Again $$\frac{1}{i} = \frac{1 \times i}{i \times i} = \frac{i}{i^2} = \frac{i}{-1} = -i \quad \text{(ii)}$$

So from (i) and (ii), $i = -i = \frac{1}{i}$

(The fallacy is in writing $\sqrt{\frac{1}{-1}} = \sqrt{-1}$, as $\frac{\sqrt{a}}{\sqrt{b}} = \sqrt{\frac{a}{b}}$ only if a, b are both positive.)

20. Zeno's Paradoxes on Infinity

There are two paradoxes related to the concept of infinity which are often attributed to Greek philosopher Zeno.

One of these is a little modification of the well known story of hare and tortoise. In the story of hare and tortoise, tortoise wins because hare falls asleep midway.

According to Zeno tortoise would have still won even if the hare had not fallen asleep provided tortoise had some initial head start. His argument is like this. Suppose to start with the hare was at A and tortoise at B, a little ahead of A. During the race when hare reaches B, tortoise will have moved some distance ahead to C. When hare reaches C, tortoise will have moved ahead to D and so on. So how can hare overtake the tortoise?

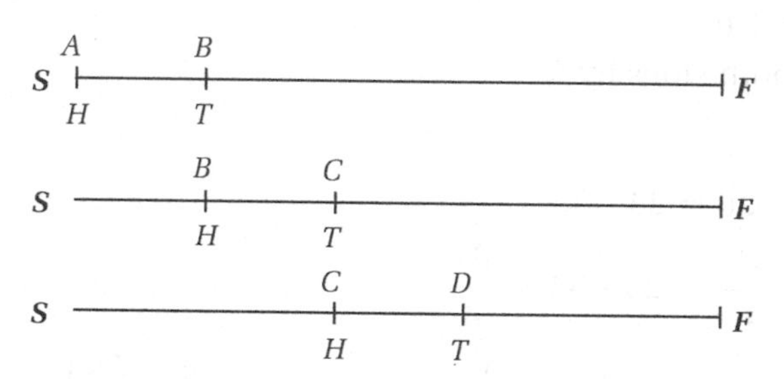

Another similar paradox credited to Zeno goes like this.

Suppose a runner wishes to cover a distance of one kilometer and runs in such a way that at each stage he covers half of the remaining distance (speed gradually decreasing). The question is will he be ever able to reach the goal post? Following argument shows that he will not be able to do so.

In the first stage he covers half a kilometer. In the second stage he covers ¼ kilometer. In the next stage 1/8 kilometers and so on. If he continues like this it will take him infinite time to cover 1 kilometer as at each stage there will be some distance left to cover, howsoever small that be.

What is wrong with these reasonings? Apparently nothing. But we know that in reality hare will overtake tortoise if initial head start A is not very large and the runner will reach the goal post in finite time. Such like situations have lead us to the concept of limit and limiting value which we shall have a chance to discuss in some detail in a later chapter.

Mind Teasers

1. Integers 1 to 10 are written in a row. Is it possible to place '+' and '–' signs between these integers so that the final answer is zero.
2. Find some two digit positive numbers which are divisible by both of their digits (examples 24, 39).
3. Find the smallest natural number which is four times smaller than the number with same digits written in reverse order.
4. Is there a natural number such that the product of its digits is 1980?
5. A natural number n is not divisible by 3. Is it possible for $2n$ to be divisible by 3?

6. Given $15n$ is divisible by 6. Is n also divisible by 6?

7. Find the smallest natural number which leaves remainder 1 when divided by 2, remainder 2 when divided by 3, remainder 3 when divided by 4 and remainder 4 when divided by 5.

8. Show that the product of three consecutive natural numbers is always divisible by 3.

9. Two natural numbers have their sum 100. Can their product be 3000?

10. A number ends in digit 2. If we move this 2 from the last place in the number to the first place, the number becomes twice the original number. What is the smallest such number?

11. A book has 750 pages numbered 1 to 750. What is the total number of digits used in printing these page numbers.

12. In order to number pages of a book, a printer used 1890 digits. How many pages this book has?

13. Ram has a book of 96 pages with pages numbered on both sides from 1 to 192. His younger brother Shyam tore out 25 consecutive pages from this book. Can the sum of these torn pages be 1990?

14. Obtain geometrically the lengths which may be used to represent irrational numbers $\sqrt{3}$ and $\sqrt{5}$ on real line.

15. Perform the following arithmetic operations using binary arithmetic
(i) $81 + 129$, (ii) $129 - 81$, (iii) 129×81, (iv) $129 \div 18$.

16. Multiply 129 with 34 using multiplication and division with 2 only.

17. Following is a half erased addition problem. Find the missing digits and the number base in which this operation is valid.

$$\begin{array}{ccccc} 2 & 3 & - & 5 & - \\ 1 & - & 6 & 4 & 2 \\ \hline 4 & 2 & 4 & 2 & 3 \end{array}$$

18. Find number base if such a base exists in which the following results are true
(i) $3 + 4 = 10$ and $3 - 5 = 15$ (ii) $2 + 3 = 5$ and $2 \cdot 3 = 10$
(iii) $3 \cdot 4 = 10$

19. Perform the following operations on imaginary and complex numbers
(i) $\dfrac{1+i}{1-i}$ (ii) $(2 + 3i) \times (4 - 2i)$ (iii) $i^3 \times i^4$

20. Show that real number system is a subset of complex number system by giving an operation on complex numbers which results in a real number.

21. In section 16 operations of real numbers with zero and infinity were defined. Is it possible to perform basic arithmetic operations amongst these two numbers themselves? Which of the following operations on these have unique values?

    (i) $\infty + 0$ (ii) $\infty - 0$ (iii) $\infty \times 0$ (iv) $\frac{\infty}{0}$ (v) $\frac{0}{0}$ (vi) $\frac{\infty}{\infty}$

Chapter 2
Beauty in Numbers

1. Introduction

In the previous chapter we considered number systems and arithmetic operations on numbers. Many natural numbers have certain interesting unusual built-in features which normally skip a casual reader and are generally not highlighted in schools. In fact beauty lies in certain operations on numbers and this needs to be searched. Whereas some of this has been searched through experimentation, in some cases people stumbled upon certain beautiful and interesting results. Carl Friedrich (1777–1855), a famous mathematician of eighteenth century discovered many such interesting results. Some of these he later proved as Theorems. There is no end to discovery of such results. Discovering such results can be fun. Even you can try.

In this chapter we present some such results which highlight the beauty of numbers for the benefit of reader with the hope that he/she will enjoy these and may feel motivated to discover more such results.

2. Surprising Number Patterns

(i) Following arithmetic symmetries are quite fascinating.

$1 \times 1 = 1$ $\qquad$ $111 \times 111 = 12321$

$11 \times 11 = 121$ $\qquad$ $1111 \times 1111 = 1234321$

and so on.

(ii) $1 \times 8 + 1 = 9$ $\qquad$ $123 \times 8 + 3 = 987$

$12 \times 8 + 2 = 98$ $\qquad$ $1234 \times 8 + 4 = 9876$

(iii) $142857 \times 2 = 285714$ $\qquad$ $142857 \times 5 = 714285$

$148257 \times 3 = 428571$ $\qquad$ $142857 \times 6 = 857142$

$148257 \times 4 = 571428$

(digits of 142857 appear in different orders when this number is multiplied with 2, 3, 4, 5 and 6.)

(iv)

$0 \times 9 + 1 = 1$
$1 \times 9 + 2 = 11$
$12 \times 9 + 3 = 111$
$123 \times 9 + 4 = 1111$
$1234 \times 9 + 5 = 11111$
$12345 \times 9 + 6 = 111111$
$123456 \times 9 + 7 = 1111111$
$1234567 \times 9 + 8 = 11111111$
$12345678 \times 9 + 9 = 111111111$

(in other words as many ones appear in the result as is the digit which is added to the product of nine with number starting with one and ending with digit one less than that digit)

We have another similar result

$0 \times 9 + 8 = 8$
$9 \times 9 + 7 = 88$
$98 \times 9 + 6 = 888$
$987 \times 9 + 5 = 8888$
$9876 \times 9 + 4 = 88888$
$98765 \times 9 + 3 = 888888$
$987654 \times 9 + 2 = 8888888$
$9876543 \times 9 + 1 = 88888888$

(v) **Repeated digit 9**

$999,999 \times 1 = 0,999,999$
$999,999 \times 2 = 1,999,998$
$999,999 \times 3 = 2,999,997$
$999,999 \times 4 = 3,999,996$
$999,999 \times 5 = 4,999,995$
$999,999 \times 6 = 5,999,994$
$999,999 \times 7 = 6,999,993$
$999,999 \times 8 = 7,999,992$
$999,999 \times 9 = 8,999,991$
$999,999 \times 10 = 9,999,990$

(it is of interest to note the pattern in the results of products. Sum of the first and last digits is 9. All middle digits are 9).

Similarly we notice

$9 \times 9 = 81$
$99 \times 99 = 9801$
$999 \times 999 = 998001$
$9999 \times 9999 = 99980001$
$99999 \times 99999 = 9999800001$
$999999 \times 999999 = 999998000001$, etc.

(vi)

$1 = 1 = 1.1 = 1^2$
$1 + 2 + 1 = 2 + 2 = 2.2 = 2^2$
$1 + 2 + 3 + 2 + 1 = 3 + 3 + 3 = 3.3 = 3^2$
$1 + 2 + 3 + 4 + 3 + 2 + 1 = 4 + 4 + 4 + 4 = 4.4 = 4^2$

... ...

... ...

$1 + 2 + 3 + 4 + 5 + 6 + 7 + 8 + 9 + 8 + 7 + 6 + 5 + 4 + 3 + 2 + 1$
$= 9 + 9 + 9 + 9 + 9 + 9 + 9 + 9 + 9 = 9.9 = 9^2$

(vii) Again $81 = (8 + 1)^2 = 9^2$
$512 = (5 + 1 + 2)^3 = 8^3$
$4913 = (4 + 9 + 1 + 3)^3 = 17^3$

(Try to find some more numbers with similar properties)

(viii) In some cases it is possible for the number to be equal to sum of its digits raised to consecutive exponents. For example

$135 = 1^1 + 3^2 + 5^3$ $\quad$ $1306 = 1^1 + 3^2 + 0^3 + 6^4$

$175 = 1^1 + 7^2 + 5^3$

In some cases a number equals sum of its digits raised to the power of the digit itself. For example

$3435 = 3^3 + 4^4 + 3^3 + 5^5$

Who says mathematics is dull and boring. It has its beauty which has to be discovered. Proceeding further we observe

(ix) **Some unusual relationships**

Numbers	Their Product	Their Sum
9, 9	$9 \times 9 = 81$	$9 + 9 = 18$
3, 24	$3 \times 24 = 72$	$3 + 24 = 27$
2, 47	$2 \times 47 = 94$	$2 + 47 = 49$
2, 497	$2 \times 497 = 994$	$2 + 497 = 499$

(in these cases digits appearing in sums of numbers are in reverse order of digits appearing in their products)

(x) **Numbers which when divided by *n* (= 2, 3, 4, 5, 6, 7, 8, 9) leave *n* – 1 as remainder**

$2519 \div 2 = 1259$ remainder 1
$2519 \div 3 = 839$ remainder 2
$2519 \div 4 = 629$ remainder 3
$2519 \div 5 = 503$ remainder 4
$2519 \div 6 = 419$ remainder 5
$2519 \div 7 = 359$ remainder 6
$2519 \div 8 = 314$ remainder 7
$2519 \div 9 = 279$ remainder 8
$2519 \div 10 = 251$ remainder 9

We notice that when 2519 is divided by 2 remainder is 1, when divided by 3 remainder is 2, when divided by 4 remainder is 3, when divided by 5 remainder is 4, when divided by 6 remainder is 5, when divided by 7 remainder is 6, when divided by 8 remainder is 7, and when divided by 9 remainder is 8.

In fact 2519 is not only such number. Others exist (try to find some). However this is the smallest number exhibiting this property.

(xi) **Numbers whose product remains unchanged when digits are reversed**

It can be checked that product of 203313 and 657624 is same as the product of 313302 with 426756.

Are there more such numbers?

(xii) **Numbers that can be expressed as sums of consecutive integers**

For example $3 = 2 + 1$

$5 = 2 + 3$

$6 = 1 + 2 + 3$

$7 = 3 + 4$

$9 = 5 + 4$

$10 = 1 + 2 + 3 + 4$

$11 = 5 + 6$

$12 = 3 + 4 + 5$

$13 = 6 + 7$

$14 = 2 + 3 + 4 + 5$

$15 = 4 + 5 + 6$ (or $7 + 8$)

$17 = 8 + 9$

$18 = 5 + 6 + 7$

$19 = 9 + 10$

$20 = 2 + 3 + 4 + 5 + 6$

and so on.

If we continue like this we observe a pattern.

For instance numbers which are multiples of 3 and are of type $3n$ can be expressed as

$$3n = (n - 1) + (n) + (n + 1)$$

For example $30 = 9 + 10 + 11$. Also $30 = 6 + 7 + 8 + 9$

However numbers such as 2, 4, 8, 16 (which are powers of 2) can not be expressed in this form.

Try to discover if there are more such patterns.

(xiii) **Numbers which multiplied with an integer from 2 to 9, yield reversal of digits**

1089 when multiplied with 9 yields 9801 which is reversal of digits of the original number 1089. Similarly 2178 when multiplied with 4 yields 8712 which is again reversal of digits of 2178. These are the smallest four digit numbers which show this property. Infact no two or three digit numbers exist which exhibit this property. Numbers having digits more than four exist which have this property. (Try to find some).

3. Strange Cancellations

It is well known that in the case of a fraction a factor which is common to both the numerator and the denominator can be cancelled.

For example

$$\frac{128}{176} = \frac{16 \times 8}{16 \times 11} = \frac{8}{11}$$

However certain cancellations from numerator and denominator which are technically incorrect sometimes yield correct answers. Some of these we present here.

$$\frac{2\not{6}}{\not{6}5} = \frac{2}{5} \text{ which is correct as } \frac{26}{65} = \frac{13 \times 2}{13 \times 5} = \frac{2}{5}$$

Similarly $\frac{1\not{6}}{\not{6}4} = \frac{1}{4}$ is also correct as $\frac{16}{64} = \frac{16 \times 1}{16 \times 4} = \frac{1}{4}$

Again $\frac{499}{998} = \frac{4}{8} = \frac{1}{2}$ is correct as $\frac{499}{998} = \frac{499 \times 1}{499 \times 2} = \frac{1}{2}$

Also $\frac{484}{847} = \frac{4}{7}$ is true because $\frac{484}{847} = \frac{121 \times 4}{121 \times 7} = \frac{4}{7}$

Even $\frac{1428571}{4285713} = \frac{1}{3}$ is true because $\frac{1428571}{4285713} = \frac{1428571 \times 1}{1428571 \times 3} = \frac{1}{3}$

However it will not always be true. For example

$\frac{73}{67} = \frac{3}{6} = \frac{1}{2}$ is not correct as one can easily see.

In fact conditions have been established under which such invalid cancellations will yield correct results. For instance whenever x, y, z satisfy the relation

$$z = \frac{10xy}{(9x + y)} \text{ cancellation } \frac{xy}{yz} = \frac{x}{z} \text{ will be correct.}$$

4. Generating all Natural Numbers from a Given Natural Number

It is possible to generate all natural numbers from a given natural number using basic arithmetic operations of addition, subtraction, multiplication and division.

Consider for example the number 4. We can use it to generate other natural numbers. We present here few for illustration

$$4 - 4 = 0, \frac{4}{4} = 1, \frac{4}{4} + \frac{4}{4} = 2, \frac{4}{4} + \frac{4}{4} + \frac{4}{4} = 3, 4 + \frac{4}{4} + \frac{4}{4} = 6$$

$$4 \times 4 - \frac{4}{4} = 15 \text{ and so on.}$$

To construct 85 we may write $4 \times 4 \times 4 + 4 \times 4 + 4 + \frac{4}{4}$

It may be noted that these representations are not unique.

For example 3 could also have been generated using $4 - \frac{4}{4}$

Reader can try and generate more such numbers. The procedure becomes still more simple if one is permitted to use other arithmetic operations such as square root. For instance to generate 14 we could have used $4 \times 4 - \sqrt{4}$.

5. Numbers with Interesting Properties

Some numbers show interesting behaviour. We present here some of these numbers.

Number 1089

Start with any three digit number whose unit and hundred digits are not same and perform the following operations in succession in the sequence prescribed.

(i) Reverse its digits

(ii) Subtract smaller from the larger

(iii) Again reverse the digits

(iv) Add last two numbers

We will always end up with the number 1089.

Example: Suppose we chose number 825, then

(i) On reversing its digits we get 528.

(ii) Subtracting smaller from the larger we get

$$825 - 528 = 297$$

(iii) Reversing its digits we get 792

(iv) Adding numbers obtained in (ii) and (iii) we get

$$297 + 792 = 1089$$

Start with any other three digit number you will get same answer.

Number 1089 has another interesting property called **reversal property**. Consider its multiplication with numbers from 1 to 9.

$1089 \times 1 = 1089$ $\quad$ $1089 \times 6 = 6534$

$1089 \times 2 = 2178$ $\quad$ $1089 \times 7 = 7623$

$1089 \times 3 = 3267$ $\quad$ $1089 \times 8 = 8712$

$1089 \times 4 = 4356$ $\quad$ $1089 \times 9 = 9801$

$1089 \times 5 = 5445$

It will be of interest to note that multiple of 9 is reverse of multiple of 1, multiple of 8 is reverse of multiple of 2, multiple of 7 is reverse of multiple of 3 and multiple of 6 is reverse of multiple of 4.

Similar results hold for numbers 10989, 109989, 1099989 and so on which are obtained by inserting one, two, three nines between 10 and 89.

Another number which shows reversal property when multiplied with an integer is 2178.

For example $2178 \times 4 = 8712$. However unlike 1089 multiple of 9 is not reverse of multiple of 1, multiple of 8 is not reversal of multiple of 2, etc.

Are there some more numbers having such reversal properties? (Try to find).

Number 142857

Number 142857 has some interesting properties. If we multiply it with 1, 2, 3, 4, 5 or 6, digits in the number change cyclically

$142857 \times 1 = 142857$ $\quad$ $142857 \times 4 = 571428$

$142857 \times 2 = 285714$ $\quad$ $142857 \times 5 = 714285$

$142857 \times 3 = 428571$ $\quad$ $142857 \times 6 = 857142$

and that is not all. See what happens if these obtained numbers are each first multiplied with 7 and then divided by 9

$$142857 \times 7 = 999999/9 = 111111$$

$$285714 \times 7 = 1999998/9 = 222222$$

$$428571 \times 7 = 2999997/9 = 333333$$
$$571428 \times 7 = 3999996/9 = 444444$$
$$714285 \times 7 = 4999995/9 = 555555$$
$$857142 \times 7 = 5999994/9 = 666666$$

Indeed quite an interesting behaviour!

Number 2519

As mentioned earlier in 2(x) number 2519 has the property that when it is divided by n(2, 3, 4, 5, 6, 7, 8, 9 or 10) remainder is $n - 1$.

Number 3435

As mentioned earlier in 2 (viii) number 3435 has another type of interesting property

$$3435 = 33 + 44 + 33 + 55$$

Are there some other numbers having such a property?

6. Special Types of Numbers

In this section we present certain types of numbers which have some special properties.

6.1 Cyclic Numbers

Numbers created by moving its digits cyclically are called cyclic numbers. For example in the case of number 2387 numbers which can be generated cyclically from it are: 7238, 8723 and 3872. Such numbers have an interesting property about their sum. For instance in the present case sum of digits at each place will be 2 + 3 + 8 + 7 = 20. So in the sum zero will appear at unit place and there will be carryover of two. Since sum at each other place will also be twenty to which a carryover of two will have to be added, so sum of all these cyclic numbers will be 22220. We could also achieve this sum as

$$(2 + 3 + 8 + 7) \times 1111 = 20 \times 1110 = 22220$$

Same approach can be used to find sum of all cyclic numbers generated from a given number.

As mentioned in section 5, number 142857 has additional cyclic properties.

6.2 Palindromic Numbers

Palindrome is a word or a sentence which reads same from both directions. For example RADAR, ROTATOR are palindromic words and MADAM I'M ADAM and STEP NOT ON PETS are palindromic sentences.

We similarly have palindromic numbers which read same both ways. For example 1991 or 2002 were palindromic years (which next year will be palindromic?) Similarly date 20–02–2002 was palindromic.

Interestingly not only 11 but its first four powers are also palindromic

$11^1 = 11$ $11^3 = 1331$

$11^2 = 121$ $11^4 = 14641$

Number 1347007431 is a large palindromic number.

We can generate a palindromic number from a given number by adding to the number a number obtained by reversing its digits once or more than once.

For example $32 + 23 = 55$ which is palindromic

$75 + 57 = 132$ which is not palindromic

However repeating on 132

$132 + 231 = 363$ is palindromic.

Similarly in the case of 86

$86 + 68 = 154$ is not palindromic

again $154 + 451 = 605$ is also not palindromic

but $605 + 506 = 1111$ is palindromic.

In fact it can be checked that whereas by starting with 97 we reach palindromic number in six steps, by starting with 98 we shall reach palindromic number in 24 steps!

Palindromic numbers have a property that each of these numbers is divisible by 11 (why?)

Another interesting observation about palindromic numbers is that numbers that yield palindromic cubes are themselves palindromic. For example cube of 11 is 1331 which is palindromic and so is 11. However cube of every palindromic number need not be palindromic. For instance 363 is a palindromic number but its cube 47832147 is not palindromic.

6.3 Friendly Numbers

Two numbers are considered friendly (or amicable) if sum of the proper divisors of one number equals second number and sum of proper divisors of second number equals first number.

For example 220 and 284 are friendly numbers since proper divisors of 220 are 1, 2, 4, 5, 10, 11, 20, 22, 44, 55 and 110 and their sum is

$$1 + 2 + 4 + 5 + 10 + 11 + 20 + 22 + 44 + 55 + 110 = 284$$

which is the second number

Again proper divisors of 284 are 1, 2, 4, 71 and 142 and their sum is

$$1 + 2 + 4 + 71 + 142 = 220$$ which is the first number

Similarly it can be checked that 17296 and 18416; 1184 and 12105; 2620 and 2942; 5020 and 5564; and 6232 and 6368 are pairs of friendly numbers.

A question naturally arises how may it be checked whether a given pair of numbers is friendly or not without determining proper divisors of the two numbers. It can be done as follows

Let $a = 3 \times 2^n - 1, b = 3 \times 2^{n-1} - 1$, and $c = 3^2 \times 2^{2n-1} - 1$

Where a, b, c are prime numbers and n an integer greater than or equal to 2. (A natural number is prime if it is divisible by only 1 and itself). Then it can be checked that $2^n ab$ and $2^n c$ will be friendly numbers.

Taking $n = 2, a = 3 \times 2^2 - 1 = 11, b = 3 \times 2^1 - 1 = 5$ and $c = 3^2 \times 2^3 - 1 = 71$ are all prime.

Therefore $2^n ab = 2^2 \times 11 \times 5 = 220$ and $2^n c = 2^2 \times 71 = 284$.

We have already seen that 220 and 284 are friendly numbers. It may be of interest to note that in the case of friendly numbers 5965 and 7706, each of these can be expressed as sum of squares of two numbers

$$5965 = 77^2 + 6^2 \text{ and } 7706 = 59^2 + 65^2$$

Check if this property is universal and possessed by every pair of friendly numbers.

It may be of some interest to note that 4522265534545208537974785 and 4539801326233928286140415 are two very large 25 digit friendly numbers!

6.4 Fibonacci Numbers

Fibonacci's manuscript on Hindu-Arabic numerals also contained a problem on regeneration of rabbits. It was the solution of this problem which has produced numbers which are now popularly known as Fibonacci numbers.

The problem was this. **How many rabbits will be produced in a year beginning with a single pair, if each month each pair gives birth to a baby pair which becomes reproductive the very next month. The process is shown below diagrammatically. A pair is denoted by letter A and non-productive baby pair by B.**

Month		Number of reproductive pairs	Baby pair	Total
January	A	1	0	1
February	AB	1	1	2
March	ABA	2	1	3
April	ABAAB	3	2	5
May	ABAABABA	5	3	8
June	ABAABABAABAAB	8	5	13
July	... ...	13	8	21
August	... ...	21	13	34
September	... ...	34	21	55
October	... ...	55	34	89
November	... ...	89	55	144
December	... ...	144	89	233

Number of reproductive pairs yield sequence of numbers 1, 1, 2, 3, 5, 8, 13, 21, 34, 55, 89, 144, 233..... This sequence is known as **Fibonacci sequence** and numbers appearing in this sequence **Fibonacci numbers.**

This sequence has an interesting property. Let F_n denote n^{th} number of this sequence so that $F_0 = 1, F_1 = 1, F_2 = 2, F_3 = 3, F_4 = 5, F_5 = 8, F_6 = 13, F_7 = 21$, etc. Then it may be of interest to observe that

$$F_{n+1} = F_n + F_{n-1} \qquad n = 1, 2, 3, 4, 5, ...$$

Starting with $F_0 = 1, F_1 = 1$, all numbers of Fibonacci sequence can be generated using this recurrence relation.

Another interesting property of these numbers is that ratio $\frac{F_{n+1}}{F_n}$ approaches golden ratio as n becomes larger and larger. Fibonacci numbers now find several applications in different fields of applied mathematics. Fibonacci believed (and this is also observed) that nature prefers Fibonacci numbers. Number of petals in flowers and leaves in plants are generally observed to be some Fibonacci number (check).

6.5 Perfect Numbers

A natural number is called a perfect number if it equals the sum of its perfect divisors. Smallest perfect number is 6. It has 1, 2, 3 as its divisors and 1 + 2 + 3 = 6. Next perfect number is 28 (28 = 1 + 2 + 4 + 7 + 14). After this the next perfect number is 496

$$496 = 1 + 2 + 4 + 8 + 16 + 31 + 62 + 124 + 248$$

and after this it is 8128. These were the first four perfect numbers known to the Greeks. Fifth perfect number is 33,550,336. It appeared for the first time in fifteenth century in a German manuscript.

Pythagoras had defined perfect numbers in 6th century B.C. In fact Greeks had been very much fascinated by perfect numbers. In the book 'Creation of World (III)', the first century Hebrew philosopher Philo Judaes claimed that God created the world in six days (and rested on seventh day, Sunday) as six was the smallest perfect number (In the City of God (XI, 30)) and Augustine agreed with it!

Euclid came up with a theorem to generalize search of perfect numbers. According to this theorem, if $2^k - 1$ is a prime number, then $2^{k-1}(2^k - 1)$ will be a perfect number. Here k can be any natural number.

Using this theorem

k	$2^k - 1$	Perfect Number $2^{k-1}(2^k - 1)$
2	3 (prime)	6
3	7 (prime)	28
4	15 (not prime)	
5	31 (prime)	496
6	63 (not prime)	
7	127 (prime)	8128
8, 9, 10, 11, 12	(not prime)	
13	(prime)	33, 550, 336

Perfect numbers are thus related to prime numbers of type $2^k - 1$ (known as **Mersenne primes**). It may also be of interest to note that perfect numbers after 6 are partial sums of series $1^3 + 3^3 + 5^3 + 7^3 + 9^3 + 11^3$. For example $1^3 + 3^3 + 5^3 + 73^3 = 496$. However it is not easy to show that all perfect numbers are of this type. By third quarter of twentieth century (1978 to be precise) 24 perfect numbers had been found and all of these are even. The largest being $2^{19936}(2^{19937} - 1)$ which has 12003 digits! Can a perfect number be odd? None has been found thus far.

6.6 Composite Numbers

A natural number which is divisible by numbers other than one and itself is called a composite number. On the contrary a number which is not composite (i.e., it is divisible only by itself and one) is called a prime number. For example 42 is a composite number as it is divisible by 2, 3, 6, 7, 21 besides 1 and 42. However 41 is a prime number as it is divisible by 1 and 41 only. In fact prime numbers are regarded as bricks with which entire edifice of natural numbers can be built because every composite number can be expressed in a unique way as product of prime numbers. For example $42 = 2 \times 3 \times 7$. This is in fact known as '**fundamental theorem of arithmetic**'. We shall learn more about prime numbers in the next section.

7. Prime Numbers

As mentioned earlier, a natural number which is divisible only by itself and one is called a prime number. For example 1, 2, 3, 5, 7, 11, 13, 17, 23, 29, 31, 37, 41, ... are prime numbers. Occurrence of prime numbers amongst natural numbers decreases as we go to higher natural numbers. Starting with 1, 2, 3, 5, 7,..., prime number at serial number 3000 is 27449 and prime number at serial number 4000 is 37813 and at serial number 5000 is 48611. In fact it has been calculated that number of primes having digits less than 65 is of the order of 2.12×1063. At the end of 20th century the largest known prime number was $2^{6972593} - 1$.

An interesting fact that has been observed about prime numbers is that if P(N) denotes the prime number which appears at serial number N of primes, then

$$(N \log_6 P(N))/P(N) \to 1 \text{ as } N \to \infty.$$

Surprisingly eminent mathematician Gauss had conjectured that it should be so at the young age of 14 years!

In general it is conjectured that number of primes upto n approaches $n/(\log_6 n)$ as $n \to \infty$. A question naturally arises whether total number of primes is finite or infinite.

Prime numbers have been extensively studied by mathematicians and several of their properties established. For example as mentioned in earlier section every composite number can be uniquely expressed as product of prime numbers leading to prime number being called building blocks of natural numbers. Similarly it has been shown that prime numbers other than 2 and 3 can be expressed either in form $6n + 1$ or $6n - 1$, where n is a positive integer. However converse need not be true. In other words, a natural number which can be expressed in the form $6n + 1$ or $6n - 1$ need not be prime. For instance 25 is $6n + 1$ for $n = 4$ but is not prime.

Some other well known results which have been established in the case of prime numbers are:

(i) If p is a prime number and n is prime to p, then $n^{p-1} - 1$ is a multiple of p (This result is known as **Fermat's theorem**).

(ii) If p is prime then $1 + (p - 1)!$ is divisible by p (This is known as **Wilson's theorem**).

However there are still some unanswered questions and results which have neither been proved nor disproved. For example

(i) Every even number greater than 2 can be expressed as sum of two primes.
Examples: $4 = 2 + 2$, $6 = 3 + 3$, $18 = 7 + 11$, etc.
(This is known as **Goldbach's first conjecture**).

(ii) Every odd number greater than 3 can be expressed as sum of three prime numbers.
Examples: $7 = 2 + 2 + 3$, $17 = 5 + 5 + 7$, $51 = 3 + 17 + 31$, etc.
(This is known as **Goldbach's second conjecture**).

There is also a conjecture due to eminent Indian mathematician Ramanujan on prime numbers.

Interestingly as yet no rational algebraic expression has been found which can exclusively express prime numbers.

7.1 Checking a Number to be Prime

A commonly held belief is that in order to check whether a given natural number n is prime or not one should check if it is divisible by any of the numbers 2, 3, 4, ... upto $n - 1$. This may be alright if the number n is small. The task becomes labourious if n is a large number. However one need not feel disheartened on this account. Mathematicians have shown that in order to check whether n is prime or not one need not check if it is divisible by 2, 3, ... upto $n - 1$ but check its divisibility by 2, 3, ... upto $\sqrt{n}$. For example in order to check whether 103 is prime or not, one need not check its divisibility by 2, 3, ... upto 102 but only by 2, 3, ... upto $\sqrt{103}$, i.e., only divisibility by 2, 3, 4, 5, 6, 7, 8, 9 and 10. Clearly if it is not divisible by 2, it will not be divisible by 4, 6 and 8 and if it is not divisible by 3 it will not be divisible by 9 and if it is not divisible by 5 it will not be divisible by 10. So we need check its divisibility by just 2, 3, 5 and 7 only. Clearly it is not divisible by any of these and so is a prime.

7.2 Generating a Prime Number

Euler had discovered an efficient method to check whether $2^m - 1$ is prime or not (m positive integer). Euler's test is used in computers in search of large prime numbers. In fact computer programs are now available which can be used to generate prime numbers.

7.3 Total Number of Primes

A question that naturally comes to the mind is that whether total number of primes is finite or infinite as their number keeps on diminishing as we search for them amongst higher natural numbers.

As far back as 300 B.C., Euclid had demonstrated that the total number of primes can not be finite. His argument was like this. Suppose total number of primes is finite, then there will be some largest prime number say N_l. This implies that every natural number greater than N_l is composite and has factors. Let $N = (2 \times 3 \times 5 \times 7 \times 11 \times \ldots \times N_l) + 1$. Then N is a number greater than product of all primes upto N_l. Obviously N is a prime number as it is not divisible by any of primes upto N_l but is greater than N_l. This contradicts the assumption that N_l is the largest prime number.

A systematic way to compute the total number of primes in the set $\{1, 2, 3, \ldots, n\}$ is also available. It is known as Inclusion-Exclusion principle. (Interested reader may look up chapter on prime numbers in a book on number theory).

7.4 Number RSA129

A one hundred and twenty nine digits number which is product of two prime numbers is known as RSA129 number. Existence of such a number was announced by computer scientists Ron Rivest Adi Shamir and Leonard Adelman in connection with their work on encryption of messages. They had challenged mathematicians and computer enthusiasts to determine two prime numbers whose product is a 129 digit number which they decided to call number RSA 129.

In 1993, a group of more than 600 computer enthusiasts and academicians used internet to coordinate their efforts for solving this challenge puzzle. It took them more than a year to determine two such prime numbers. One of these has 64 digits and the other 65 digits (This is from the book 'Road Ahead' by Bill Gates). Possibility of using such numbers in development of unbreakable codes for encryption has revived interest in such types of large prime numbers.

7.5 Twin Primes

Pairs of consecutive primes which differ from each other by two are known as twin primes. For example 5, 7; 11, 13; 17, 19; 29, 31; 41, 43; 59, 61 and 71, 73. In fact these are the only 7 pairs of twin primes from 1 to 100. There are again 7 pairs of twin primes from 101 to 200 (check). However it is not always so. From 201 to 300 there are only 4 pairs of twin primes and 2 pairs only between 301 and 400. But again there are 3 pairs of twin primes between 401 and 501. In fact this number keeps on varying.

8. Factorial Loop

We define factorial of a natural number n (written as $\lfloor n$ or $n!$) as a natural number which is product of all natural numbers starting from one upto n. In other words $n! = 1 \cdot 2 \cdot 3 \cdot 4 \cdot ..... \cdot (n-1) \cdot n$.

Thus $1! = 1$, $2! = 1 \cdot 2 = 2$, $3! = 1 \cdot 2 \cdot 3 = 6$, $4! = 1 \cdot 2 \cdot 3 \cdot 4 = 24$ and so on.

It may be noted that $0! = 1$ and not zero.

Let us observe the sum of the factorials of the digits of a natural number, say 145. This is

$$1! + 4! + 5! = 1 + 24 + 120 = 145$$

which is same as number 145. A question naturally arises is it true in some particular cases only or is it true in general? The answer is that it is not always so. For example in the case of 871

$$8! + 7! + 1! = 40320 + 5040 + 1 = 45361 \neq 871$$

However if we repeatedly apply this rule, we ultimately get the number. For instance if we sum factorials of digits of 45361 which was sum of factorials of digits of 871, we get

$$4! + 5! + 3! + 6! + 1! = 24 + 120 + 6 + 720 + 1 = 871$$

Consider another number say 872. In its case

$$8! + 7! + 2! = 40320 + 5040 + 2 = 45362$$
$$4! + 5! + 3! + 6! + 2! = 24 + 120 + 6 + 720 + 2 = 872$$

Again we get back number in two steps. However in the case of 169

$$1! + 6! + 9! = 1 + 720 + 362880 = 363601$$
$$3! + 6! + 3! + 6! + 0! + 1! = 6 + 720 + 6 + 720 + 1 + 1 = 1454$$
$$1! + 4! + 5! + 4! = 1 + 24 + 120 + 24 = 169$$

So we get back 169 in three steps.

A question which naturally arises is that will it be true in case of all numbers? In other words if we add factorials of digits of a given natural number repeatedly shall we get back the number? Answer is no. It is true only for a few numbers. It has been noticed that upto 2,00,000 this property is observed only in the case of a few numbers. Numbers 1, 2, 145, 40585 return back to the number in just one cycle, numbers 871, 872, 45361 and 45362 return back in two cycles, numbers 169, 1454, 363601 return back to the original number after 3 cycles. Try to find some more such numbers.

9. Results Involving Sums of Squares and Cubes

In this section we present some interesting results where squares or cubes of numbers are being used.

9.1 Sums of Squares and Cubes of Numbers

In general it is not always possible to find natural numbers a, b and c such that $a^2 = b^2 + c^2$. It is possible only in some cases. For example $5^2 = 3^2 + 4^2$ but we can not find b and c such that $b^2 + c^2 = 6^2$.

A point to be noted is that if the sum of squares of two natural numbers is a perfect square then their product will be divisible by 12. For example $3^2 + 4^2 = 5^2$ and $3 \times 4 = 12$ which is divisible by 12.

Similarly $6^2 + 8^2 = 36 + 64 = 100 = 10^2$ and $6 \times 8 = 48$ is divisible by 12.

In case of cubes, it is not possible at all to find natural numbers a, b and c such that $a^3 = b^3 + c^3$ $(a \neq b \neq c)$.

However it may be of interest to note that sum of the cubes of the first n integers equals the square of the sum of these integers.

Example: $1^3 + 2^3 = 1 + 8 = 9 = (1 + 2)^2$

$1^3 + 2^3 + 3^3 = 1 + 8 + 27 = 36 = (1 + 2 + 3)^2$

$1^3 + 2^3 + 3^3 + 4^3 = 1 + 8 + 27 + 64 = 100 = (1 + 2 + 3 + 4)^2$

$1^3 + 2^3 + 3^3 + 4^3 + 5^3 = 1 + 8 + 27 + 64 + 125 = 225 = (1 + 2 + 3 + 4 + 5)^2$

and so on. An interesting result. Isn't it?

9.2 Ending in a Loop

Start with any natural number. Add the cubes of its digits. Again add the cubes of the digits of this new number obtained. Keep on repeating this. One will always end up in a loop.

Example 1: Consider the number 352. In this case

$3^3 + 5^3 + 2^3 = 27 + 125 + 8 = 160$

$1^3 + 6^3 + 0^3 = 1 + 216 + 0 = 217$

$2^3 + 1^3 + 7^3 = 8 + 1 + 343 = 352$

and this is same number with which we started.

Example 2: Again consider the number 123.

$1^3 + 2^3 + 3^3 = 1+ 8 + 27 = 36$

$3^3 + 6^3 = 27 + 216 = 243$

$2^3 + 4^3 + 3^3 = 8 + 64 + 27 = 99$

$9^3 + 9^3 = 729 + 729 = 1458$

$1^3 + 4^3 + 5^3 + 8^3 = 1 + 64 + 125 + 512 = 702$

$7^3 + 0^3 + 2^3 = 343 + 0 + 8 = 351$

$3^3 + 5^3 + 1^3 = 27+125 + 1 = 153$

$1^3 + 5^3 + 3^3 = 1 + 125 + 27 = 351$

So we end in a loop between 153 and 351. However it does not involve the number 123 with which we started.

A question that naturally arises is that is this property true for powers other than 3 also. The answer is that it is not so in general.

Consider for example squares of 123

$1^2 + 2^2 + 3^2 =1+ 4 + 9 = 14$

$1^2 + 4^2 = 1 + 16 = 17$

$1^2 + 7^2 = 1 + 49 = 50$

$5^2 + 0^2 = 25 + 0 = 25$

$2^2 + 5^2 = 4 + 25 = 29$

$2^2 + 9^2 = 4 + 81 = 85$

$8^2 + 5^2 = 64 + 25 = 89$

$8^2 + 9^2 = 64 + 81 = 145$

$1^2 + 4^2 + 5^2 =1 + 16 + 25 = 42$

$4^2 + 2^2 =16 + 4 = 20$

$2^2 + 0^2 = 4 + 0 = 4$

$4^2 = 16$

$1^2 + 6^2 = 1 + 36 = 37$

$3^2 + 7^2 = 9 + 49 = 58$, etc.

One can continue as long as one likes, a loop will not be found. However in certain numbers it is true as is shown in next subsection.

9.3 An Interesting Pattern in the Sum of Squares of Digits of Certain Numbers

Consider number 30

$3^2 + 0^2 = 9 + 0 = 9,\ 9^2 = 81,\ 8^2 + 1^2 = 64 + 1 = 65,\ 6^2 + 5^2 = 36 + 25 = 61$

$6^2 + 1^2 = 36 + 1 = 37,\ 3^2 + 7^2 = 9 + 49 = 58,\ 5^2 + 8^2 = 25 + 64 = 89$

$8^2 + 9^2 = 64 + 81 = 145,\ 1^2 + 4^2 + 5^2 = 1 + 16 + 25 = 42,$

$4^2 + 2^2 = 16 + 4 = 20,\ 2^2 + 0^2 = 4,\ 4^2 = 16,\ 1^2 + 6^2 = 37$

It is to be noticed that after this pattern starts repeating.

Similarly in the case of 31

$3^2 + 1^2 = 9 + 1 = 10,\ 1^2 + 0^2 = 1,\ 1^2 = 1$. Now pattern starts repeating.

In the case of 32

$3^2 + 2^2 = 9 + 4 = 13,\ 1^2 + 3^2 = 1 + 9 = 10,\ 1^2 + 0^2 = 1,\ 1^2 = 1$

Pattern starts repeating after this.

In the case of 33

$3^2 + 3^2 = 18,\ 1^2 + 8^2 = 65,\ 6^2 + 5^2 = 61,\ 6^2 + 1^2 = 37$

Now pattern will repeat as happened in the case of 30 when 37 is reached.

In the case of 34

$3^2 + 4^2 = 9 + 16 = 25$, $2^2 + 5^2 = 4 + 25 = 29$, $2^2 + 9^2 = 4 + 81 = 85$,

$8^2 + 5^2 = 64 + 25 = 89$, $8^2 + 9^2 = 64 + 81 = 145$,

$1^2 + 4^2 + 5^2 = 1 + 16 + 25 = 42$, $4^2 + 2^2 = 16 + 4 = 20$, $2^2 + 0^2 = 4$,

$4^2 = 16$, $1^2 + 6^2 = 37$

and now pattern repeats as in the case of 30 when 37 is reached.

In the case of 35

$3^2 + 5^2 = 9 + 25 = 34$

and pattern will repeat after 37 is obtained as happened in case of 34.

In the case of 36

$3^2 + 6^2 = 9 + 36 = 45$ $\quad$ $5^2 + 0^2 = 25 + 0 = 25$

$4^2 + 5^2 = 16 + 25 = 41$ $\quad$ $2^2 + 5^2 = 4 + 25 = 29$

$4^2 + 1^2 = 16 + 1 = 17$ $\quad$ $2^2 + 9^2 = 4 + 81 = 85$

$1^2 + 7^2 = 1 + 49 = 50$

now pattern will repeat as happened in case of 34 after 85 was obtain.

A question naturally arises does it happen for every number between 30 and 40 or for that matter any number from 0 to 40 or still further any natural number. Try for some other numbers. It will be a fun.

9.4 Numbers the Sum of Cubes of whose Digits Equals the Number

In case of certain numbers, the sum of the cubes of their digits equals the number.

Example: $1 = 1^3 = 1$ $\quad$ $371 = 3^3 + 7^3 + 1^3 = 371$

$153 = 1^3 + 5^3 + 3^3 = 153$ $\quad$ $407 = 43 + 03 + 73 = 407$

$370 = 3^3 + 7^3 + 0^3 = 370$

Are there more such numbers? Try.

(In fact these are the only known numbers for which this property holds).

10. Very Large and Very Small Numbers

Numbers can be very big as well as very small. Number of hair on a person's head, number of stars in the sky are very large numbers. However in the structure of an atom we deal with very small numbers where we have to deal with the dimensions and masses of protons and electrons. Exponents are very helpful in expressing very big and very small numbers. For example one million can be expressed as 10^6, one billion 10^{12} and so on. Similarly 10^{-6} is just one millionth part of one (0.000001). Cambridge mathematician S. Skewes while working on prime numbers noted that these become less and less frequent as we consider bigger numbers.

His number known as Skewe's number is $10^{10^{10^{10^{34}}}}$. Indeed a very very big number.

This number despite its rather harmless appearance is actually quite ferocious. It will need billions of notepads to write this number in full in decimal form. Not only numbers in base 10, exponents of numbers with binary base 2 can be even still bigger. For example $2^{199937} - 1$ is a very large number having 6002 digits in decimal form.

In the case of large numbers written in exponent form, last digit is always zero if base is ten and this digit is always an even number if base is two. However if a number is written in some base other than 2 or 10, it may not be always obvious as to what the last digit of the number will be. For example to find the last digit of 1989^{1989} it can not be decided off hand. However it can be determined. For example the last digit of this number will be same as that of 9^{1989}. Now last digits of first few powers of 9 are 9, 1, 9, 1, 9, ... Thus odd powers end with digit 9 and even with digit one. Hence last digit of 1989^{1989} will be nine.

11. Some Well Known Large Number Problems

There are some interesting real life large number problems. We present here some of these.

11.1 Grains of Rice on a Chess Board

Long time ago there lived a king. He was very wealthy and powerful. He had a young beautiful daughter of marriageable age. One day while bathing in a river with her friends, a crocodile caught her leg. She started crying. Her friends ran out of the river in panic. A young man was passing by. He heard her cries for help and jumped into the river. Through his valour he saved the princess from the clutches of certain death at the hands of the crocodile.

When the king came to know of it, he was very much pleased with the youngman and asked him to name his reward which he (king) would gladly grant. The youngman was a shrewed person well versed in numbers. He replied, 'Sir, I am not a greedy person. With your blessings I have what I need in life. However if you insist upon rewarding me, then I just ask for grains of rice that will cover a chess board in such a manner that whereas one grain is placed in the first square, two are placed in the second square, 4 in the third, 8 in fourth, 16 in fifth and so on (in other words number of grains in next square is double of number of grains in the previous square).

The king and his ministers thought that this person was a fool. He had asked for such an insignificant reward. Anyway, king's word is a word. A big size chess board was prepared and some sacks of rice grains brought out from the godown thinking that grains in these sacks would more than suffice. It was noticed that eight squares in the first row needed $1 + 2 + 2^2 + 2^3 + 2^4 + 2^5 + 2^6 + 2^7 = 255$ grains, squares in the second row needed $2^8 + 2^9 + ... + 2^{15} = 65,535$ grains (around 1.45 kg). Grains needed in the third row were $2^{16} + 2^{17} + ... + 2^{23}$ (around 373 kgs). Similarly grains in the fourth row need 95400 kg! By this time the king could visualize the shrewdness of the youngman. The king realized that the entire stock of rice not only in his godowns but even in his entire kingdom would not suffice to cover the chess board the way it was desired by the youngman. Pleased with his ingenuity, the king married his daughter to him and made him the heir of his throne as he had no other issue.

This example illustrates how sometimes things which look small and trivial may not really be so.

11.2 Brahma's Tower in Benaras

Long ago in a temple on the banks of the river Ganges in Benaras, the following problem was posed.

There are three vertical rods of small thickness. Left most rod has 64 rings of varying diameter placed one over the other in such a manner that the largest diameter ring is at the bottom. The rings of ever decreasing diameter are stacked one above the other so that top most ring has the smallest diameter. The other two rods have no rings. The problem is to transfer all these 64 rings in the same pattern on the right most rod so that the ring with the largest diameter is again at the bottom and ring with smallest diameter at the top. Middle rod can be used as a stop gap arrangement. However at no time a ring of bigger diameter is to be put on the top of a ring of smaller diameter in the right most rod.

The answer to the problem is yes it is possible but in $2^{64} - 1$ moves. At the rate of one move per second it will need 585 billion years to finish $2^{64} - 1$ moves! A too long period of time. Life on earth is estimated to be about 10 billion years. Even a superfast computer will take very-very long time to solve it. That is why this problem is nicknamed 'Brahma's tower' as the time needed for its completion is almost equivalent to the life of the universe! So it is only Brahma (the creator of universe) who will be able to complete it when he starts making his next universe!

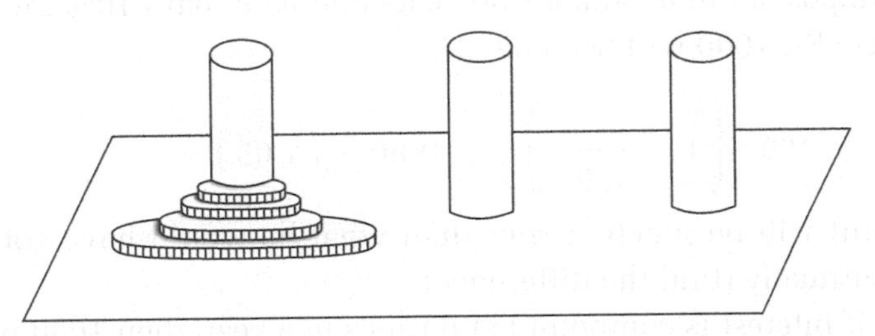

We shall consider in subsequent chapters simpler versions of this problem when the number of discs is less and transfer of discs can be made in a finite number of moves.

11.3 Compound Interest

There are two types of interest which are paid on money deposited in a bank. One is simple interest and the other compound interest. Whereas simple interest is calculated each year on the initial amount of money deposited in the bank irrespective of whether the interest accrued on it in the preceding year is withdrawn or not, the compound interest is computed on the total amount including principal plus interest accrued if interest is not withdrawn. If principal amount P is deposited in a bank for n years at annual rate r % of interest, then at the end of n years it will become

$$A_S = P + \frac{P \times r \times n}{100} = P\left(1 + \frac{nr}{100}\right)$$

if interest is simple and

$$A_C = P\left(1 + \frac{r}{100}\right)^n$$

if interest paid is compound.

It may be of interest to observe how vastly A_S and A_C vary as n (the number of years) becomes large. Suppose initial amount invested is 1000 at 10% rate of interest. Then at the end of five years it will become

$$A_S = 1000\left(1 + \frac{5 \times 10}{100}\right) = 1500$$

at simple interest.

and $$A_C = 1000\left(1 + \frac{10}{100}\right)^5 = 1000 \times (1.1)^5 \approx 1610.51$$

at compound interest.

There being a difference of more than Rs. 110 between the two. Similarly at the end of 50 years it will be only Rs. 6000 if interest is simple and around Rs. 117,400 if interest is compound (a huge difference of around Rs. 111,400!)

Some banks credit interest half yearly (and some even quarterly) to attract customers. If Rs. 1000 is deposited in a bank for 50 years and bank pays 10% interest half yearly then in this case Rs. 1000 will become

$$1000\left(1 + \frac{5}{100}\right)^{100} = 1000 \times (1.05)^{100}$$

This amount will be much greater than what he would have got if interest was compounded annually (find the difference).

In general if interest is compounded n times in a year, then 1000 will become

$$1000\left(1 + \frac{10/n}{100}\right)^n = 1000 \times \left(1 + \frac{0.1}{n}\right)^n \text{ in an year.}$$

This may lead one to think that if interest is credited everyday (or every hour or every minute or every second) one may become even a millionaire within a year by depositing a not very large amount. However this is not so.

We have in mathematics a result according to which

$$\left(1+\frac{x}{n}\right)^n \to e^x \text{ as } n \to \infty, \text{ where } e \text{ is transcendental number } (e \approx 2.7178281).$$

Hence maximum that we can expect by depositing Rs. 1000 for a year at 10% is $1000 \times e^{0.1} \approx 1{,}105.17$ or just Rs. 5.17 more than what we would have got at simple interest.

One is often interested in knowing in how many years money put in a bank will double or triple itself.

In this case we want to find n such that

$$P\left(1+\frac{r}{100}\right)^n = 2P \text{ or } n \text{ for which } \left(1+\frac{r}{100}\right)^n = 2$$

Roughly speaking money will double itself in $72/r$ years. In other words if rate of interest is 6%, it will double itself in 12 years. If rate of interest is 9%, it will double in 8 years. Similarly at rate of interest r it triples itself in $114/r$ years approximately. So at 6% it triples itself in 19 years and at 9% in 13 years.

11.3.1 Population Growth

Population growth is similar to compound interest. Even if population of a country grows at a paltry rate of 2% per annum, its effect on long run is manifold. A population of 100 million will become

$$10^8\left(1+\frac{2}{100}\right)^{50} = 10^8\,(1.02)^{50}$$

which is more than twice (i.e. more than 200 million) in fifty years.

11.4 Teeth Extraction Problem

A person went to a dentist. After examining his teeth, the dentist told the patient that all his teeth needed extraction and a new denture need to be fixed. When the person asked the doctor for his charges for teeth extraction, the doctor told him that there were two schemes and he could choose any one of these two schemes. According to the first scheme, he was to pay Rs. 10,000 lumpsum for extraction of all 32 teeth.

According to the second scheme, he would be charged one paisa for extraction of the first tooth, 2 paisa for extraction of second, 4 for extraction of third and so on (each time twice the amount of previous sitting). Which of the two schemes he should prefer?

At first sight the second scheme might appear a better option but it is not so. The problem is similar to that of the grains of rice on a chessboard. If we analyse cost under scheme two we notice that the cost of extraction doubles itself at every next extraction and as there are 32 teeth to be extracted, the total cost will be

$$1 + 2 + 2^2 + 2^3 + ... + 2^{31} \text{ paisa or } \frac{2^{32} - 1}{100} \text{ rupees.}$$

and this is almost Rs. 43 lacs! This amount is much more than Rs. 10,000 for the first scheme.

Of course if more than one doctor is offering second scheme and minimum number of teeth to be extracted is not specified then the customer can keep on changing dentist every eighth extraction and save a lot. His total cost will be just Rs. 10.20.

12. Certain Unsolved Problems of Number System

Do not imagine that every problem in mathematics gets solved. There still exist several problems even in number system which, even though look simple, have remained unsolved for centuries and some of these are even still unsolved. A problem which has remained unsolved thus far excites one to try and see if one can crack it particularly when the problem looks simple and is easy to understand.

Unsolved problems have excited and tantalized many mathematicians, professionals and even amateurs over the years. There are still some such problems (referred to as conjectures in literature) for which still no satisfactory proofs for or against have been possible. In some cases even with the availability of fast computers, it has still not been possible to find even a counter example. Such results are therefore expected to be true. However mathematicians do not feel satisfied till an unambiguous confirmative proof is provided. Efforts put in by mathematicians to solve such problems have in certain cases lead to some other important discoveries and results which otherwise might have remained hidden.

Even though Fermat's theorem regarding infinity of primes now seems to have been finally established there are still several unsolved problems of number system. However many of these are not easy to understand. We present here some of these which are easy to understand and appreciate. You might feel tempted in trying your hand in solving some of these.

12.1 Goldbach's First Conjecture

Every even number greater than 2 can be expressed as sum of two prime numbers.

Goldbach (1690–1764) was a Prussian mathematician. In 1742 he wrote a letter to famous Swiss mathematician Euler in which he posed this problem. However this simple looking result remains unproved to this day even though no contradiction has been reported.

Below is given a list of some illustrative even numbers expressed as sum of two prime numbers

Even Number	Prime Numbers
4	2 + 2
6	3 + 3
8	3 + 5
10	3 + 7
12	5 + 7
14	7 + 7 or 3 + 11
16	5 + 11
18	5 + 13
20	7 +13
...	...
48	19 + 29
...	...
100	3 + 97 or 11 + 89

12.2 Goldbach's Second Conjecture

Every odd number greater than 5 can be expressed as sum of three prime numbers.
Examples:

Number	Sum of primes	Number	Sum of primes
7	2 + 2 + 3	19	5 + 7 + 7
9	3 + 3 + 3	21	7 + 7 + 7
11	3 + 3 + 5	...	... ...
13	3 + 5 + 5	41	3 + 7 + 31
15	3 + 5 + 7	51	3 + 17 + 31
17	5 + 5 + 7	...	... ...
77	5 + 5 + 67	...	... ...
101	5 + 7 + 89		

12.3 Fermat's Last Theorem

In 1637, Fermat on reading Disphantus's Arithmetica (a monumental third century treatise on arithmetic) noted on a margin of one of its pages, 'Dividing a cube into two cubes or for that matter any n^{th} power of a number into sum of nth powers of two other numbers, for $n > 2$ is impossible. I have found a proof of it but am unable to write for lack of space.' This problem may be stated formally as:

Given z^n, find integers x and y such that $x^n + y^n = z^n$ for $n \geq 3$.

(According to Fermat, this problem has no solution).

A general proof of this had evaded mathematicians of the world till as late as end of twentieth century even though top mathematicians had worked on it. Some of them had been able to establish its correctness in some special cases. Euler could establish it in 1753 for $n = 3$. Dirichlet and Legander confirmed it for $n = 7$ in 1825. Lane confirmed its truth for all n less than 100 in 1839. By 1980 it had been verified for every n upto 125000. However a general proof of it was still missing. It was in 1993 that Andrew Wile, a Princeton University professor of mathematics could manage a part of its general proof. For this he was given Wolf's prize. It was only in 1999 that Brian Conard, Richard Taylor, Christopher Beruil and Fred Diamond could finally complete Wile's work. On the hind sight sometimes one wonders whether Fermat really had a proof for it or was it just a conjecture.

Mind Teasers

1. Can we use each of the digits 0, 1, 2, 3, 4, 5, 6, 7, 8 and 9 just once to create integers whose sum is one hundred?

   $$(\text{Here is an example } 4 + 5 + 9 + 13 + \frac{72}{8} + 60).$$

   Try to find some more such combinations. Can it be done without using a fraction?

2. Construct the following numbers using 4's only four times:

   (i) 516 (ii) 614 (iii) 3624

   What is the largest number which can be constructed this way?

3. Construct numbers of problem 2 using (i) 5 only, (ii) 3 only a suitable number of times.

4. Choose any number however large. Reverse its digits and subtract it from the original number. Check that remainder is divisible by 9. Can you reason why it is so?

5. Check that any prime number divided by 30 always yields remainder which is a prime number or 1.

6. Can the sum of squares of three consecutive numbers be a perfect square?

7. Show that a number written with hundred zeros, hundred ones and hundred two's can not be a perfect square.

8. A certain number k is a multiple of 9. Add all its digits. If the resulting number has still more than one digit, again add these. Continue till it has one digit. What will this digit be?

9. Find distinct integers m and n (in case these exist) such that their sum equals their product.
10. Determine how many zeros 100! has at its end.
11. Does there exist a natural number n such that $n!$ has exactly 5 zeros at the end of its decimal representation?
12. How many zeros are there at the end of (i) 200! (ii) 2^{300}, 5^{600}, 4^{400}
13. What is the last digit of 3^{4798}?
14. What is the last digit of (i) 7^{77} (ii) $(777)^{777}$?
15. Which is greater: (i) 2^{300} or 3^{200} (ii) 50^{99} or 99!
16. Show that $n^3 + 2n$ is always divisible by 3 for all integer values of n.
17. Do there exist natural numbers a and b such that $ab\,(a - b) = 45045$? If yes, find these.
18. Denote the sum of three consecutive numbers by a. The sum of next three consecutive numbers b. Can their product ab be 11,11,11,11?
19. Suppose a muslin napkin which is 1 mm thick is folded half of its existing size repeatedly 32 times. How much thick will it become at the end?
20. World's population crossed 6 billion in 1999. Suppose it is increasing annually at the rate of 2%. What will be world's population in 2020? When will it reach a stage when each individual will just get 1 m^2 space to exist (assume that habitable land area of earth is 150 million km^2)?

Chapter 3

Numerical Computations

1. Introduction

As human civilization progressed and trade and commerce grew, it led to increase in the use of arithmetical computations. A necessity was therefore felt of developing techniques and computational aids which could help in performing these computations fast. In this chapter we review some of these computing techniques and computing aids which have been developed over the ages.

2. Speed Computations

Most of us are familiar with rules commonly used for performing basic arithmetic operations of addition, subtraction, multiplication, division, squaring, cubing and finding square root, etc. However in some special cases these operations can be speeded up. We present in this section some such techniques.

2.1 Clever Addition

It is said that a well known mathematician Carl Friedrich Gauss while at school could mentally add 100 big numbers and write nothing but answer whereas his other classmates worked for a long time on paper to arrive at the answer. Suppose we are to add the following one hundred numbers

$$81297 + 81495 + 81693 \ldots\ldots + 100701 + 100899$$

At first sight it looks a formidable task. However a closer look at the series to be summed shows that this series has 100 terms, the first being 81297 and the last 100899. Moreover successive terms of this series have a common difference of 198. What Gauss used to do in summing such a series was to add the first and the last term which yields $(81297 + 100899 = 182196)$, then second term and the second from the last term which yields $(81495 + 100701 = 182196)$ and so on. Each pair yielding the same sum. There being 50 such pairs, the result obviously is

$$182196 \times 50 = 9109800$$

The result is obtained fast. If we start adding these hundred numbers in the normal way it will take unduly long time.

In fact if we want to add a sequence of numbers whose consecutive terms differ from each other by a fixed number, then the sum of such a sequence is just half the number of terms multiplied with the sum of the first and the last terms. Similar rules for faster addition are available for summing a sequence of terms whose each succeeding term is a constant multiple of its preceding term. We learn about such techniques in summation of series in chapter on algebra.

2.2 Faster Multiplication

In certain cases multiplication can also be speeded up. We present in this sub section some such techniques.

Multiplication with 11: There is a simple rule to multiply a two digit number with eleven. The rule is: add the two digits of the number to be multiplied with eleven and place this sum between the two digits of the number. For example

$$45 \times 11 = 4\ (4 + 5)\ 5 = 495$$

Similarly $$87 \times 11 = 8\ (8 + 7)\ 7 = 957$$

It can also be used in succession when there are more than two digits in the number.

For example $$6789 \times 11 = 6\ (6 + 7)\ (7 + 8)\ (8 + 9)\ 9 = 6\ (13)\ (15)\ (17)\ 9 = 74679$$
(making carryovers when sum exceeds nine)

An alternative rule is to put a zero on the right of the number and add the number to it.

For example $$387 \times 11 = 3870 + 387 = 4257$$

The reason for validity of this rule is obvious. It is essentially multiplying the number first with ten and then adding the number to it to make multiplier 11.

Multiplication with 9: To multiply a number with 9, write zero to the right of the number and subtract the number from it. For example

$$387 \times 9 = 3870 - 387 = 3483$$

Multiplication with 21: For multiplying a number with 21, double the number, write a zero to its right and add the original number

Example $$184 \times 21 = 3680 + 184 = 3864$$

Multiplication with 31: Multiply the number with three, write a zero to its right and add the original number.

Example $$184 \times 31 = (184 \times 3)\ 0 + 184 = 5520 + 184 = 5704$$

Justification of these rules is obvious. We can make similar rules for multiplication with 41, 51, 61, as well as 19, 29, 39, ...

Multiplication with 99: To multiply a number with 99 append two zeroes to its right and subtract the original number from it.

The reason is obvious. However in certain cases a more simple and interesting rule is also possible. For example **to multiply 21 digit number 112359550561797752809 with 99 simply append one at both ends**. This is the smallest number which displays this curious property.

2.3 Lattice Multiplication

Multiplication of two numbers containing several digits can be conventionally performed in lattice form as under

(i) Form a table (lattice) having as many rows as the number of digits in one number and as many columns as number of digits in the other number. Now draw in each cell diagonal lines starting from top right and moving to bottom left so that each cell of the table gets divided into two halves as shown in the figure below.

(ii) Now write digits of one number on the left of the table and that of the other on top of the table so that only one digit appears along each left most cell and one digit at the top of each topmost cell.

(iii) Now in each cell multiply the digit against its row with digit at the top of table, writing unit part of the product in lower half and tens part in upper half of appropriate cell (product of digit in i^{th} row with digit at top of jth column is to be written in cell (i, j).

(iv) Now add the entries in the cells diagonal wise starting from right and moving to left. Writing sum at bottom of the table, (carryovers if any of sum being greater than 9 are to be added to sum of entries of the next diagonal).

The product of two numbers is the number we get at the bottom of the table.

We illustrate the method by multiplying 2348 with 289. We choose 3 rows for the second number and 4 columns for the first number. Carryovers are shown circled at top of next diagonal.

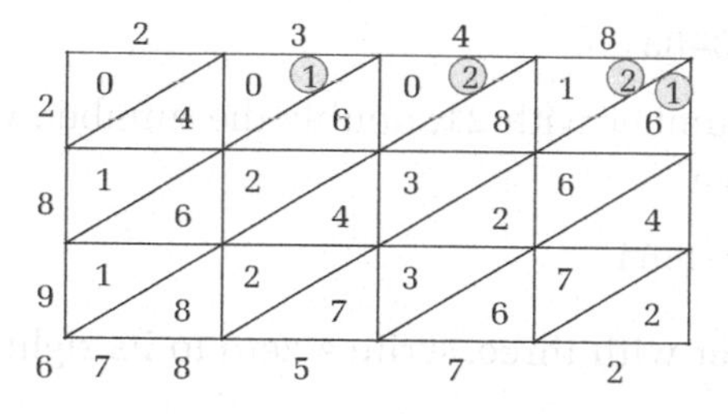

So, $2348 \times 289 = 678572$.
It can be verified that it is indeed so.

2.4 Russian Method of Multiplication

It is said that in Russia peasants used a special method to multiply two numbers. The method is somewhat similar to the method of multiplication of two numbers using multiplication and division with 2 only as given in chapter one (section 12.2). The method is as follows.

Write the two numbers to be multiplied in two separate columns. Keep on doubling one number and halving the other (keeping only integral part) till the number being halved reduces to one. Now add the numbers in the other column which correspond to odd numbers of this column. This will be the result of the product of two numbers.

Example: Multiply 43×92

Column One (double)	Column Two (halve)
43	92
86	46
172*	23
344*	11
688*	5
1376	2
2752*	1

We get the same result even if we interchange entries.

Column One (double)	Column Two (halve)
92*	43
184*	21
368	10
736*	5
1472	2
2944*	1

Product = 92 + 184 + 736 + 2944 = 3956

The product is

172 + 344 + 688 + 2752 = 3956

Working of the method can be explained using binary representation.

Writing in binary form one of the two terms

$$43 \times 92 = (1 \times 2^5 + 0 \times 2^4 + 1 \times 2^3 + 1 \times 2^1 + 1 \times 2^0) \times 92$$
$$= 92 \times 2^5 + 92 \times 2^3 + 92 \times 2^1 + 92 \times 2^06$$
$$= 2944 + 736 + 184 + 92 = 3956$$

Clearly non-zero terms only arise corresponding to odd powers of 2 of number written in binary format.

2.5 Checking Divisibility without Performing Division

We are familiar with formal method of division. Sometimes we are interested in checking whether a given number is divisible by another given number or not before performing actual division process. In this sub-section we present some such methods:

(i) **Divisibility by 2:** A number is divisible by 2 if its last digit is even. Reason is obvious.

(ii) **Divisibility by 3:** A number is divisible by 3 if the sum of its digits is divisible by 3.

Example: 296357: Sum of the digits is 2 + 9 + 6 + 3 + 5 + 7 = 32 which is not divisible by 3. So the number is not divisible by 3.

However in the case of number 457875, the sum of the digits is 4 + 5 + 7 + 8 + 7 + 5 = 36 which is divisible by 3. So 457875 is divisible by 3.

(iii) **Divisibility by 5:** A number is divisible by 5 if its last digit is either zero or five. Reason is obvious.

(iv) **Divisibility by 7:** To check divisibility of a number by 7 delete the last digit of the number and then subtract twice of this deleted digit from the remaining number. If the remainder is divisible by 7, then so is the original number (In case the remainder is too large, the process can be repeated more than once).

Example: Is 876547 divisible by 7?

Applying the rule: 876547 : 87654 – 14 = 87640,

8764 : 876 – 8 = 868

868 : 86 – 16 = 70

and 70 is divisible by 7.

Hence original number 876547 is also divisible by 7 (check).

The reason why this rule works is that we are subtracting in each case a number which is divisible by 7 as shown below:

Terminal digit dropped	Number subtracted from original number	Divisibility of subtracted number by 7
1	20 + 1 = 21	Yes
2	40 + 2 = 42	Yes
3	60 + 3 = 63	Yes
4	80 + 4 = 84	Yes
5	100 + 5 = 105	Yes
6	120 + 6 = 126	Yes
7	140 + 7 = 147	Yes
8	160 + 8 = 168	Yes
9	180 + 9 = 189	Yes

So essentially in each case we are taking out bundles of multiples of 7 from the original number. Hence the original number will be divisible by 7 if final remainder is divisible by 7.

Similar rules are available for checking divisibility by 13 and 17.

Divisibility by 13: In this case delete the last digit and subtract 9 times the deleted digit from the remaining number.

Example: Is 5616 divisible by 13?

5616 : 561 – 6 × 9 = 561 – 54 = 507

507 : 50 – 7 × 9 = 50 – 63 = –13 which is divisible by 13.

So 5616 is divisible by 13.

Divisibility by 17: Delete the extreme digit on right and subtract 5 times the deleted digit from the left over number. Number will be divisible by 17 if remainder is divisible by 17.

Example: Is 5616 divisible by 17?

$5616 : 561 - 5 \times 6 = 561 - 30 = 531$

$531 : 53 - 5 \times 1 = 53 - 5 = 48$ which is not divisible by 17.

So original number 5616 is not divisible by 17.

Next consider divisibility of 56168 by 17.

In this case $56168 : 5616 - 5 \times 8 = 5616 - 40 = 5576$

$5576 : 557 - 5 \times 6 = 557 - 30 = 527$

$527 : 52 - 5 \times 7 = 52 - 38 = 17$ divisible by 17.

Hence, 56168 is divisible by 17.

Divisibility by 11: A number is divisible by 11 if the difference of the sum of its alternate digits is divisible by 11.

Example: Is 918082 divisible by 11?

In this case difference of sum of its alternate digits is

$(9 + 8 + 8) - (1 + 0 + 2) = 25 - 3 = 22$ which is divisible by 11.

Hence, 918082 is divisible by 11.

However in the case of number 86245

Difference of sum of its alternate digits is

$(8 + 2 + 5) - (6 + 4) = 15 - 10 = 5$ which is not divisible by 11.

So 86245 is not divisible by 11.

Divisibility by 9: Rule for checking divisibility by 9 is similar to the rule for checking divisibility by 3. If the sum of the digits of the number is divisible by 9, then the number is divisible by 9.

Example: Is 296357 divisible by 9?

In this case sum of digits is

$2 + 9 + 6 + 3 + 5 + 7 = 32$ which is not divisible by 9.

So number is not divisible by 9.

Again in the case of number 457875, sum of its digits is

$4 + 5 + 7 + 8 + 7 + 5 = 36$ which is divisible by 9.

So, 457875 is divisible by 9.

Some More Tests for Checking Divisibility by 7, 11 and 13.

(i) Every six digit number comprising of a 3 digit repeated sequence of the type (*abcabc*) is divisible by 7, 11 as well as 13.

Example: 647647 is of type *abcabc*. It can be checked that it is divisible by 7, 11 as well as 13.

(ii) Every palindromic number having even number of digits is divisible by 11.

Example: 14641 is a palindromic number. It can be checked that it is divisible by 11.

Divisibility by 4 and 8

To check whether a given large number is divisible by 8 or not just check if number formed by its digits at hundred, ten and unit place is divisible by 8 or not. Similarly, to check if a large number is divisible by 4 or not just check divisibility by 4 of number formed by its digits at tens and units places.

Example: Is 8765168 divisible by 8?

Since 168 is divisible by 8 so 8765168 is also divisible by 8. It is also divisible by 4 as 68 is divisible by 4.

However 8765924 is not divisible by 8 as 924 is not divisible by 8. But it is divisible by 4 as 24 is divisible by 4.

Divisibility by a Composite Number

A given number will be divisible by a composite number if it is divisible by each of its relatively prime factors. Thus for a number to be divisible by 4, it must be divisible by 2, to be divisible by 6 it must be divisible by 2 and 3, to be divisible by 8 it must be divisible by 2 and 4, to be divisible by 26 it must be divisible by 2 and 13 and so on.

Example: Is 647647 divisible by 91?

Now $91 = 13 \times 7$ and as seen above 647647 is divisible by both 7 and 13. Hence 647647 is divisible by 91.

2.6 Checking Whether a given Number is a Perfect Square or Not

We are often required to find the square root of a given number. Standard methods for extracting square root of a number are available. However before starting to find the square root of a number we often want to check whether it is a perfect square or not because exact square root of a number can only be obtained if it is a perfect square.

We present here two simple guidelines which can help in deciding whether a given number is a perfect square or not.

(i) Since the squares of first nine natural numbers 1, 2, 3, 4, 5, 6, 7, 8 and 9 are 1, 4, 9, 16, 25, 36, 49, 64 and 81 respectively, a number ending in digits 2, 3, 7, or 8 can not be a perfect square.

(ii) When a number on division by 4 leaves remainder 2 or 3, it can not be a perfect square. However if the remainder is 1 or 0 it may or may not be. For example 1274 when divided by 4 leaves remainder 2. It can be checked that it is not a perfect square. Similarly number 8761 when divided by 4 leaves remainder 1 but is still not a perfect square. On the contrary 81 when divided by 4 leaves remainder 1 and is a perfect square.

2.7 Checking Correctness of Performed Multiplication or Division Operation

Fibonacci in his book Liberabacci (in which he also introduced Fibonacci numbers) proposed a method named 'casting out bundles of 9' to check whether the performed operation of multiplication carried out on two large numbers is correct or not. The procedure consists in taking out bundles of 9 from the sums of digits of the two numbers being multiplied and their product. For example suppose we want to check whether the product

$734 \times 879 = 645186$ is correct or not.
In the case of 734 the sum of its digits is $7 + 3 + 4 = 14$
Summing the digits of this sum we get $1 + 4 = 5$.
Next in the case of 879 the sum of its digits is $8 + 7 + 9 = 24$
which when again summed yields $2 + 4 = 6$.
Now product of $5 \times 6 = 30$.
Casting out bundles of 9 from this yields $30 - 3 \times 9 = 3$.

Now in case of the product number 645186, sum of its digits is $6 + 4 + 5 + 1 + 8 + 6 = 30$, which when summed again yields $3 + 0 = 3$. This is same as result obtained after casting out bundles of 9 from the numbers being multiplied. Hence answer must be correct (check).

Same procedure can be used for checking the correctness of performed division operation. For example to check whether $645186 \div 879 = 734$ is correct or not, we may rewrite it as $734 \times 879 = 645186$ and check the correctness of this multiplication.

Caution. Suppose we claim that

$734 \times 879 = 654168$ and want to check if product is correct or not.

Here since numbers being multiplied are same as earlier so result after casting out bundles of 9 from the sum of digits of multiplicands is again 3.

Now in case of the product 654168, sum of its digits is $6 + 5 + 4 + 1 + 6 + 8 = 30$ which when again summed yields 3. So the answer of 654168 of the product again seems to be correct but this is not. In fact any cyclic interchange of digits in the product will yield same value 3. Thus the test alone may not suffice. This test is a necessary condition but not sufficient.

3. Vedic Mathematics

Over the ages different civilizations developed and perfected their own techniques for performing arithmetic operations in an efficient manner. We have already seen Russian peasants method of multiplication. In the past Aryans in India of Vedic times also perfected methods for efficient performance of basic arithmetic operations. These appear at different places in the Vedic literature and are therefore often referred to as 'Vedic Mathematics'. Some of these techniques are totally unconventional and difficult to understand on first reading.

There are four Vedas: Rigveda, Samveda, Yajurveda and Atharvaveda. These are the earliest books known to human civilization. Besides these four Vedas there are also four Upavedas and six Vedangas. All this forms vedic literature. The main techniques of arithmetic operations appear in the form of sixteen 'sutras'. These are not located at a single place but appear scattered at different places in vedic literature. Moreover none of these appears in the basic four Vedas (that is why some persons have reservation in calling these techniques as 'vedic mathematics'). These sutras appear in 'Parisista' (Appendix portion) of Atharvaveda. These occur there as aids for fast mental calculations of highly intricate and complex arithmetic operations. It must be kept in mind that in the ancient past during vedic period rishi munis depended more on oral communication. In fact for generations Vedas had been communicated from guru to shishyas orally. The pundits who were well versed in all the four Vedas were called 'Chaturvedis', those who were well versed in three Vedas were called 'Trivedis' and those who were well versed in two Vedas were called 'Dwivedis'. Since there was very little dependence on written literature, efforts were made to communicate the knowledge in an easy to remember compact form. This resulted in formulation of easy to remember compact sutras in verse form.

Jagadguru Swami Shri Bharti Krishna Tirathji Maharaj has compiled these computational sutras at one place in his well known book on Vedic Mathematics which has been published by Motilal Banarsi Das Publishers, Delhi in 1992 (It has been reprinted several times since then). The subject of late has become popular especially amongst school going children and teachers and as a result several more books have appeared on the subject.

3.1 Vedic Numeral Code

In Vedic times knowledge was preserved and transmitted from Guru to Shishya orally. Therefore it was necessary that rules framed for arithmetic operations be compact and in an easy to learn format. For this a numeral code was developed. This Vedic numeral code associated specific devenagri script letters with numbers. This code has been frequently used in the past to convey rules of arithmetic operations and numerical results in easy to remember verse form. Roughly the code is as under

	ka (क), ta (त), pa (प) and ya (य)	all denote numeral 1
	kha (ख), tha (थ), pha (फ) and ra (र)	each stand for numeral 2
	ga (ग), da (द), ba (ब) and la (ल)	each stand for numeral 3
	gha (घ), dha (ध), bha (भ) and va (व)	each stand for numeral 4
	gna (यं), na (न), ma (म) and sa (स)	each stand for numeral 5
	ca (क), ta (ता), and sa (स)	each stand for numeral 6
	cha (च), tha (थ), and saa (सा)	each stand for numeral 7
	ja (ज), da (द), and ha (ह)	each stand for numeral 8
	jha (झ) and dha (ध)	stand for numeral 9
and	ksa (क्ष)	stands for numeral 0

Thus papa is 11, mama is 55, tata is again 11 and mara 52 etc. Vowels are not included. It is generally left to the writer to select appropriate consonant or vowel which he prefers. Poets generally availed of this latitude to frame the numerical data and desired arithmetic operations to be formed in easy to remember verse form. For example मुझे मामा चाहिए (I want mama) will mean I want 55. मुझे पापा चाहिए (I want papa) will mean I want 11. Basic sutras of Vedic Mathematics make use of these codes.

4. Arithmetic Computations in Vedic Mathematics

Following the book on vedic mathematics of Jagadguru Swami Shri Bharti Krishna Tirathji Maharaj, we present in this section some techniques of arithmetic computations based on the basic sutras of Vedic Mathematics. For more details the interested reader can refer to the original book. Most of these techniques are situation specific and that is why we often have more than one rule for the same arithmetic operation, each being effective in a specific situation. There is no doubt that in situations where applicable, most of these techniques are superior to the conventional methods but these are unlikely to prove effective in situations for which these are not intended. However with the advent of computers such fast computational techniques now do not have that much significance and attraction as these used to have in the earlier days.

4.1 Multiplication

Multiplication rule is based on a sutra which reads निखिलं नवतश्चरम दशतः. This literally means 'all from nine and the last from 10.'

The rule is as follows: (i) Select as the base of calculations that power of 10 which is nearest to the numbers to be multiplied, (ii) subtract each of the numbers from this base and write down the remainder along side on the right with a minus connecting symbol '-' in between.

The product will have two parts one on the left side and the other on right side. Left side digit can be arrived at in any of the four ways, each yielding same result. Two of these are: (a) Subtract the chosen base from the sum of the numbers on the left, or (b) subtract the sum of deficiencies of the two from the base. This is the left part of product. To obtain the right part of the product multiply the deficit figures.

1. 9×7 Choosing base 10 and writing

$$\begin{array}{l} \qquad\qquad 9 - 1 \\ \qquad\qquad 7 - 3 \\ \hline (9 + 7) - 10 \,/\, 1 \times 3 \\ \text{i.e.} \qquad\quad 6 \,/\, 3 \end{array}$$

So the answer of the product is 63.

2. 93×97

Choosing base 100 (this is that power of 10 which is nearest to the numbers to the multiplied), we write these numbers as

$$\begin{array}{r}93 - 7\\97 - 3\\\hline (93 + 97) - 100 \ / \ 7 \times 3\\ \text{i.e.} \quad 90 \ / \ 21\end{array}$$

So the answer of the product is 9021.

A difficulty arises when product of deficit digits yields a digit of the same order as the base. In such a case surplus is to be carried over to the left portion as is done in usual multiplications

3. 7×6 Choosing base 10 we write

$$\begin{array}{r}7 - 3\\6 - 4\\\hline \text{we have} \quad 13 - 10 \ / \ 3 \times 4\\ \text{i.e.} \quad 3 \ / \ 12\\ \text{or} \quad 3 + 1 = 4 \ / \ 2\end{array}$$

So the answer of the product is 42.

4. 25×98 Choosing base 100 we write

$$\begin{array}{r}25 - 75\\98 - 02\\\hline \text{we have} \quad 123 - 100 \ / \ 75 \times 2\\ \text{i.e.} \quad 23 \ / \ 150\\ \text{or} \quad 23 + 1 = 24 \ / \ 50\\ 24 \ / \ 50\end{array}$$

So the answer of the product is 2450.

(*Note: We have to keep only as many digits in the right hand side of answer as are there in the deficits*)

5. 112×998 Choosing base 1000

$$\begin{array}{r}112 - 888\\998 - 002\\\hline \text{we have} \quad 1110 - 1000 \ / \ 888 \times 2\\ \text{i.e.} \quad 110 \ / \ 1776\\ \text{or} \quad 110 + 1 = 111 \ / \ 776\\ 111 \ / \ 776\end{array}$$

So the answer of the product is 111776.

When the multiplicand and the multiplier are little over a certain power of 10.
In such cases we follow similar procedure using excess in place of deficit.

Example: Multiply 111 × 109 Choosing base 100 we have

$$\begin{array}{r|l} & 111 + 11 \\ & 109 + 9 \\ \hline 220 - 100 & 11 \times 9 \\ \text{i.e. } 120 & 99 \end{array}$$

So answer of the product is 12099.

If one of the two numbers exceeds the base and the other is in deficit.
In such case subtract the product on right hand side from the base by transferring one from the left hand side.

Example: Multiply 1033 × 997 Choosing base 1000 we have

$$\begin{array}{r|l} & 1033 + 33 \\ & 997 - 3 \\ \hline 2030 - 1000 & 33 \times (-3) = -99 \\ \text{i.e. } 1030 - 1 & 1000 - 99 \\ 1029 & 901 \end{array}$$

So answer of the product is 1029901.

It is permissible to choose a base which is some multiple of 10 if that seems more suited to the problem. However in that case the left hand side answer will have to be multiplied with multiplier and the base.

Example: Multiply 41 × 41 Choosing base 40 (40 = 10 × 4) we have

$$\begin{array}{r|l} & 41 + 1 \\ & 41 + 1 \\ \hline (82 - 40) & 1 \times 1 \\ \text{i.e. } 42 \times 4 & 1 \\ 168 & 1 \end{array}$$

So the answer of the product is 1681.

Effect of the choice of base
It is important to note that choice of base will have no effect on the result if rules are followed correctly. Choice of any two different bases will yield the same result. (However base should not be chosen as zero). We illustrate this with the following example

Example: Multiply 62 × 48 Choosing base 40 (40 = 10 × 4) we have

$$\begin{array}{llcl}
 & 62 + 22 & & \\
 & 48 + 8 & & \\
\hline
\text{we have} & (110 - 40) \times 4 & / & 22 \times 8 \\
\text{i.e.} & 70 \times 4 & / & 176 \\
\text{or} & 280 + 17 & / & 6 \\
 & 297 & / & 6
\end{array}$$

(as base is multiple of 10) So the answer of the product is 2976.

(*Note: we could not retain 76 on deficit side. Number on deficit side has to be less than the chosen base*)

Again choosing base 60 (60 = 10 × 6)

$$\begin{array}{llcl}
 & 62 + 2 & & \\
 & 48 - 12 & & \\
\hline
\text{we have} & (110 - 60) \times 6 & / & 2 \times (-12) \\
\text{i.e.} & 300 & / & -24 \\
\text{or} & 300 - 3 & / & 30 - 24 \\
 & 297 & / & 6
\end{array}$$

So the answer of the product is 2976.

Again choosing base 50 (50 = 10 × 5)

$$\begin{array}{llcl}
 & 62 + 12 & & \\
 & 48 - 2 & & \\
\hline
\text{we have} & (110 - 50) \times 5 & / & 12 \times (-2) \\
\text{i.e.} & 300 & / & -24 \\
\text{or} & 300 - 3 & / & 30 - 24 \\
 & 297 & / & 6
\end{array}$$

Again the answer is 2976.

Some more Examples

1. Multiply 231 × 582 Choosing working base as 600 (600 = 6 × 100)

$$\begin{array}{llcl}
 & 231 - 369 & & \\
 & 582 - 18 & & \\
\hline
\text{we have} & (813 - 600) \times 6 & / & -369 \times (-18) \\
\text{i.e.} & 213 \times 6 & / & 6642 \\
\text{or} & 1278 + 66 & / & 42 \\
 & 1344 & / & 42
\end{array}$$

So the answer is 134442.

2. Multiply 389×516 Choosing working base as 500 $(500 = 5 \times 100)$

$389 - 111$

$516 + 16$

we have	$(905 - 500) \times 5$ /	$-111 \times (+16)$
i.e.	2025 /	-1776
or	$2025 - 18$ /	$1800 - 1776$
	2007 /	24

So the answer is 200724.

It is thus obvious that the method works well only when both numbers are closer to the common chosen base. For instance in the above examples where it is not so multiplying numbers on the right hand side is itself a problem. Similarly it is not easy to write the product of 369 with 18 mentally in case of example (1).

4.1.1 An alternate method of multiplication

Another method of multiplication which is rather more general than the previous one when translated into English language reads as: 'Vertically and cross wise'. Its working is illustrated with the following examples

Example 1: Multiply 12×13.

Multiplying left digits in the present case yields $1 \times 1 = 1$.
This is the left digit of the answer.
Next cross multiply left digit of each number with right digit of the other and add.
This is middle digit of answer. In this example it is $1 \times 3 + 1 \times 2 = 5$.
Next multiply the right digits. This is right digit of the answer. In this example it is $2 \times 3 = 6$.

Hence $12 \times 13 = 156$.

Example 2: Multiply 37×33

Applying the rules

Left digit is $3 \times 3 = 9$
Middle digits are $3 \times 3 + 7 \times 3 = 9 + 21 = 30$
Right digit is $7 \times 3 = 21$

Therefore the product is: 9 (30) (21) = 9 (32)1 = 1221 (Carryover where digits exceed nine)

Multiplication of numbers with three digits

For multiplication of three digit numbers this rule works as under:

- Left digit of the answer is product of the left digits of two numbers.
- Next digit from left is sum of the cross products of the left two digits of the two numbers.
- Next digit is sum of the product of the left most digit of one number with the right most digit of other, middle digit of one with middle digit of other and right most digit of one with left most digit of other.

- Next digit is the sum of the products of middle digit of first with the right most digit of the other and right most digit of first with middle digit of the other.
- Right most digit is product of the right most digits of two numbers (carryover to the left digit is to be carried out in the usual way when digit exceeds 9).

Example: Multiply 795×362.

Using the rule

Left most digit of answer is $7 \times 3 = 21$

	Next digit is	$7 \times 6 + 3 \times 9 = 42 + 27 = 69$
	Next digit is	$7 \times 2 + 9 \times 6 + 5 \times 3 = 14 + 54 + 15 = 83$
	Next digit is	$9 \times 2 + 6 \times 5 = 18 + 30 = 48$
	Right most digits	$5 \times 2 = 10$.
	So the desired product is	(21) (69) (83) (48) (10)
	Making usual carryovers	(21 + 6) (9 + 8) (3 + 4) (8 + 1) (0)
or	(27 + 1) (7) (7) (9) (0)	= 287790

In order to multiply a three digit number with a two digit number make two digit number three digit number by putting a zero on its left.

Example: Multiply 795×62

For this we multiply 795 with 062 using rules of three digit number.

Left most digit of the answer is $7 \times 0 = 0$.

	Next digit is	$7 \times 6 + 0 \times 9 = 42$
	Next digit is	$7 \times 2 + 9 \times 6 + 0 \times 5 = 14 + 54 + 0 = 68$
	Next digit is	$9 \times 2 + 5 \times 6 = 18 + 30 = 48$
	Right most digit is	$5 \times 2 = 10$
	So the desired product is	(0) (42) (68) (48) (10)
i.e.	(4) (2 + 6) (8 + 4) (8 + 1) (0) = 49290.	

4.2 Division

There are several methods which have been recommended for division, each being efficient in a specific situation. We present here some of these

Division by 9: In case a two digit number is to be divided by 9, partition the dividend into its left and right digits. Then left digit is quotient and the sum of the two digits remainder.

Examples:	$21 \div 9$	Quotient 2, remainder $2 + 1 = 3$
	$80 \div 9$	Quotient 8, remainder $8 + 0 = 8$.

In case remainder becomes nine or more subtract 9 from the remainder and add one to quotient.

	$87 \div 9$	Quotient 8 remainder $8 + 7 = 15$
i.e.	quotient $8 + 1$ remainder $15 - 9 = 6$	
i.e.	quotient 9 remainder 6.	

Incase of division by 9 of numbers having more than two digits, number of digits in quotient will be one less than the number of digits in dividend. First digit of the quotient is again the first digit of dividend. The second digit of the quotient is to be obtained by adding the first digit of dividend to the second and next by adding this sum to the next digit of the dividend and so on. Remainder is the sum of the digits of this number. Remainder is to be treated as in case of division of two digit number when this last sum is nine or more.

Example 1: 311 ÷ 9 Quotient 3 (3 + 1), remainder 3 + 1 + 1

i.e. Quotient 34, remainder 5.

Example 2: 120021 ÷ 9

Quotient 1(1 + 2) (1 + 2 + 0) (1 + 2 + 0 + 0) (1 + 2 + 0 + 0 + 2)

Remainder (1 + 2 + 0 + 0 + 2 + 1)

i.e. Quotient 13335, remainder 6.

Example 3: 1011649 ÷ 9

Quotient: 1 (1 + 0) (1 + 0 + 1) (1 + 0 + 1 + 1) (1 + 0 + 1 + 1 + 6) (1 + 0 + 1 + 1 + 6 + 4)

Remainder 1 + 0 + 1 + 1 + 6 + 4 + 9

i.e., Quotient 1 1 2 3 (9) (13) remainder (22)

or Quotient 112(3 + 1) (0) (1 + 4), remainder (2 × 9 + 4)

i.e., Quotient 112(4 + 0) (1 + 4) and remainder (22 – 9 – 9)

i.e., Quotient 112405 and remainder 4.

Division by 8, 7 or 6

Rules for division by 8, 7 and 6 are similar to the rules for division by 9. However in the case of division by 8, we have to first multiply each dividend digit with (10 – 8), i.e., two and then add it to next (In case of 9 it was 10 – 9 i.e. one). Similarly in case of 7 each digit has to be multiplied with 10 – 7 i.e., 3 before adding to next and so on.

Examples:

(i) 23 ÷ 8 Quotient 2 remainder 2 × 2 + 3

i.e. Quotient 2, remainder 7.

(ii) 69 ÷ 8 Quotient 6 remainder 6 × 2 + 9 = 21 = 8 × 2 + 5

i.e. Quotient 6 + 2, remainder 5.

or Quotient 8, remainder 5.

(iii) 34 ÷ 7 Quotient 3, remainder 3 × 3 + 4

i.e. Quotient 3, remainder 13

i.e. Quotient 3, remainder 1 × 7 + 6

or Quotient 4, remainder 6.

This method is however not very convenient when a three or more digit number is to be divided. In such problems sums will exceed divisors and therefore suitable carryovers have to be performed. There is another method of division called 'Nikhlim Method of Division' which is recommended in such cases as well as in problems in which divisor has more than one digits.

4.2.1 Nikhlim method of division

This method of division is very effective when all the digits of the divisor exceed 5. Simply stated the method works as under.

(i) Subtract all digits of the divisor, except the last, from nine and subtract the last digit of the divisor from ten.

(ii) Multiply thus obtained number with the left most digit of the dividend and write it below the dividend starting from one place to the right of the left most digit and add. The left out left most digit of dividend is the quotient and the resulting sum the remainder. In case the remainder exceeds divisor subtract a suitable multiple of the divisor from this remainder to get remainder less than divisor and add this multiple to the quotient.

Examples:

(i) Divide 111 by 73

73	1 ¦	1	1	Quotient 1
27		2	7	Remainder 38
		3	8	

(ii) Divide 1234 by 888

888	1 ¦	2	3	4	Quotient 1
112		1	1	2	Remainder 346
		3	4	6	

(iii) Divide 210012 by 8997

8997	2 ¦	1	0	0	1	2	Quotient 23
1003		2	0	0	6		Remainder 3081
		3 ¦	0	0	7	2	
		¦	3	0	0	9	
			3	0	8	1	

(iv) Divide 1030007 by 98987

98987	1 ¦	0	3	0	0	0	7	Quotient 10
01013		0	1	0	1	3		Remainder 40137
		0 ¦	4	0	1	3	7	
		¦	0	0	0	0	0	
			4	0	1	3	7	

(v) Divide 101020 by 8888

```
8888 | 1 ¦ 0  1  0  2  0      Quotient 11
1112     1  1  1  2           Remainder 3252
       ----------------
       1 ¦ 2  1  4  0
         ¦ 1  1  1  2
         -------------
           3  2  5  2
```

(vi) Divide 1294567 by 89997

```
89997 | 1 ¦ 2  9  4  5  6  7   Quotient 13 + 1 = 14
10003       1  0  0  0  3      Remainder 34609
        ----------------------
            3 ¦ 9  4  5  9  7
              ¦ 3  0  0  0  9
            ------------------
            1 ¦ 2  4  6  0  6
              ¦ 1  0  0  0  3
              ----------------
                3  4  6  0  9
```

If required division process can be carried forward to the decimal stage by writing an additional zero to the right of remainder at each stage as in normal division process.

Supposing in example (i) we want quotient upto 4 decimal places then the procedure will be as under

```
73 | 1 ¦ 1  1                     Quotient is 1 (3 + 2) (1 + 1) 0 (4 + 1)
27     2  7                       i.e. 1.5205
    ----------                    Remainder is 35.
     3 ¦ 8  0
       ¦ 8  1
    ----------
     1  6  1
    -1  4  6                      (–73 × 2)
    -----------
        1 ¦ 5  0
          ¦ 2  7
          -------
           7  7
          -7  3                   (–73 × 1)
        -----------
           0 ¦ 4  0
             ¦ 0  0
           ----------
              4 ¦ 0  0  0
                ¦ 1  0  8
                ---------
                  1  0  8
                    -7  3         (–73 × 1)
                  -------
                     3  5
```

It may be noted that at a stage when resulting addition exceeds divisor, appropriate multiple of divisor is to be subtracted to make it less than the divisor and that multiple is to added to quotient of that stage.

Nikhlim method of division rule however is not eminently suited for division problems in which digits of divisor are 5 or less. A rule which is more eminently suited for division problems in which all the digits of divisor are less than six is known as Paravartya Yojayet Method.

4.2.2 Paravartya Yojayet method

Paravartya Yojayet Method (transpose and apply) method is as under:

Leaving the first digit from left of the divisor, write negatives of the remaining digits. Then multiply these with the first digit from the left of the dividend and add it to the dividend by writing it below the dividend from second digit from the left onwards. The left out digit of the dividend on the left is the quotient and rest of the sum remainder. If remainder is still more than divisor, repeat the process. In case remainder becomes negative at some stage add to it an appropriate multiple of the divisor to make remainder positive and subtract this multiple from the quotient.

Example:

1. Divide 1234 by 112

Divisor					
112	1 ¦	2	3	4	Quotient 11
− 1 − 2		− 1	− 2		Remainder 2
		1 ¦	1	4	
		¦	− 1	− 2	
				2	

2. Divide 12349 by 1133

Divisor						
1133	1 ¦	2	3	4	9	Quotient (11 − 1) = 10
− 1 − 3 − 3		− 1	− 3	− 3		Remainder 1019
		1 ¦	0	1	9	
		¦	− 1	− 3	− 3	
	- 1 ¦	1	1	3	3	
		1	0	1	9	

It may be noticed that in the second step, the normal procedure is yielding negative remainder so the divisor is substracted once to achieve a positive remainder and consequently one substracted from the quotient

In problems where some digits of the divisor are less than six and some more than six, a judicious use of any of the two methods can be made. A mix of two methods can also be used.

Example: Divide 4009 by 882.

By Nikhlin method

Divisor				
882	4 ¦	0	0	9
118		4	7	2
		4	8	1

Quotient 4
Remainder 481

By Pravartya Yojayet method

Divisor				
882	4 ¦	0	0	9
– 8 – 2		– 3	– 2	– 8
	¦	– 3	– 2	1
	– 1 ¦	8	8	2
	+ 1 ¦	5	6	3
	¦		– 8	– 2
		4	8	1

So quotient is (4 – 1 + 1 = 4) and remainder is 481.

Mix of the two methods

Divisor				
882	4 ¦	0	0	9
1 2 – 2		4	8	– 8
		4	8	1

Quotient 4
Remainder 481

It may be noted that Pravartya Yojayet method is not very convenient in the present case. There is still another method called the method of straight division which can be used in all cases (Interested reader may refer to the book on Vedic Mathematics by Jagadguru Swami Shri Bharti Krishna Tirathgi Maharaj).

4.3 Expressing a Fraction in Decimal form

We often need to express a fraction in decimal form. We are aware of the techniques which we ordinarily use for this purpose. Vedic mathematics has convenient to use rules for this. These are particularly effective when fraction is expressible in recurring decimal form. There are two types of methods. One which is effective in fractions whose denominators end in nine and other for fractions whose denominators do not end in 9.

4.3.1 Fractions whose denominators end in 9

Relevant sutra is 'एकाधिकेन पूर्वण', i.e., multiply or divide by one more than the preceding one. The first method in which we multiply is called 'Ekadhika Purvana'. In this method last but one digit of the denominator is increased by one to get the multiplicand. The modus operandi is illustrated below in case of 1/19.

Here multiplicand will be last but one digit of denominator plus one, i.e. 1 + 1 = 2.

1. Put down the digit in numerator or the fraction (which is 1) as the last right hand digit of decimal representation.
2. Multiply this last digit one by multiplicand 2 and put down 2 as the digit preceding the last digit.

3. Next multiply this 2 by 2 and get 4 as next digit to left.
4. Next multiply 4 with 2 and get 8 as next digit to left.
5. Now multiply 2 with 8 to get 16 as product. Since it has two digits we write 6 as the next digit to left and keep 1 as carryover.
6. Now multiply 6 with 2 to get 12. Adding carryover one we get 13. Writing 3 as next digit we get 1 as carryover.
7. Next multiply 3 with 2 to get 6. Adding 1 of carryover we write 7 as next digit.

We continue like this to achieve 18 digits after which digits start repeating (recurring decimal).

The decimal representation of 1/19 is

$$1/19 = .052631578947368421052 63.....$$
$$= .\dot{0}5263157894736842\dot{1}$$

An alternative method is division by one more than the number preceding nine in the denominator.

Taking again the example of writing 1/19 in decimal form, the number preceding 9 in denominator being 1, number one more than this is 2. So we divide 1 by 2 and get 0 quotient and 1 remainder. So the first digit from left is 0. Shifting remainder to next decimal place we have 10 which divided by 2 yields 5 as next digit. Now remainder is 0. Taking previous quotient 5 as next dividend and dividing it by 2 we get 2 as quotient and one as remainder. So the next digit in decimal form is 2 with 1 as remainder. Taking 1 suffixed with quotient 2 we get 12 which when divided by 2 yields 6 with remainder 0. Remainder 0 suffixed with quotient 6 is (06). This divided by 2 yields 3 as the next digit and so on. Process can be continued till result has been obtained upto desired number of decimal places or recurrence is observed.

Compared to Nikhlim, this method is recommended when answer has to be obtained upto desired number of decimal places. For example value of 1/19 upto 5 decimal places is .05263. In case of recurring decimal form it may be of interest to note that in case of 1/19 recurrence is after 18 digits. If we write first nine digits in one row and the next nine digits in the next row, we find that their sum consists of all nines

$$\begin{array}{r} .0\,5\,2\,6\,3\,1\,5\,7\,8 \\ 9\,4\,7\,3\,6\,8\,4\,2\,1 \\ \hline 9\,9\,9\,9\,9\,9\,9\,9\,9 \end{array}$$

This happens in case of all recurring decimals. Thus having worked half way we can obtain the remaining digits using this fact. This is called complement rule.

It can be similarly verified that 1/39 = .025641111....... = $.02564\dot{1}$

$$1/29 = .\dot{0}\,3\,4\,4\,8\,2\,7\,5\,8\,6\,2\,0\,6\,8$$
$$9\,6\,5\,5\,1\,7\,2\,4\,1\,3\,7\,9\,3\,\dot{1}$$
$$1/49 = .\dot{0}\,2\,0\,4\,0\,8\,1\,6\,3\,2\,6\,5\,3\,0\,6\,1\,2\,2\,4\,4\,8$$
$$9\,7\,9\,5\,9\,1\,8\,3\,6\,7\,3\,4\,6\,9\,3\,8\,7\,7\,5\,5\,\dot{1}$$

(*A dot on a single digit indicates its recurrence, dots on two digits indicate all digits from first dot to second appear in recurrence*).

4.3.2 Fractions with endings of denominators other than 9

Fractions whose denominators end in 7, 3 or 1 yield their decimal forms ending in 7, 3 and 1 respectively. In case of 1/7 we can write it as 7/49 and use the rule of fraction with denominator ending in 9. Similarly 1/13 may be written as 3/39, 1/11 as 9/99, 1/23 as 3/69.

In the case of 1/7 writing it as 7/49, digit preceding 9 in denominator is 4 and one more than it is 5. As the numerator of fraction is 7, so we start with 7 as the last digit of decimal representation on right and multiply it with 5 to obtain 35. We put 5 as digit next to 7 on its left and take 3 as carryover. Now we multiply 5 with 5 and adding carryover 3 get 28. We write 8 next to 5 on its right and take 2 as carry over. Multiplying 5 with 8 and adding carryover 2 we get 42. We write 2 next to 8 on its right and take 4 as carryover. Multiplying 2 with 5 and adding carry over 4 we get 14. So digit to the right of 2 is 4 and 1 is carryover. Multiply 4 with 5 adding 1 we get 21. So digit to right of 4 is 1 and 2 is carry over.

We have thus obtained 1/7 = ... 1 4 2 8 5 7

However we notice that digits of 142 are obtainable from digits 857 using compliment rule. Hence there is recurrence on these digits and therefore $1/7 = .\dot{1}4285\dot{7}$

Similarly it can be verified that

$$1/13 = 3/95 = .\dot{0}\,7\,6\,9\,2\,\dot{3}$$
$$1/11 = 9/99 = .\dot{0}\,\dot{9}$$
$$1/23 = 3/69 = .\dot{0}\,4\,3\,4\,7\,8\,2\,6\,0\,8\,6\;\;9\,5\,6\,5\,2\,1\,7\,3\,9\,1\,\dot{3}$$

Fractions whose denominators end in 2, 4, 8 or 5 always have non recurring decimal representations and are easy to write.

4.4 Squaring and Cubing

One way to square a number is to multiply the number with itself using some appropriate multiplication rule discussed earlier and for cubing multiplying the result again with the number itself. However some special rules such as 'Yavaduman sutra' for squaring and cubing numbers are also available. We discuss here some of these which are based on 'Yavaduman sutra'.

(i) For squares of numbers upto 9

Taking 10 as base whatever is deficiency of number from this base, lessen the number further by it. This is the digit at left of answer. Digit at right is deficiency square.

Example: Find 9^2

Deficiency of 9 from 10 is one. Decreasing 9 further by it we get 8. So

$$9^2 = (9 - 1)\ 1^2 = 81$$

Similarly,

$$8^2 = (8 - 2)\ (2)^2 = 84$$
$$7^2 = (7 - 3)\ (3)^2 = 49$$
$$6^2 = (6 - 4)\ (4)^2 = 2\ (16) = (2 + 1)\ 6 = 36$$
$$5^2 = (5 - 5)\ (5)^2 = (0)\ 25 = 25, \text{ etc.}$$

(ii) For numbers between 10 and 20

For numbers between 10 and 20 find excess of number above 10 and add it to the number. This is digit on left of the answer. Digit on the right is the square of excess.

Example:

$$11^2 = (11 + 1)\ (1)^2 = 121$$
$$12^2 = (12 + 2)\ (2)^2 = 144$$
$$13^2 = (13 + 3)\ (3)^2 = 169$$
$$14^2 = (14 + 4)\ (4)^2 = 18\ (16) = (18 + 1)\ 6 = 196$$
$$15^2 = (15 + 5)\ (5)^2 = 20\ (25) = 225$$
$$16^2 = (16 + 6)\ (6)^2 = 22\ (36) = 256$$
$$17^2 = (17 + 7)\ (7)^2 = 24\ (49) = 289, \text{ etc.}$$

Similar rules can be framed for finding squares of numbers between 20 and 30 choosing 20 as base and so on.

For numbers close to 100 such as 95, 89 etc., 100 may be chosen as base and deficit rule used keeping in view that result will have four digits.

Example:

$$91^2 = (91 - 9)\ (9)^2 = 8281$$
$$97^2 = (97 - 3)\ (3)^2 = 94\ (09) = 9409$$
$$89^2 = (89 - 11)\ (11)^2 = 78\ (121) = 7\ (8 + 1)\ 21 = 7921$$
$$88^2 = (88 - 12)\ (12)^2 = 76\ (144) = 7\ (6 + 1)\ 44 = 7744$$

For numbers close to 100 but exceeding hundred excess rule can be used keeping in mind that number of digits in the answer is to be 5

Example:

$$108^2 = (108 + 8)\ (8)^2 = 116\ (64) = 11664$$
$$103^2 = (103 + 3)\ (3)^2 = 106\ (09) = 10609, \text{ etc.}$$

Rule can be used for writing squares of numbers close to 1000 or 10000 etc. also using these as the base and keeping in mind number of digits expected in the answer.

Example:

$$993^2 = (993 - 7)\ (7)^2 = 986\ (049) = 986049$$
$$9989^2 = (9989 - 11)\ (11)^2 = 9978\ (0121) = 99780121$$
$$9993^2 = (9993 - 7)\ (7)^2 = 9986\ (0049) = 99860049, \text{ etc.}$$

(iii) Squaring a two or more digit number ending in 5

We know that square of 5 is 25. In order to find square of a more than one digit number ending in 5 we just need to multiply the number to left of last digit 5 with a number one more than itself and prefix the result to 25.

Example:

$$15^2 = (1 \times 2)\ 25 = 225$$
$$65^2 = (6 \times 7)\ 25 = 4225$$
$$105^2 = (10 \times 11)\ 25 = 11025$$
$$195^2 = (19 \times 20)\ 25 = 38025, \text{ etc.}$$

(iv) Using sum of first *n* odd numbers

Another method which can be used to find the square of a positive integer n is to add first n odd numbers.

For example 7^2 = sum of first 7 odd numbers

$$= 1 + 3 + 5 + 7 + 9 + 11 + 13 = 49$$

$$13^2 = 1 + 3 + 5 + 7 + 9 + 11 + 13 + 15 + 17 + 19 + 21 + 23 + 25$$
$$= 169$$

It can be proved mathematically that it should be so. The sum of first n odd numbers is the sum of arithmetic series 1 + 3 + 5 + ... terms, where first term is 1, common difference is 2 and number of terms is n.

$$\frac{n}{2}[2a + (n-1)\,d]$$

i.e.

$$\frac{n}{2}[2 \times 1 + (n-1)\,2] = \frac{n}{2} \cdot 2n = n^2$$

If we know square of n then square of $n + 1$ can be obtained using

$$(n+1)^2 = n^2 + (n+1)^{\text{th}} \text{ odd number.}$$

So knowing 13^2, 14^2 may be calculated using

$$14^2 = 13^2 + 14^{\text{th}} \text{ odd number} = 169 + 27 = 196.$$

Similarly $\quad (n-1)^2 = n^2 - n^{\text{th}}$ odd number

So, $\quad 12^2 = 13^2 - 13^{\text{th}}$ odd number $= 169 - 25 = 144.$

Method is eminently suited for finding squares of small integers. However it is not so convenient in the case of large numbers.

4.4.1 Finding cube of a number

Yavaduman Sutra can also be applied to find the cube of a number. For finding cube of a number, it has to be applied as under.

Instead of adding excess over base to the number as in case of square, add twice excess to the number. Next write to its right, thrice of excess multiplied with original excess and finally cube of excess and simplify keeping in mind number of digits expected in the answer.

Examples: $104^3 = (104 + 4 \times 2)\ [(3 \times 4) \times 4]\ 4^3 = (112)\ (48)\ (64) = 1124864$

$113^3 = (113 + 13 \times 2)\ [(3 \times 13) \times 13]\ 13^3 = 139\ (507)\ (2197)$

$= (139 + 5)\ (07 + 21)\ (97) = 1442897$

$93^3 = (93 - 7 \times 2)\ [(-7) \times (-21)]\ (-7)^3 = 79\ (147)\ (-343)$

$= 80\ (47)\ (-400 + 57) = 80\ (47 - 4)\ 57 = 804357$

$996^3 = (996 - 4 \times 2)\ [(-4 \times (-12)]\ (-4)^3 = 988\ (048)\ (-64)$

$= 998\ (048)\ (-1000 + 936) = 998\ (047)\ (936) = 998047936$

It may be noticed that number of digits has to be carefully adjusted. One must bear in mind that vedic mathematics aims at performing computations fast and efficiently in a simple manner using appropriate rules. So the user has to be sure that he is not making a mistake in applying suggested shortcut technique.

4.5 Finding square root of a number

For extracting the square root of a number, some basic points have to be kept in mind. These are:

(i) Starting from right to left arrange the number in two digit groups. If single digit is left at the end, treat it as a group. The number of digits in the square root will be the number of such groups formed (This rule is same as is used in common square root method).

(ii) Since squares of first nine natural numbers 1, 2, 3, 4, 5, 6, 7, 8 and 9 are 1, 4, 9, 16, 25, 36, 49, 64 and 81 respectively, a perfect square number can not end with digit 2, 3, 7 or 8 on its right.

(iii) If a number on division by 4 leaves 2 or 3 as remainder, it can not be perfect square. However if remainder is 1 or 0 it may or may not be a perfect square.

(iv) Square root of a number ending in 4 will end in digit 2 or 8. Square root of a number ending in 5 will end in digit 5. Square root of a number ending in 0 will end in digit 0. Square root of a number ending in 6 will end in digit 4 or 6. Square root of a number ending in 9 will end in digit 7 or 3.

Keeping these points in mind we can decide by looking at the number, whose square root is to be found, whether it is a perfect square or not and how many digits its square root will have as well as what is the likely terminating digit on the right. For example, the square root of 74562814 will have 4 digits and its right most digit will be either 2 or 8. Square root of 96306719 will have 4 digits and its right most digit will be 7 or 3. Number 963106713 will not be a perfect square and so on.

We will now illustrate the rule for finding the square root taking specific examples.

Example 1:
Suppose we want to find the square root of 119716. It can be perfect square. Its square root will have 3 digits the right most digit being 6 or 4. Now we start the process of extracting square root. For this we proceed as under:

(i) Extract the square root of first pair on left, i.e. 11. It is 3 with 2 as remainder. Write 3 below 11 and remainder 2 below 9 and take divisor as twice of the square root digit 3 (i.e. 6) and write these as shown next on the left.

6	1 1 9 7 1 6
	2
	3

(ii) Our dividend is now 29 and divisor 6. Division yields 4 as quotient and 5 as remainder. We write quotient 4 besides 3 on its right and 5 a little to left before 7 as shown.

6	1 1 9 7 1 6
	2 5
	3 4

(iii) Our next dividend is 57. From this we subtract square of 4 (the second obtained quotient), i.e., 16 to obtain 57 – 16 = 41. This when divided by 6 yields quotient 6 and remainder 5. These we append as shown.

6	1 1 9 7 1 6
	2 5 5
	3 4 6

(iv) Our next dividend is 51 from this we subtract twice the product of second and third quotient, i.e., 2 × 4 × 6 = 48 to obtain 3 as remainder. These we append as shown.

6	1 1 9 7 1 6
	2 5 5 3
	3 4 6

(v) This now gives 36 as dividend. From this subtract square of third quotient, i.e., 36. This yields remainder 0. This completes the process. We have 346 as the square root. It has desired number of digits and expected digit on its right. The whole process can be carried out in a single table appearing at the end.

6	1 1 9 7 1 6
	2 5 5 3 0
	3 4 6

Example 2: Square root of 3249.
It will have two digits with possible end digit on right as 3 or 7. Applying the technique we get square root in step (ii) itself.

10	3 2 4 9
	7 4
	5 7

Example 3: Square root of 552049.
Square root will have 3 digits. Digit on the right will be 3 or 7. Taking square root of 55 we get 7 as square root and 6 as remainder. Writing double of square root 7 as divisor and divident as remainder 6 written before 2 we get 62. This when divided by 14 yields 4 as quotient and 6 as remainder. This yields next dividend as $60 - 4^2 = 44$ which when divided by 14 yields 3 as quotient and 2 as remainder. This yields next dividend to be 24 – 2 × 4 × 3 = 0. So square root finding process is over. The square root is 743.

14	5 5 2 0 4 9
	6 6 2
	7 4 3

4.5.1 Square root of a number less than one

In case the number whose square root is to be determined is less than one and it is to be expressed in decimal form, then the following guidelines will be helpful in finding its square root.

Form pairs of digits starting from left (and not right as in case of whole numbers). If a single digit is left on the right make a pair by writing a zero on its right and then apply usual techniques for finding the square root. Each pair of zeroes that appear immediately after decimal will result in a zero in the square root.

Example: $\sqrt{.16} = .4$, $\sqrt{.9} = .948^*$, $\sqrt{.09} = .3$, $\sqrt{.0064} = .08$

$\sqrt{.000049} = .007$, $\sqrt{.00009} = .003$, (* *upto 3 decimals*)

$\sqrt{.00009} = \sqrt{.000090} = .00948^{**}$, etc. (** *upto 5 decimals*)

4.5.2 Square root of a number which is not a perfect square

It is generally helpful if one knows before hand whether a number whose square root is being determined is a perfect square or not. Square roots of numbers which are perfect squares can be determined exactly. For numbers which are not perfect squares, their square roots can be determined accurate upto desired number of places after decimal.

The following guidelines can help in deciding whether a given number whose square root is to be determined is a perfect square or not.

A number can not be a perfect square if it

(i) ends in digit 3, 7 or 8.

(ii) it ends in odd number of zeroes.

(iii) its last digit is 6 but digit preceding it is even (example 326 or 47886)

(iv) its last digit is 6 but digit preceding it is odd (example 2316 or 7456).

(v) The number is even but its last two digits are not divisible by 4 (example 3452 or 3866).

In order to find the square root of a number which is not a perfect square we may proceed as under

Example: Find square root of 732108 correct to 3 decimal places.

Proceeding in the usual manner we notice that as there are three pairs of digits its integer part will have 3 digits. To determine its decimal part upto desired accuracy we may if necessary continue process further by adding necessary number of zeroes after decimal

```
16 | 73 | 2  1  0  8  0  0  0
        | 9  12 16 14 15
   ----------------------------
     8  5  5  6  3  3
```

Ans. 855.633 upto 3D.

We have presented in this section some of the popular techniques for arithmetic operations of vedic mathematics and also illustrated their use in solving problems.

Interested reader can use these techniques to solve some more problems. Correctness of results in most cases can be easily verified. Interested reader may refer to the book of Jagadguru Swami Shri Bharti Krishna Tirathji Maharaj which contains several problems on these techniques and much more. These shortcut techniques show the versatility of the brains of ancient Aryans of vedic period.

5. Computing Aids

To begin with humans were making use of the fingers of their hands to perform simple addition and subtraction operations (and infact is still being done till date). Before invention of paper, use was being made of aids such as writing and marking on earth, wooden boards and slates for performing more complicated arithmetic operations. Some persons even developed the knack of performing such operations mentally as demonstrated in Vedic Mathematics. However, besides working on slate, board or paper computational aids have also been developed over the ages to perform more complex arithmetic operations faster. One of the earliest such aids was Abacus. It is still popular amongst small merchants and traders in certain countries. It is also being used as an aid in introducing kids to arithmetic operations of addition and subtraction in many nursery and kindergarten schools. Of course with the passage of time more and more sophisticated computing aids in the form of logarithmic tables, slide rules, calculating machines, calculators and computers and laptops have now appeared. In this section we briefly review these computing aids.

5.1 Abacus

Abacus is a calculating tool which has been used over the ages in parts of Asia as a counting aid for performing basic arithmetic operations of addition and subtraction. It is also called a counting frame. It is still used in many pre-school and elementary schools as an aid for teaching kids how to perform basic operations of addition and subtraction.

An abacus is usually a rectangular wooden frame with wires connecting the opposite longer sides at equal distances. Each wire has beads sliding on it (Originally beads or stones used to be moved in grooves made in sand or on tablets of wood, stone or metal).

Chinese are credited with being one of the earliest users of abacus. The earliest known written document referring to the use of abacus dates back to 2nd century B.C. Chinese abacus (also known as Suânpân meaning counting tray) is usually 8 inches (20 cm) tall and comes in various widths. It generally has more than seven rods, each rod denoting a position value in the number system. There is usually a horizontal bar which divides the frame into two rectangles, lower rectangle being bigger than the upper rectangle. On each wire there are two beads in the upper rectangle and five beads in the lower rectangle. The beads can be moved up and down on the rods. Each bead in the upper rectangle denotes 5 and each bead in the lower rectangle denotes one. The intended numeral in a particular position value of the decimal system is indicated by sliding appropriate number of beads close to the dividing bar. For example to represent 7, we move one bead from the upper side and two beads from the lower side of the wire close to the dividing bar. In an abacus with 10 wires numbers upto 10 digits can be represented. The position of beads in the following abacus with ten wires represents the number 6302715408.

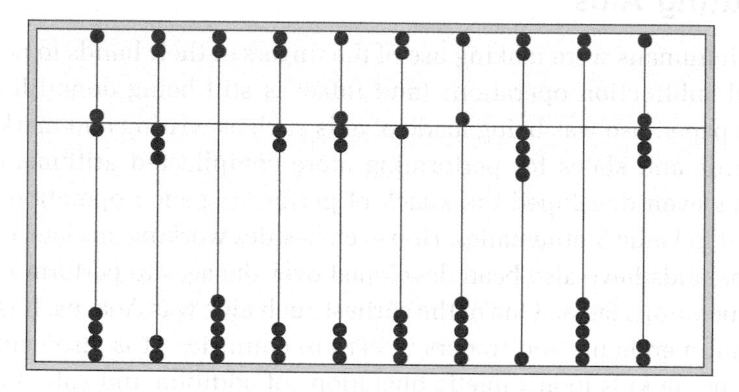

For an arithmetic operation the beads are moved up and down. When a bead is moved towards the dividing bar it means addition and when it is moved away from the dividing bar it means subtraction. When number of beads in the lower part of the rod which have to be moved close to the dividing bar have to be five or more (or beads on a wire close to the dividing bar in the upper rectangle has to be two or more), it is adjusted by moving one bead of the upper rectangle close to the dividing bar on the next wire to the left. For working in decimals suitable position for positioning of the decimal can be decided between two appropriately chosen neighbouring wires.

Working of abacus is more or less replica of counting on fingers of hands. Each bead in the lower rectangle represents one finger and each bead in upper rectangle represents one hand of five fingers.

Abacus can also be used for performing arithmetic operations other than addition and subtraction. Efficient techniques have been developed to perform multiplication division and even square root and cube root operations. A simpler version of abacus which is commonly used in elementary schools these days has ten rods with ten beads on each rod (five of one colour and five of another colour). There is no dividing rod. It is used to teach students counting from one to hundred. In such an abacus each bead on each bar has same value.

5.2 Logarithmic Tables

Napier developed logarithmic tables which proved helpful in performing multiplication, division and exponent operations fast. The basic idea behind the construction of logarithm tables was that instead of directly performing these operations on numbers as such, the numbers be first expressed in exponent form and then these operations be performed.

Suppose we want to multiply or divide two numbers say m and n. Then we first express these in exponent form with the same base. Let $m = a^x$ and $n = a^y$ be exponent form of these numbers in base a. Then

$$m \times n = a^x \times a^y = a^{x+y} \text{ and } m/n = a^x/a^y = a^{x-y}.$$

Thus operations of multiplication and division of two numbers m and n in decimal form reduce to adding or subtracting their exponents when these numbers are expressed in exponent form with the same base. Once this operation has been performed, the result can be expressed back in decimal form to get the answer. Thus operations of multiplication and division essentially reduce to operations of addition and subtraction when numbers are expressed in exponent form. Expressing a number in exponent form is called taking its logarithm and expressing back an exponent in its decimal form is called antilogarithm operation.

Not only multiplication and division, even more complex arithmetic operations can be performed using these tables. For example suppose it is desired to calculate

$$\frac{m^3 \times n^2}{k^4}, \text{ then taking base a we may express it as}$$

$$\frac{(a^x)^3 \times (a^y)^2}{(a^z)^4} \text{ which is equivalent to } \frac{a^{3x} \times a^{2y}}{a^{4z}}, \text{ i.e., } a^{3x + 2y - 4z}.$$

Thus after expressing, m, n and k in exponent form (i.e. taking log) if the exponents are x, y and z respectively, then we need only find $3x + 2y - 4z$ and then express this result back in decimal form taking antilog using the same base a.

Napier prepared tables which could be used to write logarithms and antilogarithms of numbers and rules have been framed to effectively use them.

When we write a number m in exponent form as a^x, then x is called the logarithm of m to base a and we write $x = \log_a m$. Normally two bases are in common use. These are 10 and transcendental number e. Whereas base 10 is more convenient for normal arithmetic operations, base e is used when operations of calculus such as differentiation or integration are to be performed. Another base which is now also used is 2. This is helpful in binary operations. It is possible to shift from one base to another.

It may be of interest to note that logarithm of a negative number is not defined. This is because suppose a is $-ve$ number say $-m$. Let its log to base 10 be b. Then $\log(-m) = b$. Hence $10^b = -m$. Now there can not be any power of 10 (positive or negative) for which 10^b is negative. Moreover $\log_{10} 1 = \log_e 1 = 0$, as $e^0 = 10^0 = 1$. Also $\log_{10} 10 = \log_e e = 1$, $\log_{10} 100 = \log_e e^2 = 2$, $\log_{10} 1000 = \log_e e^3 = 3$ and so on. In fact whereas $\log_{10} 0 = \log_e 0 \to -\infty$; $\log_{10} \infty$ as well as $\log_e \infty \to \infty$. Graph of logarithm to base 10 is shown in the following figure.

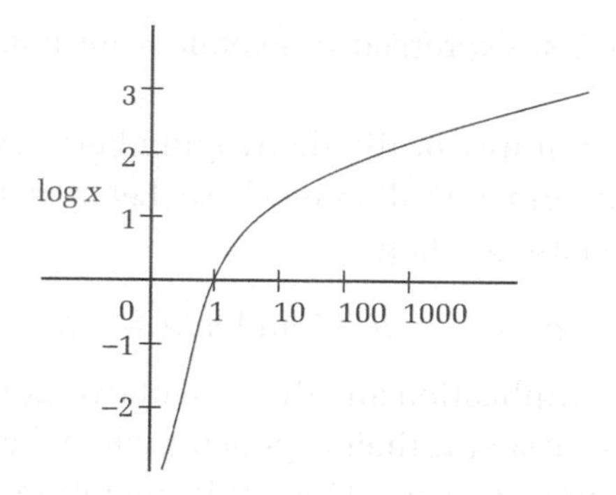

In case of log to the base e, 10, 100, 1000 etc. get replaced by e, e^2 and e^3 on *x-axis* without any change on y-axis.

5.3 Slide Rule

Slide rule was designed to aid engineers in performing complex arithmetic operations in a fast and convenient way. It was compact and convenient to carry and on earlier days engineers and overseers usually carried it with them for fast calculations during survey and construction work on sites. It made its appearance in late nineteenth century and was popular in early twentieth century. It was in vogue till middle of twentieth century when calculators gradually replaced it.

A slide rule consists of a long strip of metal on which numbers are printed on both of its edges. A central strip edged with numbers can slide in its middle. (That is why it was called a slide rule). It also has a pointer which indicates the result of arithmetic operation. It essentially mimicks the working of logarithm tables in compact and easy to carry form. In fact its use was considered so essential for engineers that in the very first year of an engineering course students were made to purchase a slide rule and trained in its use.

5.4 Calculating Machines

Calculating Machine is a mechanical device which was invented to aid scientists, mathematicians and engineers in performing arithmetic operations mechanically while working in offices or labs. It has a drum with a handle which is encased in a cylindrical cover. The cover has a number of slits from each of which digits from 0 to 9 printed on the drum are visible. Each slit has also a pointer. By suitable adjustment of the pointers desired number can be stored in the machine. Next the number with which this number had to be operated upon (added, subtracted, multiplied, divided) is set. Giving one rotation to the cylinder in the forward direction yields sum of the two numbers. A rotation in the backward direction means subtraction. The machine can be manipulated to perform all the four basic arithmetic operations of addition, subtraction, multiplication and division. It also has facility for fixing position of decimal. Suitable variants of such machines were also in use earlier in departmental stores and shops.

Initially these machines were hand operated. Later on these got fitted with electric motors. Subsequently electric versions of these machines also appeared which worked through electric circuits built in them. These had on their panels buttons for numerals from 1 to 10 and different arithmetic operations. By suitable pressing of these buttons various arithmetic operations can be performed efficiently and fast.

5.5 Calculators

Calculators made their appearance towards the second half of twentieth century. These can be regarded as battery operated compact electric calculating machines based on circuits. These are quite handy and gradually replaced slide rule and calculating machines. With the passage of time their circuitry become more and more complex and they could perform several intricate and even sequential arithmetical and statistical calculations fast and correct. They are still in use. In fact besides students, engineers and scientists, even grocers and shopkeepers now make use of these.

5.6 Computers

Computers made their appearance in the middle of twentieth century. Their working is based on electric circuits and therefore they perform arithmetic operations in binary mode. Initially computers used to be quite big machines which used valves in circuits. These valves generated lot of heat. As a result such computers had to be housed in air-conditioned rooms. At Roorkee we received a computer in mid sixties of the last century. It was an IBM 1620 machine. It was one of the earliest computers that came to India. It was housed in Central Building Research Institute (CBRI) of Roorkee and was meant for the use of engineers and scientists working in the university as well as the research institutes such as CBRI and Irrigation Research Institute located in Roorkee. I was amongst the earliest users of this computer. For working on this computer a program had to be written by the user in a language called 'Forgo' and then punched on cards using punching machines. These punching machines punched holes at appropriate places on the card through which electric current could pass to create the desired circuits. The deck of cards on which the computer program of the problem to be solved was punched was then converted into another pack of cards on which program got punched in what was then called machine language. This pack was first checked for accuracy. It was then fed to the main computer which carried out the desired computations and finally punched out the results on another set of cards. These punched cards were then put in another machine (called printer) which printed the results on sheets of paper.

Even though now it seems to be a very cumbersome process, those days these were the fastest computational facilities available. In fact I performed my entire Ph.D. related computational work on this machine. One of my research problems needed around eight hours of non-stop computation time on this machine (these days this much computation can now be executed on a PC in less than 5 minutes!). Later on for feeding a program into the computer punch cards were replaced by paper tapes and still later by magnetic tapes. In fact now we directly type the program to be executed into the computer or PC and it prints out the results directly on screen or a sheet of paper.

With the passage of time the capabilities and speeds of computers have fast increased. Big computing machines have now been replaced by more compact and sophisticated computers which have the capabilities of performing computations, much faster and even in parallel. In fact modern PC's and laptop's have computing capabilities that far exceed the capabilities of earlier big computing machines.

6. Rapid Advances in Computing Aids

There have been rapid advances in the availability of computing aids in the last few decades. I joined as a lecturer in mathematics in Roorkee University in 1964. At that time we used to teach first year engineering students the use of slide rule and calculating machines. At that time I was not Ph.D. and had enrolled myself for Ph.D. in Roorkee University. My research work required a lot of numerical computations. Initially I used to work with mechanical and electrical calculating machines. However gradually computers appeared on the scene and I shifted to working on computers. I recall an incident connected with my early experience of working on IBM 1620 machine in 1960's. One of my research problem required continuous non stop operation for about 8 hours. I was not getting that much time as a number of users had to be served each day and so the time was allotted in slots of 15 minutes, 30 minutes or at the most an hour depending upon the requirement of the user. However there was no provision for allotting the entire working day to a single user. When I explained my difficulty to the operating staff of the computer centre, they were kind enough to agree to run the computer exclusively for my use for the entire day on a Sunday (those days it was not possible to leave machine unattended and let it work on its own as temperature had to be checked manually and maintained with the help of AC's and so the staff of computer centre had to be there).

As in the case of computers, calculators are also becoming more and more sophisticated. With the availability of PC's and laptops, very few use computers and calculators now. Besides performing day to day computations, laptops also provide internet facilities, which is an additional benefit to the user. And this is not the end. Computation scenario is changing fast. Every other day we find new technological innovations being reported.

6.1 Seamy Side of Computing Aids

No doubt with rapid advances in computer technology, computations can be performed fast and these are generally reliable. However, this has its seamy side also. Through disuse, humans are fast loosing their mental capabilities. If a calculator or a PC says

$5 + 7 = 35$, we immediately accept it, little realizing that some error might have been committed due to some manual or technical snag or oversight. Most of us will accept the result provided by the computer or PC as a Gospel. On account of too much reliance upon computers, in fact many of us are loosing our faculties of critical analysis and judgment through its disuse. In this connection I recall an incident which happened to me as far back as 1973.

I was in England those days on a visit to universities in U.K. under British Commonwealth exchange of younger scientists programme. England of those days was ahead of India in the availability of computing aids. Use of calculating machines in the departmental stores was common. Once I had gone to a shopping mall to make some purchases. When I presented the purchased items to the lady at the counter, she made some quick calculations on the calculating machine which prepared the bill. She asked me to pay £5.75. Meanwhile I had made some rough mental calculations and expected the amount to be £15.75 and not £5.75. I asked the lady to recheck the amount of the bill. She virtually screamed 'How can it be wrong when it has been prepared by a calculating machine' (She might have been thinking that I expected the bill to be still less). When I told her that it was up to her and in fact according to my calculations she was charging £10 less, she became alert and rechecked the totals to find that what I was saying was correct. She thanked me for this and asked how I could make such long calculations mentally fast and that too correctly. In fact those days even £10 was a big amount (this was almost her weekly pay package).

No doubt faster computational facilities are necessary but those must be used when absolutely necessary. Their use should not be at the cost of human faculties. We must retain our capabilities of critical judgment and analysis and not become slaves to the machines and get cheated through their dubious miss use by shrewd persons.

Exercises

In case you feel excited with techniques of fast arithmetic operations and Vedic mathematics presented in this chapter, then, for practice perform arithmetic operations of multiplication, division, squaring, cubing, finding square root and expressing a fraction in decimal form using appropriate techniques of fast computations and Vedic mathematics presented in this chapter on numbers arbitrarily chosen. Also verify the correctness of the obtained results.

Chapter 4
Mathematics and Logical Reasoning

1. Introduction

Mathematics prides itself in being a subject based on logical reasoning. But what is logic after all? Dictionary defines logic as a subject of reasoning. Logic lays down precise rules on the basis of which one should argue to establish whether a stated proposition is true or false. Logic lays emphasis on having a sequence of statements, each of which follows as a deduction from the preceding one till one reaches the statement which is to be proved true or false. Starting statements of such a deductive process are known as 'axioms'. Axioms are basically self evident truth statements whose truth can be accepted on their face value. For example, 'two halves make a full' or 'a part can not be greater than the whole', etc. Axioms are in a way universal truths requiring no proof.

Logic is now studied as a full fledged subject in advanced courses on mathematics. Not only that we now have a subject called 'Fuzzy logic' which is generalization of conventional logic. Whereas in conventional logic a statement may be either true or false, in fuzzy logic it is also permissible to assume that the statement may be partially true and partially false or the concerned person is unable to say whether it is true or false.

Logic finds extensive applications in different branches of mathematics in particular Euclidean geometry. It is not only in mathematics alone but even in our day to day real life that logical reasoning helps in arriving at correct decision regarding the truth or falsity of a statement.

In this chapter we shall primarily restrict ourselves to the use of logic in its application to day to day practical life problems and demonstrate how logical reasoning can enable one to arrive at correct decisions. Its applications to mathematics based problems will follow in subsequent chapters.

2. Examples of Logical Reasoning

In this section we present some simple but interesting examples of everyday life which demonstrate how logical reasoning can help in arriving at correct decisions.

1. Countries of birth

Example: Ram, Rahim and John were born in India, Bangladesh and Malaysia, one in each country but which one is not known. However it is known that John has never been to India and Rahim was not born in Malaysia. Moreover Rahim's birth place is also not India. What are the names of the countries of their birth?

Solution: We notice that there are three persons: Ram, Rahim and John and three countries: India, Bangladesh and Malaysia. Only one person is born in each country. Since Rahim was neither born in Malaysia nor his birth place is India, so he must have been born in Bangladesh. Since John has never been to India and he can not have been born in Bangladesh (as Rahim was born there), so he must have been born in Malaysia. Now Ram is the only person left and India is the only country left. So Ram must have been born in India.

2. Which Job to Accept

Example: Krishna has offers of two jobs. Company A has offered him a salary of rupees 10 lacs per annum with an annual increase of 50 thousand per year. Company B has offered him a salary of rupees 5 lacs per six months with six monthly increase of rupees 20 thousand every six months. Which of the two jobs is more lucrative that Krishna should accept?

Solution: Let us compute Krishna's yearly income from the two companies for the first three years. From company A he will get Rs. 10 lacs in the first year, Rs. 10.5 lacs in the second year and Rs. 11 lacs in the third year. From company B he will get 5 lacs in the first six months and 5 lacs 20 thousand in next six months of first year. This makes a total of Rs. 10 lacs 20 thousand in the first year (This is 20 thousand more than what he will get from company A). In the second year he will get from company B Rs. 5 lacs 40 thousand in the first six months and Rs. 5 lacs and 60 thousand in the next six months making a total of Rs. 11 lacs (This is 50 thousand more than what he would get from company A). Again in the third year he will get from company B Rs. 5 lacs 80 thousand in the first six months and 6 lacs in next six months making a total of 11 lacs 80 thousand (This is also 80 thousand more than what he will get from company A in the third year). So the money that he will get from company B each year will be more than money that he will get from company A. So he should prefer company B.

3. Correcting Incorrectly Labelled Baskets

Example: There are three baskets containing fruits. One basket has only mangoes in it, other only apples and the third a mix of these two fruits. The baskets are covered and labeled. Unfortunately, all the baskets are labeled incorrectly. A basket labeled mangoes does not have mangoes in it, basket labeled apples does not have apples in it and so on. Is it possible to identify correctly the contents of all the three baskets by just looking at a fruit taken out from just one of these baskets?

Solution: At first sight it might appear difficult. However it is not that difficult. Just take out a fruit from the basket labeled mixed. Since the label is wrong it has either only mangoes or only apples. Say the taken out fruit is mango. Then this basket has mangoes and the basket labeled mangoes has apples in it and the basket labeled apples has mixed fruits in it.

4. More Milk or More Water

Example: There is a jug containing milk and another similar jug containing same amount of water. Some water is transferred from the water jug to the milk jug and thoroughly mixed. Next the same amount of liquid mix from this milk jug is transferred back to the water jug and contents thoroughly mixed. Now milk jug contains some amount of water and water jug some amount of milk. This process of transferring contents back and forth is repeated 10 times. After this which is greater, the amount of water in the milk jug or amount of the milk in the water jug?

Solution: After pouring back and forth same amount of the contents of two jugs alternatively 10 times, the jug of milk will contain same amount of water as jug of water will contain the amount of milk. This can be appreciated making use of a little bit of algebra. Let the amount of milk and water in the two jugs initially be one unit each. Now suppose after 10 transfers, the water jug has x units of milk and $1 - x$ units of water. Originally it had one unit of water. Where has this x units of water gone? Obviously to the jug of milk. So jug of milk now has x units of water and $1 - x$ units of milk.

5. Weight of Counterfeit Coin

Example: Out of 101 coins, one coin is counterfeit. Its weight is different from the weight of genuine coins (whether less or more is not known). Suppose a coin is picked. Given a balance, can it be decided just in one weighing whether the picked coin is genuine or counterfeit?

Solution: When a coin is picked we are left with 100 coins. Place 50 of these in one pan and rest fifty in the other. If the two pans balance, the picked coin is counterfeit. On the contrary if the weights in two pans do not balance, the picked coin is genuine.

6. How Far the Fly Flies

Example: Two trains are approaching each other on parallel tracks from opposite directions. They are running at the speed of 100 km per hour and are 20 km apart. At this instant a fly takes off from the front of one train and flies towards the other train. On reaching the front of the opposite running train, it turns around and flies back towards the first train. It continues flying back and forth like this till the two trains meet each other. What is the total distance traveled by the fly if it flies at the speed of 15 km per hour?

Solution: At first sight it appears to be quite a complex problem. However it is not so. As both the trains are traveling at the same speed, each will travel 10 km before they meet. For this distance traveling at 100 km/hr, they just need six minutes. During these 6 minutes, fly moving at the speed of 15 km per hour will have traveled 1.5 km. So simple, isn't it.

7. What Colour Hat

Example: Three persons are standing in a circle with their eyes blind folded. A hat is placed on the head of each person. They are told that the colour of the hat is either white or black and that all hats are not of same colour. Next their blindfolds are removed and each person who sees a white hat is asked to raise his hand and all of them raise their hands. After this, they are asked to identify the colour of their own hat. Can they do it correctly?

Solution: Call these persons A, B and C. Suppose A and B have white hats and C a black (all hats are not of the same colour). Then on opening their eyes each will see a white hat and raise his hand (If only one hat is white all three can not raise their hands when blindfolds are removed). Player A sees B wearing white and C black, player B sees A wearing white and C black and player C sees both A and B wearing white hats. C can argue that as A and C are both wearing white hats and all hats are not of the same colour, he must be wearing black hat. Now since all of them have seen at least one white hat and all hats are not of the same colour obviously colour of hats of B and A has to be white.

8. Tribes of truth Tellers and Liars

Example: A village in a remote place in a desert of Africa is inhabited by two tribes. Members of tribe I always speak truth whereas the members of tribe II are habitual liars. They always lie. An outsider happens to visit this village and knows about these tribes. He comes across three persons (say A, B and C) walking together. He asks A as to which tribe he belongs to. A's reply was not audible to the visitor. So he asked B as to what did A say. B replied that A said that he belonged to tribe I. However C contradicted him saying that what A said was that he belonged to tribe II. Based on this information, decide the tribes of A, B and C.

Solution: At first sight the supplied information seems inadequate. However if analysed logically the given information has sufficient clues for arriving at the correct answer.

There are only two possibilities. A either belongs to tribe I or tribe II. Suppose A belongs to tribe I. This means that he is a truth teller. So to the tourist's question he would have replied that he belonged to tribe I. However if he belonged to tribe II, he would have again said that he belonged to tribe I, as he is then a liar.

Now B said that A had said that he belonged to tribe I. So he is a truth teller and belongs to tribe I. C said that he belonged to tribe II. So he is a liar and belongs to tribe II. So we know the tribes of B and C for sure. However the supplied information is insufficient to decide the tribe of A. For this tourist may ask B as to which tribe A belongs in place of what he asked earlier as to what did A say.

9. Sisters at T Junction

Example: A motorist traveling from city A to city B came to a T junction where the sign board was missing. Of the two roads one leads to city B and the other somewhere else. At the T junction there was a cottage where two sisters live. Motorist knocked at the door. A lady came out of the house. When asked as to which road leads to city B, she pointed to one road. Just at that time her sister also came out and said that her sister was not telling truth and actually the other road lead to city B. The tourist was in a fix. Obviously one of them was lying, but who? He had earlier heard that of the two sisters, one was a habitual liar and the other a habitual truth teller. Which single question should the traveler ask the two sisters to be sure as to who was telling truth and who was lying?

Solution: Think for a while. The answer is simple. The motorist should ask the sisters as to which road leads to city A. Truth teller will point to the correct direction while her sister will point to some other road. As the traveler himself has come from city A, he will come to know which of the two sisters is a truth teller. The direction of B informed by that sister is the correct direction.

10. Who Lives Where

Example: There are three houses on each side of a small lane. Six professionals live in these six houses. Lawyer has a doctor and Mr. Desai as his next door neighbours. Professor lives next door to an accountant. Mr. Hari lives in a house at one end of the lane facing Mr. Balakrishna. Mr. Singh does not live opposite to accountant. Engineer and Charanjeet are both neighbours of Kukldeep. Identify the names of professionals with their professions and the houses in which they live.

Solution: Since lawyer has a doctor and Mr. Desai as his next door neighbours, so they all live on one side of the lane. As professor lives next door to accountant, so both live on same side of the lane. Charanjeet and engineer are neighbours of Kuldeep. So all are on the same side of the street. Hari lives opposite Balakrishna. This information leads us to conclude that Kuldeep is a professor, Desai an architect and Hari a doctor and all live on one side of the lane, whereas Charanjeet is an accountant, Balakrishna an engineer and Singh a lawyer and they live on the opposite side of the lane.

11. Who Stole the Mango

Example: A family had five children. One day mother bought six mangoes and gave one to each child and put sixth mango in the cupboard for herself. After sometime when she thought of eating her mango, she found that it was missing from the cupboard. Obviously one of the children had stolen it, but who? She called her children and enquired about it. They made the following statements

Amit: **Babli did not steal it.**
Eklavya: **Deepti has not stolen.**

Chinki: **Eklavya stole it.**

Deepti: **I know that Amit has not stolen.**

Babli: **It is Deepti who has stolen.**

Based on these statements, can you help the mother in identifying the culprit given that out of these four statements three are correct and one is false.

Solution: Eklavya's statement that Deepti has not stolen and Chinki's statement that Eklavya stole it are both consistent. Deepti's statement that Amit has not stolen is also in conformity with these. However Babli's statement that Deepti stole it contradicts Eklavya's statement. Based on these it is clear that either Eklavya or Deepti has stolen. If it is concluded that Eklavya has stolen then Babli is lying. If it is assumed that Deepti has stolen then it contradicts statements of both Eklavya and Chinki. As there are three correct and one incorrect statement, therefore it is Eklavya who has stolen it. The statement of Deepti that Amit has stolen is false.

The following problem is similar to the problem of who stole mango but a bit more elaborate.

12. Who Stole the Purse

Example: One day a teacher's purse was stolen. Needle of suspicion fell on five children: Leela, Juli, David, Thomas and Margret. It was sure that one of these five children had stolen the purse but who? When questioned, they made the following statements:

Leela: **I did not steal. I never steal. Thomas might have stolen.**

Juli: **I did not steal. Why should I? I come from a rich family. Margret knows the real culprit.**

David: **I did not steal. I was not even acquainted with Margret before joining this school. Thomas might have done it.**

Thomas: **I did not steal. Leela is lying when she says that I stole it. Actually Margret stole it.**

Margret: **I did not steal. Juli might have stolen. David can vouch for me. He knows me for the last so many years.**

Using inducements the students confessed that out of the three statements made by each one of them only two are correct. Based on this information can you identify the real culprit?

Solution: The first and third statements of Thomas are either both true or both false. As only one statement of each student is false, so both of these must be true. So he is not the culprit. Next examine the statements of David. If his first two statements that I did not steal and I was not acquainted with Margret earlier are correct then his third statement that Thomas might have done it is false. From this we conclude that Margret is lying when she says that David knew her for several years. So her other two statements are correct. Hence Juli is the real culprit. It is also in conformity with correctness of the first two statements of Leela. Her third statement that Thomas might have done it is false.

13. Who is Who in the Committee

Example: A managing committee has five members. Of these Archna, Bharti and Charulata are females while Deepak and Arvind are males. They hold offices of President, Vice-President, Secretary, Deputy Secretary and Treasurer. However not in that order. Secretary is a spinster. President and Treasurer shared same dormitory in a boarding house of a school for boys. Archna and Arvind are twins. Charulata's husband is treasurer's brother. President and deputy secretary are to be married shortly. Bharti will act as bride's friend in this marriage. Can you identify who is holding which office?

Solution: In this problem only one correct solution will be consistent with all the statements. Six statements have been made. These are:

(i) Secretary is unmarried
(ii) President and treasurer are both males
(iii) Archna and Arvind are twins (hence they cannot be husband wife)
(iv) Charulata's husband is brother of the treasurer.
(v) President and deputy secretary are of opposite sexes and not brother and sister.
(vi) Bharti is neither president nor deputy secretary.

From statements (ii) and (v) it is obvious that deputy secretary is a female and since she is unmarried, she cannot be Charulata. Bharti is neither president nor deputy secretary. So Archna has to be deputy secretary and she is to be married to the president. Now Archna's twin Arvind is a male. The only other male is Deepak. So Deepak must be president and Arvind treasurer. The other two portfolios of vice-president and secretary must be held by Charulata and Bharti respectively.

14. Whom to Appoint the Chief Minister

Example: A king wanted to appoint a chief minister. Three senior ministers (say A, B and C) in his council of ministers seemed suitable for the post. King had to select the most suitable amongst them for the post. For this he devised the following plan.

He seated these three ministers at the vertices of an equilateral triangle facing one another and got them blindfolded. Next he got red tilaks put on the forehead of each person. Blind folds were then removed so that each one of them could see the tilaks on the foreheads of the other two persons but not on his own. Ministers were then informed that red or white colour tilaks have been applied on their foreheads. They were then asked to raise hand if they saw at least one red tilak. At this all raised their hands. They were then asked to guess the colours of their own tilak. King said that the first to guess correctly would be made the chief minister. After thinking for a while second minister B replied that the colour of his tilak was red and that was correct. So he was made chief minister. Can you explain how he could correctly guess the colour of his own tilak.

Solution: Second minister B saw that colours of tilak of both first and third ministers (A and C) were red. He argued supposing the colour of his own tilak was white. Then A must have seen one red tilak on C's head and one white tilak on his forehead. C raised his hand as he saw white tilak on B's forehead and red on A's forehead. Thus A should have immediately deduced that his tilak was red and similarly was the case with C. However both were still thinking and undecided, that the colour of his own tilak must have also been red. In fact by this very argument, the other two could have also won but one who was more quick witted was the first to arrive at the correct conclusion and won.

15. To be Hanged Some Day Next Week

Example: A king got annoyed with his minister on some account and roared, 'You will be hanged someday next week and you will not be able to guess the day of hanging in advance.' Hearing this the shrewd minister smiled and said, 'Your majesty, am I allowed to speak?' The king said, 'O.K. go ahead.' Minister said, 'Sir, in that case you will not be able to hang me.' The king asked 'How?' Minister replied, 'Suppose you decide to hang me on the seventh day of next week. Then I will be surely able to know of it on 6th day of the next week as I shall not have been hanged by 6th day. So you will not hang me on 7th day. Next suppose you decide to hang me on 6th day, then I shall be knowing it on 5th day as it cannot be on 7th. So you can not hang me on 6th day also.' Arguing like this he convinced the king that under this condition he would not be able to hang him on any day of the next week. The king got confused. He was under the impression that he could order execution any day next week. Is there some fallacy in minister's arguments or can he get away like this?

Solution: There is a flaw in minister's reasoning. His argument that since it can not be on 7th day, it can also be not on 6th day as he will know of it on 5th day is flawed. Actually on 5th day he can not be sure whether he will be hanged on 6th or 7th day. He presumes that his ruling out hanging on 7th day is once for all. This is not so. In fact he can be hanged on any of first six days without his knowing about it on the previous day. However not on 7th day.

15.1 To he hanged or shot dead

Example: A prisoner was presented before the king. After hearing his case, the king remarked, 'Young man you have committed a grave crime and therefore deserve to be put to death. You will be either shot dead or hanged. I have already decided which way it is going to be. If you guess correctly you will be hanged. However if you guess wrongly, you will be shot. Now make your guess. On hearing this the prisoner's eyes lit up. Can you tell why?

Solution: The prisoner will answer, 'Sir, I would rather be shot.' Now if his answer is correct, he has to be hanged. But in that case he did not guess correctly and so he should have been shot. So the king was bound by the shrewd prisoner in a logical knot.

16. Superfluous data

Some times there may be a lot of data given in a problem which is not relevant to the desired solution of the problem. Here is an example.

Example: Imagine yourself to be a bus driver who is driving a bus from city A to city B. In between there are five bus stops. Bus has capacity of 40 passengers. It started with 20 passengers from city A. At the first stop no one alighted and 10 boarded the bus. At the second stop 7 alighted and 6 boarded the bus. At the third stop 14 alighted and 3 boarded the bus. At the 4th stop 6 alighted and 2 boarded. At the fifth stop 4 alighted and none boarded. On reaching the city B all the remaining passengers alighted from the bus. How many passengers were there in the bus when it reached city B and what is the age of the driver?

Solution: As regards the first part of the question, the answer is

$$20 + (10 - 0) + (6 - 7) + (3 - 14) + (2 - 6) + (0 - 4)$$
$$= 20 + 10 - 1 - 11 - 4 - 4$$
$$= 30 - 20 = 10$$

As regards the answer to the second part of this problem, the provided data has no relevance to it. It is superfluous. The age of the driver is your own age as you yourself are the driver!

17. What price to quote

Example: Two persons were traveling by same plane from London to Delhi. In London both had bought identical mementos which they had booked as accompanied luggage. On reaching Delhi when they got back their accompanied luggage, mementos of both of them were missing. The two persons were unacquainted. Each of them had been asked by the Airline to apply for reimbursement. None of them carried with him the proof of its price. Airline officer asked them to quote separately its price upto 200 dollars. They were told that the person who submits the lower claim will be awarded the amount claimed plus 10 dollars extra for honesty whereas the person who quotes higher price will be considered to have lied and will be paid 10 dollars less than the amount quoted by the other person. In case of tie, both will be paid the quoted amount. What price should each passenger quote to maximize the amount of compensation he receives (They are not allowed to consult each other).

Solution: Suppose passengers are A and B. If A puts claim of 200 and the other less, then he will get 20 dollars less than what other gets. If both submit claim of 200 each will get 200. So A may think of quoting 190 so that if B quotes 200, he will get 200 and B 180. However arguing the same way B may also think of quoting 190. Arguing like this each may end up quoting exact price if he does not want to get less than the other.

18. A Young Chess Champion

Example: A village chess champion was invited to play against two visiting chess masters. Unfortunately he lost to both and was feeling sad on this account. Looking at him his nine year old daughter taunted him saying, 'Daddy you need not have lost both the games. I can do better than what you managed.' Her father remarked, 'You barely know the basics of the game. How could you have done better?' Yet the girl persisted in her argument. As the chess masters were still around, girl's father persuaded them to play against his young daughter. She played with both of them simultaneously and performed better than her father. How did she manage it?

Solution: She played with black pieces on one board and white pieces on the other. Whatever move the first master played with white pieces, she played with her white pieces on the second board and whatever moves second master made with his black pieces on the second board she played with her black pieces on the first board. Thus she was virtually making one master play the other. Naturally one was to loose and one to win or both could draw. In each case she had done better than her father.

19. Students coming to college without helmets

Example: Some students were coming to college on motorbikes without wearing helmets. Both the college and their parents wanted them to wear helmets. However it was observed that most of them wore helmets near homes when their parents were watching but took them off later. It was a serious matter. One day the principal was informed that as many as 17 students had come to the college on bikes without wearing helmets. All of them were from different families. Principal, instead of just writing to the parents of these 17 students, wrote a general letter to the parents of all the students who were coming to the college on bikes that some students had been observed coming to the college on bikes without wearing helmets. Parents of such students should themselves punish their wards by getting the heads of their wards shaved off before sending them to the college. Teachers were perplexed as to why had the principal written such a letter to the parents of all the students coming on bikes instead of writing such letters to the parents of culprits only. Principal asked them to be patient and watch the fun. He was working on the premise that the parents of all the students will behave logically and each of them knew which of the other children were offenders. However they were not sure about their own ward. Nothing happened to begin with. After seventeen days had passed, on the eighteenth morning the seventeen culprits came to the college with their shaven heads covered with helmets. Can you explain how was it achieved and why nothing happened for the first 17 days.

Solution: It was like this. Each parent knew about other students except his own ward. If there had been only one guilty student his parents would have come to know of it the very first day as the principal had said that only some (and not all) students were breaking the rule. Since no other student had broken the rule, it must have been their ward. Similarly if two students had broken the rule, their parents would have become aware of it on the second day and so on. As there were 17 guilty students, their parents became sure of it on the 17th day. Therefore, on the 18th day they got the heads of their wards shaved before sending them to college.

20. Match the persons with their cities and professions

Example: Six persons Ashok, Bhatia, Charanpreet, Deepak, Gaurav and Kamal are traveling together in the same compartment of Rajdhani train from Bombay to Delhi. Two of them are doctors, two engineers and two professors. They are one each from Ahmedabad, Bombay, Chandigarh, Delhi, Jaipur and Kota but not necessarily in that order. The following facts are known about them:

(i) Ashok and person from Delhi are doctors.
(ii) Gaurav and person from Chandigarh are professors.
(iii) Person from Bombay and Charanpreet are engineers.
(iv) At Vadodara station Deepak and doctor from Ahmedabad got off.
(v) At Kota, Gaurav and engineer from Jaipur got off.
(vi) Deepak and person from Chandigarh are not of same profession.
(vii) Person from Ahmedabad and Bhatia are also not from the same profession.
(viii) However Ashok and Kamal belong to the same profession.

Based on this information match the persons with their professions and cities to which they belong.

Solution: At first sight it may appear a bit difficult as six persons have to be matched with six cities and three professions based on the above eight sets of information. However if we proceed logically things start falling in place step by step. Let us see how.

According to (i) Ashok is a doctor but not from Delhi. Person from Delhi is also a doctor. According to (ii) Gaurav is a professor but not from Chandigarh. Person from Chandigarh is also a professor (iii) implies that Charanpreet is an engineer but not from Bombay. However person from Bombay is also an engineer. According to (iv), Deepak is not from Ahmedabad but person from Ahmedabad is a doctor. According to (v) Gaurav is not from Jaipur. The person from Jaipur is an engineer. According to (vi) profession of Deepak is different from profession of person from Chandigarh. Using (ii) it means Deepak is not a professor. According to (vii) Bhatia is not from Ahmedabad. His profession is also different from profession of person from Ahmedabad. According to (viii), Ashok and Kamal are from same profession. Since by (i) Ashok is a doctor, Kamal is also a doctor and is from Delhi. We may summarise obtained information in the form of a table matching places and professions with persons.

Name	Profession	Place
Ashok	Doctor	Not from Delhi
Bhatia	Different from that of person from Ahmedabad	Not from Ahmedabad
Charanpreet	Engineer	Not from Bombay but person from Bombay engineer
Deepak	Not a professor	Not from Ahmedabad, not from Chandigarh
Gaurav	Professor	Not from Chandigarh
Kamal	Doctor	Delhi

Now Ashok and Kamal are doctors and Deepak is not a professor. So he has to be an engineer as there are to be only two doctors. So Charanpreet and Deepak are the two engineers. Since Charanpreet is not from Bombay but person from Bombay has to be an engineer. So Deepak is from Bombay. Now only Bhatia is left. So he has to be a professor.

Now we have sufficient information to match persons against professions and cities. The needed information is tabulated as under.

Name	Profession	Place
Ashok	Doctor	Ahmedabad
Bhatia	Professor	Chandigarh
Charanpreet	Engineer	Jaipur
Deepak	Engineer	Bombay
Gaurav	Professor	Kota
Kamal	Doctor	Delhi

3. What is Logically Wrong

In certain problems, even though all the stated facts in successive steps appear to be logically correct yet the final result seems to be incorrect and illogical. We know that a correct sequence of logical steps can not lead to an incorrect conclusion. It is therefore obvious that there is something logically wrong somewhere in the stated sequence of arguments and we have to detect it. In this section we present some such types of examples.

Example 1: Celebrated Hungarian mathematician Paul Erdos used to say that when he was a child scientists used to say that our earth was 2 billion years old. Now they say it is 4 billion years old. Therefore my present age is more than 2 billion years! What is wrong with this argument?

Solution: His assumption that in both cases the statements of the scientists regarding the age of the earth were absolute truths is wrong.

Example: 2. Peter's cat always sneezes before it rains. His cat sneezed today. So it will rain today.

Solution: It may or may not rain. Cat sneezes when it rains, but it can also sneeze when it does not rain.

3. Peter's Mom's Statement

Example: Peter's mom said, 'All champions are good at mathematics.' Peter replied, 'I am good at mathematics, therefore I am a champion.' What is wrong with Peter's argument?

Solution: Mom had said that 'all champions are good at mathematics'. It does not automatically imply the other way around that a person who is good in mathematics will also be a champion.

4. Three Friends in a Restaurant

Example: Three friends went to a restaurant for dinner. The bill for the dinner came to be Rs. 250. Each one gave a hundred rupees note. Waiter took Rs. 300 to the cashier and brought back 5 notes of 10 rupees each. Each of them took back one ten rupee note and left Rs. 20 as tip for the waiter. However on their way back home one of them argued like this, 'Each one of us paid Rs. 100 and got back Rs. 10 each. So each one of us has contributed Rs. 90 for the dinner making a total of Rs. 270. If Rs. 20 given to the waiter are also added, it makes a total of Rs. 290. But initially we had contributed Rs. 100 each making a total of Rs. 300 Where have the missing 10 rupees gone?' Can you help.

Solution: The calculation should have been done as follows :

Since each paid Rs. 100 and got back Rs. 10 contribution of each was Rs. 90 making a total of Rs. $3 \times 90 =$ Rs. 270. Of these Rs. 250 were paid as the bill of the restaurant and remaining Rs. 20 as tip to the waiter making a total of Rs. 270. In fact Rs. 20 given to the waiter should have been subtracted from Rs. 270 and not added to it.

4. How is it Possible

In this section we present some instances where certain facts which apparently seem to be logically incorrect are in fact correct.

Example 1: The son of a professor's father is talking to the father of professor's son but the professor is not taking part in this conversation. How is it possible?

Solution: This is possible if professor is a lady.

2. Why lift always appears to move in the opposite direction

Example: A man while working in second floor of a high rise 20 storey building used to take lift to the topmost 20th floor for lunch where office cafeteria was located. He generally observed that the lift arrived on the second floor from wrong direction (i.e., for going down). Later on he was promoted and got his office on 16th floor. Here he noticed that when he wanted to go down in the evening at 5 P.M. lift again generally came from the wrong direction (i.e., for going up). If the lift was going up and down all day long at regular intervals, why it seemed to the officer to behave erratically in both the cases?

Solution: As the lift is going up and down at regular intervals for a person on the second floor, the lift will be at a higher level (2 to 20 floors) for a duration which is 18 times more than the duration for which it is at a lower level. So when on the second floor he observed lift going down more times than going up. When he shifted to 16th floor the reverse happened.

3. Kant's Clock

Example: Immanual Kant (1724–1804) was a celebrated philosopher and mathematician who lived in Germany in 18th century. He was born and brought up in Konigsberg and reputed to have never gone out of the town. Each day he went for a walk at the same time. He always followed the same route. His pace was so regular that he always completed his walk in the same duration of time. Kant had a clock. One day it stopped. He had no other watch or clock. At a certain time he walked to his friend's house who had a clock. He stayed there for a while to chat. Then he left his friend's house and came back home following the same route. Arriving at his house he was able to set the time of his clock. How did he do it?

Solution: He counted his steps to his friend's house and counted the time of walk to his friend's house by chatting for the same amount of time which was needed for those many paces. Then he noted time on his friend's clock and returned back to his house at the same pace as he had gone. He then set time on his clock as much ahead of the time on his friend's clock for which he had sat in his friend's house.

4. Three Turtles Crawling along a Road

Example: Three turtles were crawling along a road in the same direction in a straight line. The turtle at the head said, 'Two turtles are behind me'. Turtle in the middle said, 'One turtle is ahead of me and one behind me'. Turtle at the back said, 'Two turtles are ahead of me and one at the back.' How is it possible?

Solution: The third turtle is telling a lie.

5. Pitfalls of Logic

Logic has its pitfalls. If used indiscriminately in an incorrect way, it may lead to erroneous conclusions. In this section we present some instances of such a nature.

1. Bertrand Russel's Statement

Example: Bertrand Russel was an eminent British mathematician and philosopher of twentieth century. He used to say that he had ruined himself by the study of logic and mathematics by brooding over the following simple problem.

Take a piece of paper and write on its both sides: 'Statement written on the other side of this paper is false'. Now do the truth and falsity analysis of these two statements to decide which of these is true and which false.

Solution: Statement on both sides is same. The same statement cannot be simultaneously true as well as false.

2. Russel's Paradox

There is a barber in a small village. He shaves every one who does not shave himself. Now what about the barber himself? If he shaves himself then it contradicts the statement that he shaves those who do not shave themselves. If he does not shave himself it again contradicts the statement that barber shaves all those who do not shave themselves.

Mind Teasers

1. Paul is taller than John, William is shorter than Paul. Eric is taller than William but shorter than John. Rank them from tallest to shortest.
2. Three urns contain balls. One urn contains only white balls, another only black balls and third a mix of the balls of two colours. Labels W, B and WB are pasted on the urns. However labels are incorrectly put. How can the contents of the urns be determined by just looking at the colour of one ball taken out from one of these three urns?
3. While children were playing with ball, it struck Chinky's back. Three suspected children were called by the teacher. Shyam said, 'Mam. I did not do it.' Anil said, 'I did not do it but I saw Shyam kick a shot.' 'Anil is lying.' said Jasdeep. Only one of them is telling truth. Can you tell who kicked the ball that hit Chinky?
4. Vikram, Farug and John met for dinner. One of them is a dentist, one a teacher, and one a clerk. Farug is older than the clerk, Vikram and teacher are not of the same age. Teacher is younger than John. What is the occupation of each one of them?

5. Juli, John, Krishnan and Paul were standing in a straight line. Krishnan was 3 metres ahead of John. Paul was 4 metres behind Juli. Juli was 5 metres behind Krishnan. Who was at the head of the queue and what was the length of the queue?

6. Assuming that the following statements are true:
   (i) Amongst persons having T.V. sets, there are some who are non-mathematicians,
   (ii) Non-mathematicians who swim in the swimming pool of the society everyday do not have T.V. sets.

   Do these two statements imply that not all persons having T.V. sets swim everyday?

7. Besides normal types of switches, there are two additional switches in an elevator located in a 20 storey building. Elevator goes up 13 floors when first such switch is pressed and it comes down 8 floors when second special switch is pressed (these switches however do not operate if adequate number of floors to go up or down is not possible). How can we use these switches to go from 13th floor to 8th floor?

8. There are three types of fruits on the dining table, namely apples, peaches and bananas. A family of three consisting of Prem and his mummy and papa are sitting on the dining table. Each one of them has liking for one of these fruits. Daddy does not like bananas. Mummy does not like bananas and apples. What is the favourite fruit of Prem?

9. Cake meant for dessert was already half eaten. When children were asked about it, their replies were: 'Sonia ate cake', said Ramesh. 'Ramesh is lying', said Sonia. 'I did not even open the fridge', said Tarun. Only one of them is telling truth. Can you tell who ate the cake?

10. In the problem 4 of section 2 about two flasks containing water and milk, suppose we continue pouring same amount of liquid from one flask to the other alternatively. Is it possible that after a finite number of such alternate pourings, each flask will have same amount of milk?

11. Three persons are seated in a row, one behind the other. The last person can see the other two seated in front of him. The middle person can only see the person in front of him. The person seated in front can see no one. Each one of them is asked to close his eyes and a cap is put on his head. The cap is either black or white. It is known to them that these three caps have come from a lot of 5 caps of which 3 are white and two black. After putting the caps on their heads the remaining two are hidden from their sight. The last person is asked if he knows the colour of his cap. After he has given his answer, the person in the middle is asked the same question. After he has answered, the same question is put to the person in front. What are their possible answers?

12. An island has two tribes. Members of one tribe always speak truth while members of other tribe always lie. Answer the following:
    (i) A person says, 'I am a liar.' Can he be an inhabitant of the land?
    (ii) What one question be asked to find whether a road leads to the village of truth tellers or village of liars?
    (iii) A says in the presence of B, 'At least one of us is a liar.' Who is truth teller and who is liar?
    (iv) There are three persons A, Band C. One belongs to the tribe of truth tellers and other to the tribe of liars. Third is a normal person from outside who occasionally lies. A said, 'I am outsider.' B said 'A and C sometimes speak truth.' C said 'B is an outsider.' Who is who?
    (v) If you are a truth teller how would you answer the question, 'Are you a liar?'
13. In problem 9 of section 2 regarding sisters at T junction, suppose the motorist is to ask any one of the sisters (he does not know who tells truth) a question to know correct direction. What single question should he ask?
14. In the society of five members problem 13 of section 2) given the following statements (i) Deputy secretary is unmarried lady, (ii) President and treasurer are men, (iii) Archna and Arvind are not husband and wife, (iv) Charulata's husband is treasurer's brother, (v) President and secretary are of opposite sexes but not brother and sister, If Bharti is neither president nor secretary, then what is she?
15. In problem 14 of section 2 if instead of three, there are four contestants for the post of prime ministership and similar procedure is adopted by the king, what will be the answer?
16. Are all the eight statements in the problem 20 of section 2 of matching persons with their cities and professions absolutely necessary or are some of these statements redundant and matching is possible using lesser statements?
17. Three friends were sitting together in a cafeteria. On seeing each other, one of them remarked, 'It is remarkable that one of us is wearing white shirt, one black and one brown. However the colour of the shirts of none of us matches the colour of our trousers which are also white, black and brown. 'You are correct', said the friend wearing white shirt. What is the colour of the trousers of the person wearing brown shirt?
18. In the problem 8 of section 2 regarding island having tribes of truth tellers and liars, suppose the words yes and no mean opposite (i.e., no means yes and vice versa), then (i) what one question be asked to know whether a person belongs to the tribe of truth tellers or liars, (ii) An islander X in the presence

of another islander Y said, 'At least one of us is a liar'. Does X belong to tribe of truth tellers or tribe of liars, (iii) What single question be asked to determine whether a person is a truth teller or a liar?

19. In a box there are pencils of at least two different colours and two different sizes. Does it imply that the box has two pencils which differ in size as well as colour?

20. Distribute Rs. 127 into seven wallets so that any integer amount of money upto Rs. 127 can be paid by just handing over one or more of the wallets.

21. A survey indicated high correlation between late risers and low grades of students in a college examination. This makes one believe that there is some correlation between late rising and low grades. Suggest three reasons as to why this conclusion could be in error.

Chapter 5
Algebra

1. Introduction

In mathematics algebra is regarded as a transitional subject through which a student, having learnt the basic rudiments of arithmetic and arithmetical operations, is initiated into the study of what is generally called higher mathematics which involves abstract thinking. Once a student becomes comfortable with algebra, he/she develops capacity to think about abstract concepts and to reason in a more logical and clear headed fashion.

Algebra has also been called the language of mathematics. In algebra, instead of dealing with actual numbers, we represent them by symbols such as x, y, z, etc. and use these symbols to develop mathematical expressions to represent desired real life situations. We then try to work these expressions to obtain appropriate numerical values for the symbols for which such expressions are valid.

The object of this chapter is not to survey high school algebra. Instead we discuss and survey some of the interesting and fascinating aspects of the subject which are not normally highlighted in a normal course on algebra in schools.

2. Algebraic Equations

Expressing a real life situation in the form of an algebraic equation (or a system of algebraic equations) using symbols such as x, y, z for the unknown variables and then trying to solve these equations to obtain the values of unknowns forms an important part of algebra. This procedure proves helpful in analyzing many real life situations.

By the solution of an algebraic equation we mean finding those values of the unknown variable (or variables) for which the equation gets satisfied. Real life problem may result in a single equation or a system of equations which can be linear as well as non-linear. Whereas solution of a linear equation in one variable of the type $ax + b = c$ is simple and straightforward

i.e., $$x = \frac{c-b}{a},$$

it is not so if we have a system of linear equations in more than one variable.

For example $a_{11}x_1 + a_{12}x_2 = b_1$ (i)

$a_{21}x_1 + a_{22}x_2 = b_2$ (ii)

are two linear equations in two unknown variables x_1 and x_2.

Here a's and b's are suitable constants depending upon the real life situation which these equations are supposed to represent. To solve these we generally multiply (i) with a_{22} and (ii) with a_{12} and subtract. This eliminates x_2 from the resulting expression which then gives value of x_1. Substituting this value in (i) or (ii) we can get value of x_2. This approach can also be used when we have to solve a system of more than two equations in as many unknown variables. We illustrate with some practical examples.

Example 1: Two persons live 20 km apart. They start towards each other at the same time. One is moving at the speed of 6 km per hour and the other at 4 km per hour. Where and when they would meet each other?

Solution: Suppose the two persons travel x and y kilometers before they meet. Then clearly

$$x + y = 20 \quad \text{(i)}$$

Again let t be the time after start when they meet.

Then $t = \frac{x}{6} = \frac{y}{4}$.

In other words $4x - 6y = 0$ (ii)

From (ii), $y = \frac{2}{3}x$. Substituting in (i), $x + \frac{2}{3}x = 20$ or $\frac{5x}{3} = 20$.

In other words $x = 12$ and hence $y = \frac{2x}{3} = \frac{24}{3} = 8$ km.

Thus, x will have traveled 12 km from one end and y will have travelled 8 km from the other end before they meet. They will meet after $t = \frac{x}{6} = \frac{12}{6} = 2$, i.e., 2 hours

Example 2: Four years back age of father was seven times the age of his son. Four years hence, his age will be 3 times the age of his son. What are the present ages of the father and the son?

Solution: Let the present age of father be x and that of son y. Then four years back age of father was $x - 4$ and that of his son $y - 4$. Hence

$$x - 4 = 7(y - 4).$$

or $x - 7y = -24$ (i)

Again four years hence, father will be x + 4 years old and his son y + 4 years. So according to the problem

$$x + 4 = 3\ (y + 4)$$

or $$x - 3y = 8 \qquad \text{(ii)}$$

Subtracting (i) from (ii), $7y - 3y = 8 + 24 = 32$

or $$4y = 32,$$

i.e. $$y = 8.$$

So, the present age of son is 8 years and from (ii) the present age of father is

$$x = 8 + 3y = 8 + 3 \times 8 = 32 \text{ years.}$$

2.1 Non-linear Algebraic Equations

Equations which contain terms of the unknown variable with powers other than one are called non-linear equations.

For example $$ax^2 + bx + c = 0$$

is a second order non-linear equation in the unknown variable x. This is also called a **quadratic (second degree) equation.** Whereas in a linear equation only one value of the unknown variable satisfies it, in the case of a quadratic equation two values of x satisfy it.

For example $x = 1$ as well as $x = 2$ satisfy the quadratic equation $x^2 - 3x + 2 = 0$.

These values ($x = 1, x = 2$) are called its solutions or roots.

In general $$a_n x^n + a_{n-1} x^{n-1} + a_{n-2} x^{n-2} + ... + a_2 x^2 + a_1 x + a_0 = 0$$

is called a **polynomial equation of degree *n*.** It will have n solutions. However all the solutions of a polynomial equation need not be real. Some can be complex also. In the case of quadratic equation

$$ax^2 + bx + c = 0 \qquad \text{(i)}$$

The two solutions are $x = \dfrac{-b \pm \sqrt{b^2 - 4ac}}{2a}$

The two roots of the quadratic equation (ii) are either both real or both complex.

These roots are real if $b^2 - 4ac > 0$ and complex if $b^2 - 4ac < 0$. For $b^2 - 4ac = 0$ these roots are identical. Solving a polynomial equation of degree more than two is generally not so easy.

Example: Find a real positive number such that when three times this number is subtracted from the square of this number the result is 4.

Solution: Let the number be x.

Then we have $\quad x^2 - 3x = 4.$

or $\quad x^2 - 3x - 4 = 0$

Solving $\quad x = \dfrac{3 \pm \sqrt{9 + 16}}{2} = \dfrac{3 \pm 5}{2} = 4 \text{ or } -1.$

So the desired number is $x = 4$ as the other answer -1 is not a positive number.

3. Sets and Operations on Sets

Sets play an important role in establishing results in a general form which can be applied in different fields. In fact sets dispense with the commonly held notion that mathematics deals only with numbers. Sets introduce the reader to the next level of abstraction and enable the user to establish more general types of results which are applicable to a large variety of subjects besides conventional mathematics. Based on the concept of sets a number of new subjects such as **modern algebra** have been developed.

A set is essentially a collection of similar objects, be these students, chairs, tables, fruits or numbers. We write $S = \{a, b, c\}$ to mean a set S which has a, b and c as its elements. The number of elements in the set is called its **cardinality**. Cardinality of present S (usually denoted by $n(S)$) is 3. A set can have any number of elements finite or infinite. Set $S = \{1, 1/2, 1/3, 1/4, ...\}$ has infinite number of elements. A set may also have no element. Such a set is called **null set** and is usually denoted by ϕ. It may be noted that null set ϕ does not mean that it has zero element. Null set has no element, not even zero. Corresponding to arithmetic operations of addition and subtraction on numbers, we have union and intersection operations on sets. Let $A = \{a, b, c, d\}$ and $B = \{b, c, d, e, f\}$ be two sets, then we define

union operation on A and B as

$$A \cup B = \{a, c, d, e, f\}$$

and **intersection operation** as

$$A \cap B = \{b, c, d\}$$

In other words, the set $A \cup B$ consists of those elements which are either elements of set A or elements of set B. On the contrary $A \cap B$ has only those elements which are both in the set A as well as B. Union is usually regarded as 'or' operation (i.e., element can be either from A or from B) and intersection as 'and' operation. In other words elements of $A \cap B$ are elements of both A as well as B. Union and intersection operations are usually carried on those sets whose elements are of similar type and come from the same set called **universal set**. For example, $A = \{1, 4, 5, 7\}$ and $B = \{2, 3, 6, 7, 9\}$ can be regarded as sets whose elements belong to the universe of positive integers. We usually denote the universe by U. We could also have set A as set of students of a class who study mathematics and set B as set of students of the same class who study physics. Universal set in this case is the set of students of the entire class.

3.1 Venn Diagram

The best way to appreciate sets and set operations is by depicting them in a diagram called Venn diagram. Let A and B be two sets defined on the universal set U. Then we can depict these diagrammatically as

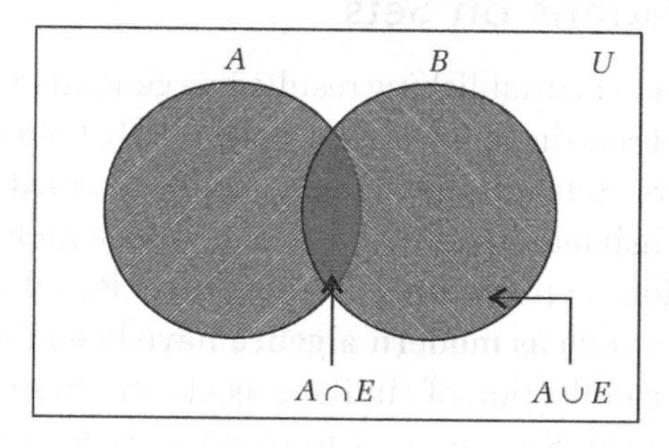

Fig. 1

In this case, $A \cup B$ is shaded as well as crossed area of diagram where as $A \cap B$ is only crossed area. The entire area of the diagram is universal set U. In case A and B have no common area (i.e., crossed portion is missing) then $A \cap B = \phi$ (an empty set).

If A and B in themselves include all the points of universal set U then $A \cup B = U$.

We also define **complement** of a set A as set C (A) which includes all those elements of U which are not elements of the set A. In the Fig. 2 the shaded portion in the diagram is C (A).

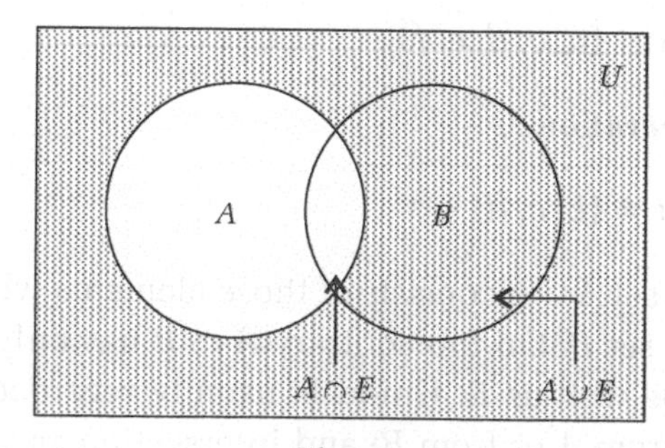

Fig. 2

Interestingly $C(U) = \phi$ and $C(\phi) = U$

Also $A \cup C(A) = U$ and $A \cap C(A) = \phi$

The number of elements in a set is called its **normality** or **cardinality**. Normality of the set A is denoted by $n(A)$. A number of useful reasults can be easily established using Venn Diagram. Some of these are:

$$n(A \cup B) = n(A) + n(B) - n(A \cap B)$$

$$n(A \cup B \cup C) = n(A) + n(B) + n(C) - n(A \cap B) - n(B \cap C) - n(C \cap A) + n(A \cap B \cap C)$$

Set theory is helpful in solving a large variety of real life problems. We illustrate this by presenting an example.

Example: 40 students took exam in English, Mathematics and Hindi. Of these 12 failed in English, 5 in Mathematics and 8 in Hindi. Two students failed in both English and Mathematics, 6 failed in both English and Hindi and 3 in both Mathematics and Hindi. One student failed in all the three subjects. How many students passed in all the three subjects?

Solution: Let the set of students who failed in English be denoted by E, the set of students who failed in Hindi by H and set of students who failed in mathematics by M. Then as per the problem

$$n(E) = 12,\ n(M) = 5 \text{ and } n(H) = 8$$

Also $n(E \cap M) = 2,\ n(E \cap H) = 6,\ n(H \cap M) = 3$

$n(E \cap M \cap H) = 1$ and $n(U) = 40$

Venn diagram for the data is shown in Fig. 3.

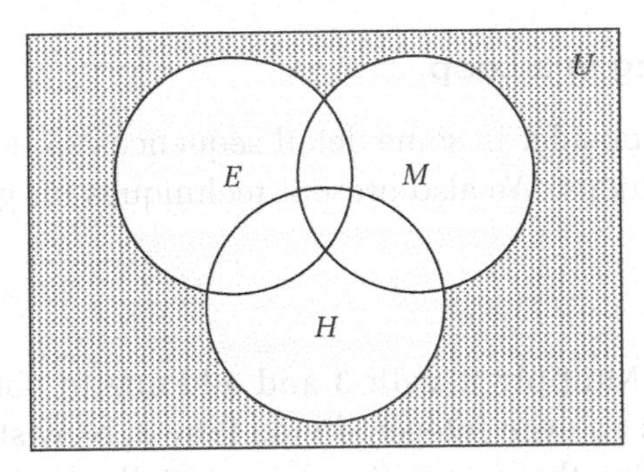

Fig. 3

We want normality or cardinality of $C(E \cup M \cup H)$. This is the normality of dotted portion of Venn diagram representing students of the class who failed in no subject. Now from the given data

$$n(E \cup M) = n(E) + n(M) - n(E \cap M) = 12 + 5 - 2 = 15$$
$$n(M \cup H) = n(M) + n(H) - n(M \cap H) = 5 + 8 - 3 = 10$$
$$n(E \cup H) = n(E) + n(H) - n(E \cap H) = 12 + 8 - 6 = 14$$

Also from the figure $n(E \cup M \cup H) = n(E \cup M) + n(H) - n(E \cap H) - n(M \cap H) + n(E \cap M \cap H)$
$$= 15 + 8 - 6 - 3 + 1 = 15$$

Hence normality of compliment of $E \cup M \cup H$ is

$$n[C(E \cup M \cup H)] = n(U) - n(E \cup M \cup H)$$
$$= 40 - 15 = 25$$

So, 25 students passed in all the three subjects.

4. Sequences

A sequence is a collection of numbers generally framed by following some specified rule.

For example $\{1, 1/2, 1/3, ..., 1/n, ...\}$ is a sequence of numbers of the type $1/n$, n being a positive integer. Sequence is said to be a finite sequence if it contains only finite number of terms and infinite sequence if the number of terms in the sequence is infinite.

If we add or subtract terms of a sequence we get a series whereas $\{1, 2, 2^2, 2^3, ... 2^n\}$ is a sequence, $1 + 2 + 2^2 + 2^3 + ... + 2^n$ is a series. Similarly $\{1, -1/2, 1/3, -1/4, ..., (1^{n-1}/n, ...\}$ is a sequence but

$$1 - \frac{1}{2} + \frac{1}{3} - \frac{1}{4}, ... + \frac{(-1)^{n-1}}{n} + ... \text{ is a series.}$$

Sequence and series play important role in mathematics and their behaviour and properties have been extensively studied. We present here some of their fascinating aspects.

4.1 Sequences Ending in a Loop

In this subsection we consider in some detail sequences $3X + 1$, $3X + 3$, (X a positive integer) which end in loops. We also present techniques for generation of such types of sequences.

1. Sequence $3n + 1$

Choose any integer n. Multiply it with 3 and add one to obtain $3n + 1$ if n is odd integer. However if n is an even integer, divide it by 2. Now starting with any integer, if we continue performing these operations sequentially on the results obtained, then regardless of any integer n with which we start, we shall ultimately end up with number one.

Example: Suppose we choose $n = 12$. Then 12 being even, we divide it by 2 to obtain 6. Now 6 is also even. So we again divide it by 2 to obtain 3 which is odd.

So we multiply it with 3 and add one to obtain $3 \times 3 + 1 = 10$. Now 10 is even. We divide it by 2 to get 5 which is odd. Multiplying it with 3 and adding one we get $3 \times 5 + 1 = 16$ which is even. Dividing it by 2 we get 8 which is again even. Dividing 8 by 2 we get 4 which is also even. Dividing 4 by 2, we get 2. 2 being even dividing it by 2 we get 1. In fact 1 being odd if we multiply it with 3 and add 1 we obtain 4. This when divided by 2 yields 2 which when divided by 2 yields 1. So we end up in a loop 4, 2, 1, 4, 2, 1, ...

The sequence generated is

{12, 6, 3, 10, 5, 16, 8, 4, 2, 1, 4, 2, 1, 4, 2, 1, ...}

Similarly, starting with 25 it can be checked that we end up with the sequence

{25, 76, 38, 19, 58, 29, 88, 44, 22, 11, 34, 17, 52, 26, 13, 40, 20, 10, 5, 16, 8, 4, 2, 1, 4, 2, 1, 4, 2, 1, ...}

Here again we have ended up in the loop 4, 2, 1 though after a larger number of iterations.

This rule is also popularly known as **irrepressible one**. Some numbers reach loop 4, 2, 1 rather fast. Others go through a large number of steps to reach this loop. For instance if we start with number 27, it goes through many ups and downs and reaches its peak 9232 in 77 steps. After this in 27 more steps it reaches loop 4, 2, 1 (try). Computers have been used to test the validity of this conjecture using starting integers upto 60,000,000! without finding an exception. However no formal mathematical proof for this exists.

2. Sequence $3n + 3$

The rule is similar to that of $3n + 1$ except that when n is odd we now add 3 in place of one after multiplying it with 3. However if n is even we again divide by 2.

Suppose we take $n = 11$. Then $3n + 3 = 33 + 3 = 36$, which is even. Dividing by 2 we get 18 which is again even. Dividing it by 2 we get 9 which is odd. Now $3 \times 9 + 3 = 30$ is even. Dividing it by 2 we get 15 which is again odd. Hence performing $3 \times 15 + 3$ we get 48 which is even. Dividing it by 2 we get 24 which is again even. Dividing it by 2 we get 12 which is again even. Its half is 6 which is again even. Dividing by 2 we get 3. This being odd when multiplied with 3 and 3 added yields $3 \times 3 + 3 = 12$. Thus in this case we end up in the loop of 12, 6, 3 in place of 4, 2, 1. The sequence generated in the present case is

{11, 36, 18, 9, 30, 15, 48, 24, 12, 6, 3, 12, 6, 3, 12, 6, 3, ...}
ending in the loop 12, 6, 3.

Again if we start with 27, we shall get the sequence

{27, 84, 42, 21, 66, 33,102, 51,156, 78, 39, 120 , 60, 30, 15,48, 24, 12, 6, 3, 12, 6, 3, 12, 6, 3, ...}

In case we start with 233 we get the sequence

{233, 702, 351, 1056, 528, 264, 132, 66, 33, 102, 51, 156, 78, 39, 120, 60, 30, 15, 48, 24, 12, 6, 3, 12, 6, 3, ...}

This again ends in loop 12, 6, 3.

Following table gives the number of steps needed to reach loop 4, 2, 1 in case of the rule $3X + 1$ and 12, 6, 3 in case of rule $3X + 3$

Starting Number	Number of steps needed to reach loop	
	rule $3X + 1$	rule $3X + 3$
27	112	30
237	35	36
3693	70	71
15292	178	60
237546	244	77
543678	209	91

Thus, for no exception to the rules has been reported even though no formal mathematical proof regarding the validity of the rules is available.

4.2 An Unexpected Sequence

Given a sequence it is generally possible to predict its subsequent terms. For example in the case of sequence {1 , 4, 7, 10, ...}, it is clear that subsequent terms are 13, 16, 19, ... Similarly in the case of the sequence {1, 2, 4, 8, 16, ...} subsequent terms will be 32, 64, 128 and so on. However there are certain sequences where it is not easy to predict subsequent terms.

Consider for instance the sequence {1, 2, 4, 8, 16, 31, ...}. At first sight it might appear as if 31 has been written by oversight and it should have been 32. However it is not so. To predict subsequent terms of this sequence, let us form the differences of its consecutive terms

Original sequence	1		2		4		8		16		31	...
First differences		1		2		4		8		15		...
Second differences			1		2		4		7		...	
Third differences				1		2		3		...		
Fourth differences					1		1		...			

So it is observed that the fourth differences of this sequence are constant, one each. Therefore to find next term of the sequence, we write one as the next 4th difference and work backwards. We find its next two terms by working backward from the fourth difference to be

Fourth differences	1	1	1	1				
Third differences	1	2	3	4	5			
Second differences	1	2	4	7	11	16		
First differences	1	2	4	8	15	26	42	
Original terms	1	2	4	8	16	31	57	99

So the sequence upto 8 terms is {1, 2, 4, 8, 16, 31, 57, 99, ...}

In fact 4th differences being one each it can be shown that the nth term of this sequence is

$$(n^4 - 6n^3 + 23n^2 - 18n + 24)/24.$$

This sequence has an interpretation in geometry. Observe the number of regions into which a circle can be partitioned by joining a specified number of points on its circumference.

Number of points on circumference	Number of regions into which circle is partitioned
1	1 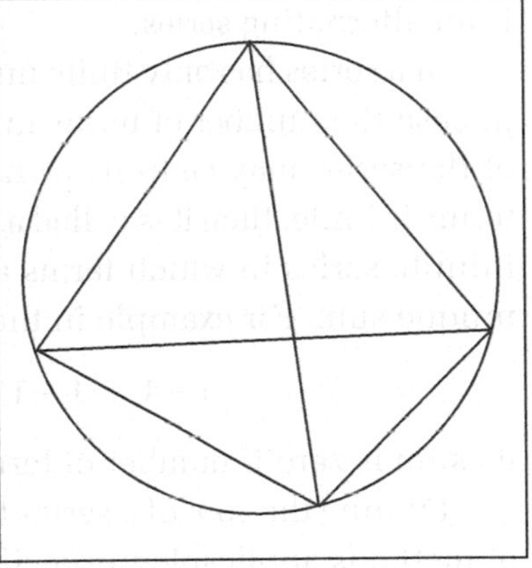
2	2
3	4
4	8
5	16
6	31
7	57
8	99

In fact number of regions are the successive terms of the sequence {1, 2, 4, 8, 16, 31, 57, 99, ...}.

4.3 Fibonacci Sequence

Another well known sequence is Fibonacci sequence {1, 1, 2, 3, 5, 8, 13, 21, ...} in which each term after the first two terms is sum of proceeding two terms. Its $(n + 1)^{th}$ term is given by

$$T_{n+1} = T_n + T_{n+1},\ n = 2, 3, 4, \ldots$$

with $T_0 = T_1 = 1$

This sequence has already been discussed in some detail in an earlier chapter.

5. Series

If $\{a_1, a_2, a_3, a_4, \dots a_n, \dots\}$ is a sequence, then $a_1 + a_2 + a_3 + a_4 + \dots + n_a + \dots$ is called a series. Like a sequence, a series can also have finite or infinite number of terms.

$$1 + 2 + 4 + 6 + 8 + \dots\dots + 120$$

$$1 + 2 + 2^2 + 2^3 + \dots\dots + 2^n$$

$$1 + \frac{1}{2} + \frac{1}{3} + \frac{1}{4} + \frac{1}{5} + \dots\dots + \frac{1}{n} + \dots\dots$$

$$1 + 2 + 2^2 + 2^3 + \dots\dots + 2^n + \dots\dots$$

Whereas the first two series have finite number of terms, the third and fourth series have infinite number of terms. Series can also have negative terms.

A series in which terms are alternatively positive and negative is called an **alternating series.**

For example $\quad 1 - 2 + 4 - 6 + 8 - 10 + \dots\dots$

is an alternating series.

If a series has only finite number of terms, then its sum is a finite number. However in case the number of terms in a series is infinite (called infinite series) then the sum of the series may be finite or infinite. If the sum of a series having infinite number of terms is finite, then it is called a **convergent series** else a **divergent series**. An alternating infinite series in which terms are alternatively positive and negative, may not have a unique sum. For example in the case of alternating series

$$1 - 1 + 1 - 1 + 1 - 1 + 1 - 1 \dots\dots$$

its sum is zero if number of terms is even and 1 if number of terms considered is odd.

Finding the sum of a series having infinite number of terms is important. A number of methods applicable in specific cases are available in the text books on algebra.

For example the sum of the infinite series

$$1 + \frac{1}{2} + \frac{1}{2^2} + \frac{1}{2^3} + \dots\dots + \frac{1}{2^n} + \dots\dots \qquad \text{(i)}$$

is $$\frac{1}{\left(1 - \frac{1}{2}\right)} = 2 \text{ and so it is a convergent series.}$$

However, the sum of the infinite series

$$1 + 2 + 2^2 + 2^3 + \dots\dots + 2^n + \dots\dots \qquad \text{(ii)}$$

is infinite and it is divergent.

For an infinite series to be convergent and have finite sum it is essential that its nth term T_n should approach zero as n approaches infinity. This happens in case of (i) but not (ii). However this is only a necessary condition but not sufficient. In other words in an infinite series T_n may approach zero as $n \to \infty$ and still it may not be convergent. For example in the case of the series

$$1 + \frac{1}{2} + \frac{1}{3} + \frac{1}{4} + \dots\dots + \frac{1}{n} + \dots\dots$$

$$n^{\text{th}} \text{ term } \frac{1}{n} \to 0 \text{ as } n \to \infty$$

Intuitively it seems the series should be convergent. However it is not so!

5.1 Arithmetic, Geometric and Harmonic Series

A series of the type $a + (a + d) + (a + 2d) + \dots\dots + [a + (n - 1)\, d] + \dots\dots$

in which difference between two consecutive terms is always same (d here) is called an **arithmetic series** or **arithmetic progression** (it is usually called progression when number of terms is finite and series when number of terms is infinite) whose first term a and common difference d which may be +ve or –ve.

The nth term of this series is $T_n = a + [a + (n - 1)]d$.

The sum of its first n terms is

$$S_n = \frac{n}{2}[2a + (n - 1)d]$$

Examples of arithmetic series are:

$1 + 2 + 3 + 4 + 5 \dots\dots$

$5 + 12 + 19 + 26 \dots\dots$

$5 + 3 + 1 - 1 - 3 - 5 - 7 \dots\dots$

whereas d is 1 in case of the first, 7 in case of second, it is –2 in the case of third series.

A series of the type $a + ar + ar^3 + ar^4 + \dots\dots + ar^{n-1} + \dots\dots$

is called a **geometric series** or **geometric progression**. In this case ratio of two consecutive terms is always same (r here). This ratio r is called common ratio. In such a series whose first term is a and common ratio r, its nth term is ar^{n-1}. The sum of first n terms of this series is

$$a\frac{(1 - r^n)}{(1 - r)}$$

In case the series has infinite number of terms, then it will be convergent and have a finite sum $\frac{a}{1 - r}$ if $|r| < 1$. Otherwise it is divergent and does not have finite sum.

Example: Discuss the convergence of the following series

$$1 + 2 + 2^2 + 2^3 + \dots\dots + 2^n + \dots\dots$$

$$1 + \frac{1}{2} + \frac{1}{2^2} + \frac{1}{2^3} + \ldots\ldots\ldots + \frac{1}{2^n} + \ldots\ldots\ldots$$

$$1 - \frac{1}{2} + \frac{1}{2^2} - \frac{1}{2^3} + \ldots\ldots\ldots + \text{“}(-1)^{n-1}\frac{1}{2^n} + \ldots\ldots\ldots$$

In case of the first series $a = 1$ and $r = 2$, so it is divergent. However if only its first n terms are added then their sum is

$$\frac{1(1-2^n)}{(1-2)} = 2^n - 1$$

In the second series $r = \frac{1}{2} < 1$ so it is convergent. Whereas its sum up to n terms is

$$\frac{1\left(1-\frac{1}{2^n}\right)}{1-\frac{1}{2}} = 2\left(1-\frac{1}{2^n}\right)$$

its sum upto infinity is $\frac{1}{1-\frac{1}{2}} = 2$ because as $n \to \infty\ \frac{1}{2^n} \to 0$

In the case of the third example $r = -\frac{1}{2}$

so its sum upto n terms is $\frac{1\left(1-\left(-\frac{1}{2}\right)^n\right)}{1-\left(-\frac{1}{2}\right)} = \frac{2}{3}\left(1-\left(-\frac{1}{2}\right)^n\right)$

and its sum upto infinity is

$$\frac{1}{1-\left(-\frac{1}{2}\right)} = \frac{2}{3}, \text{ since } \left(-\frac{1}{2}\right)^n \to 0 \text{ as } n \to \infty$$

It may be of interest to note that the series formed by the reciprocals of terms of a G.P. is also a G.P.. Whereas reciprocals of terms of a geometric series again form a geometric series, it is not so in the case of an arithmetic series. Series formed by the reciprocals of the terms of an arithmetic series

$$a + (a + d) + (a + 2d) + \ldots\ldots\ldots + (a\ (n-1)d) + \ldots\ldots$$

is $$\frac{1}{a} + \frac{1}{a+d} + \frac{1}{a+2d} + \ldots\ldots\ldots + \frac{1}{a+(n-1)d} + \ldots\ldots\ldots$$

This is called a **Harmonic series** or **Harmonic progression** (H.P.).

Example: $1 + \frac{1}{2} + \frac{1}{3} + \frac{1}{4} + \ldots\ldots + \frac{1}{n} + \ldots\ldots$

is an H.P. formed from reciprocals of the terms of A.P.

$$1 + 2 + 3 + 4 + \ldots\ldots + n + \ldots\ldots$$

Similarly $\frac{1}{5} + \frac{1}{12} + \frac{1}{19} + \frac{1}{26} + \ldots\ldots$

has been formed from the reciprocals of the terms of the A.P.

$$5 + 12 + 19 + 26 + \ldots\ldots$$

Also $\frac{1}{5} + \frac{1}{3} + \frac{1}{1} + \frac{1}{(-1)} + \frac{1}{(-3)} + \frac{1}{(-5)} + \ldots\ldots$

has been formed from reciprocals of the terms

$$5 + 3 + 1 - 1 - 3 - 5 \ldots\ldots$$

To analyse behaviour of such series help is generally taken of the fact that series has been formed by reciprocals of the terms of an arithmetic progression.

5.2 Containers Needed to Store Entire Water in the World

Here is an interesting use of the sum of geometric series. The volume V of all the water in the world (including oceans, icecaps, glaciers, ground water, river water and water in the lakes) is estimated to be $V = 1387.5 \times 10^{15}$ m^3. Assuming that a drop of water has a diameter of 3 millimeters, its volume is

$$\frac{4}{3}\pi\left(\frac{3}{2}\right)^3 = \frac{4}{3}\pi \times \frac{27}{8} = \frac{9}{2}\pi \approx \frac{99}{7} \text{ or } 14.14 \text{ mm}^3$$

i.e., 14.14×10^{-9} m^3

Hence the total number of water drops in the world is

$$\frac{1387.5 \times 10^{15}}{14.14 \times 10^{-9}} = 9.815 \times 10^{35}.$$ This indeed is a very large number.

Now suppose we want to put the entire water of the world in containers of varying sizes such that the first container can hold one drop, second 3, third 9 drops and so on. In other words, each container can hold 3 times more than what its predecessor can hold. This is a geometric progression with first term $a = 1$ and common ratio $r = 3$. Let n be the total number of such containers which will be needed to store the entire water of the world, then

$$1 + 3 + 3^2 + 3^3 + 3^n = 9.815 \times 10^{25}$$

or $$\frac{1-3^n}{1-3} = 9.815 \times 10^{25} \text{ or } \frac{1}{2}(3^n - 1) = 9.815 \times 10^{25}$$

If we solve this to find n, we shall find that n is just 55. In other words, only 55 such containers will be needed to store the entire water of the world. It may look a bit strange that just 55 containers will be needed. But visualize the dimensions of such containers. If containers are cubically shaped, then 55th container will be around 940 km in each direction!

(This problem is similar to grains of rice on chessboard problem considered in an earlier chapter)

5.3 Using Patterns to Sum Series

We have methods for summing standard types of series such as series whose terms form an A.P. or G.P. or H.P.. However, sometimes we have to find sums of series which do not fall in any of such well known standard patterns. In many such cases, manoeuvring may help in finding their sum. We illustrate this with examples

Example 1: Sum the Series $\frac{1}{1\cdot 2} + \frac{1}{2\cdot 3} + \frac{1}{3\cdot 4} + \ldots\ldots + \frac{1}{n(n+1)}$

Apparently it is neither an A.P. nor G.P. nor H.P. However we notice that

$$\frac{1}{1\cdot 2} = \frac{1}{2}$$

$$\frac{1}{1\cdot 2} + \frac{1}{2\cdot 3} = \frac{1}{2} + \frac{1}{6} = \frac{2}{3}$$

$$\frac{1}{1\cdot 2} + \frac{1}{2\cdot 3} + \frac{1}{3\cdot 4} = \frac{2}{3} + \frac{1}{3\cdot 4} = \frac{1}{3}\left(2 + \frac{1}{4}\right) = \frac{9}{12} = \frac{3}{4}$$

$$\frac{1}{1\cdot 2} + \frac{1}{2\cdot 3} + \frac{1}{3\cdot 4} + \frac{1}{4\cdot 5} = \frac{3}{4} + \frac{1}{4\cdot 5} = \frac{1}{4}\left(3 + \frac{1}{5}\right)$$

$$= \frac{1}{4} \times \frac{16}{5} = \frac{4}{5}$$

Thus the pattern is obvious.

$$\frac{1}{1\cdot 2} + \frac{1}{2\cdot 3} + \frac{1}{3\cdot 4} = \ldots\ldots + \frac{1}{n(n+1)} = \frac{n}{n+1}$$

For instance, sum of

$$\frac{1}{1\cdot 2} + \frac{1}{2\cdot 3} + \frac{1}{3\cdot 4} + \ldots\ldots + \frac{1}{49\cdot 50} = \frac{49}{50}$$

We could have done it in an alternative way also

$$\frac{1}{1\cdot 2} = \frac{1}{1} - \frac{1}{2}$$

$$\frac{1}{2\cdot 3} = \frac{1}{2} - \frac{1}{3}$$

$$\frac{1}{3\cdot 4} = \frac{1}{3} - \frac{1}{4}$$

....

$$\frac{1}{n(n+1)} = \frac{1}{n} - \frac{1}{n+1}$$

Summing $\frac{1}{1\cdot 2} + \frac{1}{2\cdot 3} + \frac{1}{3\cdot 4} + \ldots\ldots + \frac{1}{n(n+1)} = \frac{n}{n+1}$

(Identical positive and negative numbers cancel)

Example 2: Sum the Series $\frac{1}{\sqrt{1}+\sqrt{2}} + \frac{1}{\sqrt{2}+\sqrt{3}} + \ldots\ldots + \frac{1}{\sqrt{3}+\sqrt{4}} + \frac{1}{\sqrt{2001}+\sqrt{2002}}$

Here the general term is $T_n = \frac{1}{\sqrt{n}+\sqrt{n+1}}$

We can write $T_n = \frac{\sqrt{n}-\sqrt{n+1}}{(\sqrt{n}+\sqrt{n+1})(\sqrt{n}-\sqrt{n+1})}$

$$= \frac{\sqrt{n}-\sqrt{n+1}}{n-(n+1)} = \sqrt{n+1} - \sqrt{n}$$

Thus, $\frac{1}{\sqrt{1}+\sqrt{2}} = \sqrt{2} - \sqrt{1}$

$$\frac{1}{\sqrt{2}+\sqrt{3}} = \sqrt{3} - \sqrt{2}$$

$$\frac{1}{\sqrt{3}+\sqrt{4}} = \sqrt{4} - \sqrt{3}$$

....

$$\frac{1}{\sqrt{2001}+\sqrt{2002}} = \sqrt{2002} - \sqrt{2001}$$

Adding the sum of the series is $\sqrt{2002} - 1$ ($\approx$ 43.754888) as other terms being alternatively positive and negative cancel each other.

Example 3: To find the Sum of squares of first n natural numbers.

Rules of arithmetic series can be used to find the sum of first n natural numbers $1 + 2 + 3 + 4 + ... + n$. This is an arithmetic progression with first term 1, n^{th} term n and common difference one.

Therefore $$1 + 2 + 3 + \ldots\ldots\ldots + n = \frac{n}{2}[2 \times 1 + (n-1) \times 1] = \frac{n(n+1)}{2}$$

However what is the sum of squares of first n natural numbers

$$1^2 + 2^2 + 3^2 + \ldots\ldots\ldots + n^2$$

Clearly this is not an A.P. To find its sum consider the identity

$$(x+1)^3 - x^3 = 3x^2 + 3x + 1$$

Putting $x = 1, 2, 3, ..., n$ successively we have

$$2^3 - 1^3 = 3.1^2 + 3.1 + 1$$
$$3^3 - 2^3 = 3.2^2 + 3.2 + 1$$
$$4^3 - 3^3 = 3.3^2 + 3.3 + 1$$
$$\ldots\ldots \quad \ldots\ldots \quad \ldots\ldots \quad \ldots$$
$$n^3 - (n-1)^3 = 3(n-1)^2 + 3(n-1) + 1$$
$$(n+1)^3 - n^3 = 3n^2 + 3n + 1$$

Observing that 2^2, 3^3, ... etc. appear once with plus sign and once with negative sign on right hand side and so will cancel on adding, we get

$$(n+1)^3 - 1^3 = 3(1^2 + 2^2 + 3^2 + \ldots.. + n^2) + 3(1 + 2 + 3 + \ldots.. + n) + n$$

or $$n^3 + 3n^2 + 3n = 3(1^2 + 2^2 + 3^2 + \ldots.. + n^2) + 3\cdot\frac{n(n+1)}{2} + n$$

Hence, $$1^2 + 2^2 + 3^2 + \ldots.. + n^2 = \frac{n^3}{3} + n^2 + n - \frac{n(n+1)}{2} - \frac{n}{3}$$

Simplifying, $$1^2 + 2^2 + 3^2 + \ldots.. + n^2 = \frac{n(n+1)(2n+1)}{6}$$

Example 4: Sum the series $1.2 + 2.3 + 3.4 + 4.5 + \ldots\ldots\ldots + n(n+1)$

Let $$S = 1.2 + 2.3 + 3.4 + 4.5 + \ldots\ldots\ldots + n(n+1)$$

Then, $$2S = 1.2 + (1.2 + 2.3) + (2.3 + 3.4) + (3.4 + 4.5) + \ldots\ldots\ldots + ((n-1).n + n.(n+1) + n.(n+1)$$
$$= 1.2 + 2.(1 + 3) + 2.(3 + 6) + 2.(6 + 10) + \ldots\ldots\ldots + n.(n - 1 + n + 1) + n.(n+1)$$

$$= [1.2 + 2.2^2 + 2.3^2 + 2.4^2 + ... + 2.n^2] + n(n + 1)$$
$$= 2[1^2 + 2^2 + 3^2 + 4^2 + ... + n^2] + n(n + 1)$$
$$= 2 \cdot \frac{n(n + 1)(2n + 1)}{6} + n(n + 1)$$

Therefore, $$S = n(n + 1)\left(\frac{2n + 1}{6} + \frac{1}{2}\right)$$
$$= \frac{n(n + 1)}{6}[2n + 1 + 3]$$
$$= \frac{n(n + 1)(n + 2)}{3}$$

6. Arithmetic, Geometric and Harmonic Means

Given two numbers a and b,

$\frac{a + b}{2}$ is known as their **Arithmetic Mean** (A.M.),

$\sqrt{ab}$ is known as their **Geometric Mean** (G.M.)

$\frac{2ab}{a + b}$ is known as their **Harmomc Mean** (H.M.).

In fact harmonic mean is reciprocal of the means of $\frac{1}{a}$ and $\frac{1}{b}$ as can be easily verified

Arithmetic mean is usually associated with arithmetic series, geometric mean with geometric series and harmonic mean with harmonic series. Whereas arithmetic mean is simple average of numbers a and b, harmonic mean is weighted average of numbers a and b. All these averages have their own significance and are useful under specific circumstances.

Example 1: Suppose we want to calculate the average speed of a round trip going in forward direction at the speed of 60 km per hour and returning back at speed of 40 km per hour. We feel it is quite trivial and immediately come up with the answer (60 + 40)/2 = 50 km per hour. But is this answer really correct? It perhaps is not. It is not justified to give equal weightage to two different speeds. In reality if the outward journey is completed in 2 hours then the distance is 120 km. Therefore the backward journey performed at speed of 40 km per hour will be completed in 3 hours. So taking the definition of average speed to be total distance covered divided by the total time taken it should have been 240/(2+3) = 240/5 = 48 km per hour and not 50 km per hour as obtained earlier. In fact in place of finding arithmetic mean of 40 and 60, we should have obtained their harmonic mean, which is

$$\frac{2 \times 40 \times 60}{100} = 48$$

In order to appreciate this distinction further, suppose a student scores 100% marks in nine subjects and 60% marks in the 10th subject. Then his average marks are

$$\frac{(100 \times 9 + 60 \times 1)}{(9 + 1)} = \frac{960}{10} = 96 \text{ and not } \frac{(100 + 60)}{2} = 80\%$$

So while finding mean or average we should take weighted mean and not simple arithmetic mean. We define **weighted mean** of two numbers a and b with respective weights w_1 and w_2 to be.

$$\frac{(w_1 a + w_2 b)}{w_1 + w_2}$$

In fact weighted mean is nothing but harmonic mean.

Example 2: Time for a round trip by an aeroplane
An aeroplane makes a round trip between Bombay and Delhi cruising at speed of 400 km per hour. Supposing on the day under discussion there was wind blowing from Bombay towards Delhi at 50 km/hour. Because of this wind will the plane take more time, same time or less time to complete this round trip?

Solution: It may appear that since the wind is hindering motion in one direction and aiding it in the other direction at same speed, it should have no impact on the total time taken. However that is not so.

Since the wind is blowing from Bombay towards Delhi at the speed of 50 km/hour, effective speed of the plane will be 400 + 50 = 450 km/hour while flying from Bombay to Delhi and 400 – 50 = 350 km/hour while flying from Delhi to Bombay. Now suppose distance between Delhi and Bombay is d. Then time t, taken for coming from Bombay to Delhi will be $t_1 = d/450$ and time t_2 taken for return journey will be $t_2 = d/350$. Thus total time for the round trip is

$$T_1 = t_1 + t_2 = d\left(\frac{1}{450} + \frac{1}{350}\right)$$

$$= d \times \frac{35 + 45}{45 \times 350} = \frac{80d}{45 \times 350} = \frac{8d}{1575} \qquad \text{(i)}$$

However if there had been no wind, time taken would have been

$$T_2 = \frac{2d}{400} = \frac{d}{200} = \frac{8d}{1600}$$

Clearly $T_1 > T_2$. So the plane will take more time than it would have taken had there been no wind.

In fact harmonic mean could have been used to determine the time taken by the plane for the round trip. The harmonic mean of 450 and 350 being

$$\frac{2 \times 450 \times 350}{450 + 350} = \frac{1575}{4} = 393.75$$

393.75 km per hour is the mean speed for the round trip. With this speed the time for traveling distance $2d$ is

$$\frac{2d}{\frac{1575}{4}} = \frac{8d}{1575}, \text{ same as } T_1$$

6.1 Ranking Amongst Means

Arithmetic mean, geometric mean and harmonic mean find a lot of practical applications. It can be also easily verified that

$$A.M. \geq G.M. \geq H.M.$$

To justify this let a and b be two numbers.

Then their arithmetic mean is $\frac{a+b}{2}$ their geometric mean is $\sqrt{ab}$ and their harmonic mean is $\frac{2ab}{a+b}$

Let these be denoted by A, G and H respectively.

Then, $A - G = \frac{a+b}{2} - \sqrt{ab} = \frac{1}{2}[(\sqrt{a})^2 + (\sqrt{b})^2 - 2\sqrt{a}\cdot\sqrt{b}] = \frac{1}{2}[\sqrt{a} - \sqrt{b}]^2$

Now, $\frac{1}{2}[\sqrt{a} - \sqrt{b}]^2$ being a perfect square is always non-negative

Hence, $A - G \geq 0$ or $A \geq G$ (i)

Again, $G - H = \sqrt{ab} - \frac{2ab}{a+b} = \frac{\sqrt{ab}}{a+b}[a + b - 2\sqrt{ab}\,] = \frac{\sqrt{ab}}{a+b}[\sqrt{a} - \sqrt{b}\,]^2 \geq 0$

Hence, $G \geq H$ (ii)

Combining (i) and (ii), $A \geq G \geq H$

Equalities hold when $a = b$.

The fact that $A.M. \geq G.M.$ has been used to establish a technique for finding optimal (minimum) values of certain types of non-linear expressions that find a lot of practical applications in problems of engineering design.

Example: A cylindrical container open at the top and of capacity 20π m³ is to be constructed. The construction cost is Rs. 200 m² for the bottom and Rs. 400 m² for the cylindrical walls. Determine the optimum dimensions (radius and height) of the container for which construction cost will be minimum. Given that the container is to be designed for storage capacity of 400π m³ and height should not to exceed twice the diameter of the base.

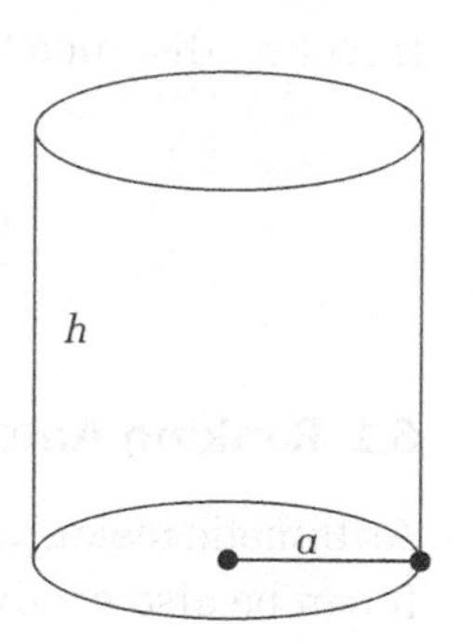

Solution: Let the radius of the base be a and height of container h. Then capacity V of the container is

$$V = \pi a^2 h$$

Areas of its base circular surface is πa^2. Also the area of its cylindrical walls is $2\pi ah$. Hence the total cost of construction is

$$C = 200 \times \pi a^2 + 400 \times 2\pi ah$$
$$= 200\pi a\,(a + 4h)$$

Thus we have to minimize

$$C = 200\,\pi a^2 + 800\pi ah$$

subject to the requirement of its capacity being $\pi a^2 h = V = 400\pi$

i.e. $$a^2 h = 400 \qquad \text{(i)}$$

and height and diameter constraint which requires that $h < 4a$ (ii)

Let $u_1 = 200\pi a^2,\ u_2 = 800\pi ah$

Then $C = u_1 + u_2$

Now using inequality $A.M. \geq G.M.$ we have

$$\frac{u_1 + u_2}{2} \geq \sqrt{u_1 u_2}$$

or $$u_1 + u_2 \geq 2\sqrt{u_1 u_2}$$

Therefore $$C \geq 2\sqrt{200\,\pi a^2 \times 800\,\pi ah}$$

i.e. $$C \geq 2 \times 400\,\pi a\sqrt{ah}$$

Hence, minimum cost is $800\,\pi a\sqrt{ah}$

Now from (i) $$ah = \frac{400}{a} \qquad \text{(iii)}$$

So minimum cost is $800\pi a\sqrt{\dfrac{400}{a}} = 800 \times 20\pi a^{1/2}$

$$= 16000\pi a^{1/2} \qquad \text{(iv)}$$

Here a is the radius of the circular base.

Cost clearly gets reduced when a is reduced but because of (i) h gets increased. However, as specified h is not to exceed twice diameter of base, thus from (ii) maximum value of h which can be chosen is $h = 2.2a = 4a$.

Writing $h = 4a$ in (iii)

$$a^3 = 100 \text{ or } a = (100)^{1/3}$$

and $$h = 4a = 4(100)^{1/3}$$

Thus for minimum cost radius should be $(100)^{1/3}$, and height $4(100)^{1/3}$. The cost will be $16000\pi \times (100)^{1/6}$. This example demonstrates the practical use of arithmetic, geometric mean inequality popularly written ($A.M. \geq G.M.$).

7. Factorial Notation

We denote factorial n for a positive integer n as $n!$ or $\underline{|n}$. It is defined as the product of first n natural numbers,

i.e. $$n! = \underline{|n} = 1 \cdot 2 \cdot 3 \cdot \ldots\ldots\ldots\ldots \cdot (n-1) \cdot n$$

Thus $$4! = 4 \cdot 3 \cdot 2 \cdot 1 = 24$$

$$6! = 6 \cdot 5 \cdot 4 \cdot 3 \cdot 2 \cdot 1 = 720, \text{ etc.}$$

and $1!=1$. But what about 0! As it will become clear shortly, this factually means placing no item in no place. Obviously it is possible in a unique way. Thus value of 0! if needed is taken to be 1, i.e., $0!=1$ and not zero.

Factorial of an integer n defined as above has a practical relevance. It is the total number of ways in which n items can be placed in n places. Suppose there are n persons who are to be seated in n chairs. We want to determine the total number of ways in which these n persons can be seated in n chairs. Clearly first person can be seated in n ways as he can be seated in any one of the n chairs. Once he has occupied a chair $n-1$ chairs are left and the second person can be seated in anyone of these $n-1$ chairs. Hence the two can be seated in $n(n-1)$ ways.

Once the second person also occupies a chair, $n-2$ chairs are left and the third person can occupy any one of these $n-2$ chairs so that three persons can be seated in n chairs in $n(n-1)(n-2)$ ways continuing this argument $(n-1)^{\text{th}}$ person will have to choose from two chairs and the last person will have only one seat available and will have no choice. Thus the total number of ways in which n persons can be seated on n chairs is:

$$n(n-1)(n-2)(n-3) \ldots\ldots\ldots\ldots 3 \cdot 2 \cdot 1 = n!$$

8. Permutations and Combinations

Each of the possible arrangements which can be made by taking some or all out of a number of objects is called a **permutation**. In our previous example, each of the $n!$ ways in which n persons could be seated on n chairs is a permutation. We could have selected r persons out of these n and seated them in n chairs. Then the number of ways would have been

$$n(n-1)(n-2) \ldots\ldots\ldots (n-(r-1))$$

This is the number of possible permutation of n distinct objects taken r at a time and is denoted by ${}^nP_r$ or $P(n, r)$.

Thus $${}^nP_r = P(n, r) = n(n-1)(n-2) \ldots\ldots\ldots (n-(r-1))$$

It is also possible to write ${}^nP_r$ as

$${}^nP_r = \frac{n!}{(n-r)!}$$

since $$\frac{n!}{(n-r)!} = \frac{n(n-1)(n-2) \ldots\ldots (n-(r-1)(n-r)(n-r-1) \ldots\ldots 3 \cdot 2 \cdot 1}{(n-r)(n-r-1)(n-r-2) \ldots\ldots 3 \cdot 2 \cdot 1}$$

$$= n(n-1)(n-2) \ldots\ldots (n-(r-1))$$

Example 1: It is required to seat 6 persons on 6 chairs. In how many ways it can be done?

Clearly in $6! = 6 . 5 . 4 . 3 . 2 . 1 = 720.$

Now suppose only three persons are to be seated in these six chairs, then it can be done in

$${}^6P_3 = \frac{6 \times 5 \times 4 \times 3 \times 2 \times 1}{3 \times 2 \times 1} = 6 \times 5 \times 4$$

$$= 240 \text{ ways.}$$

Example 2: How many four digit numbers are there? Of these how many have distinct digits.

Solution: There are 0, 1, 2, 3, 4, 5, 6, 7, 8, 9, i.e., 10 digits and we have to select any four.

So the number of four digit numbers that can be formed is

$${}^{10}P_4 = 10 \times 9 \times 8 \times 7 = 5040.$$

However these are all possible arrangements of 0, 1, 2, 3, ... upto 9 in four places. Some of these numbers will be starting with zero at thousand place and these numbers are not in reality 4 digit but three digit numbers (example 0487) and their number is ${}^9P_3 = 9 \times 8 \times 7 = 504$ since after fixing zero in thousandth place remaining 9 digits are to appear in remaining three places.

Thus actual number of 4 digit numbers is ${}^{10}P_4 - {}^9P_3 = 5040 - 504 = 4536$.

In fact in general it can be shown that if out of given n things n_1 are similar of one kind, n_2 similar of another kind and so on and n_k similar k^{th} kind, then number of their possible distinct permutations is

$$\frac{n!}{n_1!\, n_2! \ldots\ldots\ldots n_k!}$$

since n_1 things which could be permuted in $n_1!$ ways being identical can not be permuted amongst themselves and so on.

Example 3: How many distinct words can be formed from the word 'Mathematics'?

There are 11 letters in the word mathematics of which two are *m*, 2 t's and 2 a's. Rest are all distinct. Hence required number of possible arrangements is

$$\frac{11!}{2! \times 2! \times 2!} = 4989600.$$

Example 4: In how many ways can 6 persons be seated (i) in a line, (ii) in a circle?

Clearly the number of ways in which 6 persons can be seated in a line is $6! = 720$. However, when they are to be seated in a circle, the first and last position become identical and hence possible permutations are one less,

i.e. $(6 - 1)! = 5! = 120.$

Example 5: 10 persons are invited to a party and are to be seated on a circular table such that two particular persons are seated on either side of the host. In how many ways is it possible?

Clearly there are 11 persons (ten invitees plus one host) to be seated on the circular dining table. So if there had been no restriction on two specific persons being seated on either side of the host, the total number of possible ways would have been $11 - 1 = 10$ factorial. Considering two particular persons and the host as one we have 9 persons who can be seated in $9 - 1 = 8$ factorial ways. Internal arrangement of these three can be in two possible ways as each of the two persons can be on either side of the host. Hence total number of possible arrangements is $2 \times 8! = 80640$ ways.

8.1 Combinations

Suppose we have *n* items and we want to select any *r* items out of these *n*, then the numbers of possible ways in which *r* items can be selected out of *n* items is called the number of possible combinations of *r* items selected out of *n*.

Combinations differ from the permutations in that in combinations internal arrangement of these selected *r* items is immaterial whereas it is not so in permutations. Since *r* items can be arranged amongst themselves in *r*! ways, number of combinations will be $1/r!$ times number of permutations.

Now number of permutations of *r* items out of *n* is ${}^nP_r$.

Hence number of combinations of *r* items out of *n* is $\frac{1}{r!}{}^nP_r$

We denote number of combinations of *r* items out of *n* by ${}^nC_r$

Thus, $${}^nC_r = \frac{1}{r!} \times {}^nP_r = \frac{n!}{r!(n-r)!}$$

An obvious and interesting property of combinations is that ${}^nC_r = {}^nC_{n-r}$

9. Binomial Theorem

An expression of the type $a + b$, $2x + 3y$, $x - 2y$ etc. which contains two unknown variables in linear form is called a binomial expression. In fact we also consider $x + (1/y)$, $2x - (3/x)$ as binomial expressions by regarding $1/y$ in the first case and $1/x$ in the second case as second variable. We often need values of powers of binomial expressions.

We know for example

$$(x + y)^0 = 1$$
$$(x + y)^1 = x + y$$
$$(x + y)^2 = x^1 + 2xy + y^2 = x^2 + {}^2C_1xy + y^2$$
$$(x + y)^3 = x^3 + 3x^2y + 3xy^2 + y^3 = x^3 + {}^3C_1x^2y + {}^3C_2xy^2 + y^3 \text{ etc.}$$

However what about $(x + y)^{15}$ or $(x + y)^{3/2}$ or $(x + y)^{-2}$. These can not be written off hand. Binomial theorem provides us a means of writing expressions for these fast and correctly.

There are two versions of binomial theorem. One is for positive index which gives expansion of $(x + y)^n$ when n is a positive integer and the other when n is any positive or negative integer or a fraction. Without going into details of proof we present these theorems here.

9.1 Binomial Theorem for Positive Integer Index

According to this theorem

$$(x + y)^n = x^n + {}^nC_1x^{n-1}y + {}^nC_2x^{n-2}y^2 + \ldots\ldots + {}^nC_rx^{n-r}y^r + \ldots\ldots + {}^nC_{n-1}xy^{n-1} + y^n$$

keeping in view that ${}^nC_0 = {}^nC_n = 1$ it may be written in a more compact form as

$$(x + y)^n = \sum_{r=0}^{n} {}^nC_rx^{n-r}y^r \qquad \text{(i)}$$

Theorem is generally proved by induction.

Using this theorem $(1 + x)^n = {}^nC_0 + {}^nC_1x + {}^nC_2x^2 + \ldots\ldots + {}^nC_{n-1}x^{n-1} + {}^nC_n$

Putting $x = 1$, we get

$$2^n = {}^nC_0 + {}^nC_1 + {}^nC_2 + \ldots\ldots {}^nC_n \qquad \text{(ii)}$$

Thus sum of $(n + 1)$ coefficients of binomial expansion (i) is 2^n.

Again if we put $x = -1$, we get

$$0 = {}^nC_0 - {}^nC_1 + {}^nC_2 - {}^nC_3 + {}^nC_4 + \ldots\ldots - {}^nC_{n-1} + {}^nC_n$$

i.e., ${}^nC_0 + {}^nC_2 + \ldots\ldots + {}^nC_{n-1} = {}^nC_1 + {}^nC_3 + \ldots\ldots + {}^nC_n$ if n is odd

${}^nC_0 + {}^nC_2 + \ldots\ldots + {}^nC_n = {}^nC_1 + {}^nC_3 + \ldots\ldots + {}^nC_{n-1}$ if n is even (iii)

In other words sum of even coefficients in binomial expansion is equal to the sum of odd coefficients.

Results (ii) and (iii) can also be stated as

$$\sum_{r=1}^{n} {}^nC_r = 2^n \text{ and } \sum_{r=0}^{n} (-1)^r \, {}^nC_r = 0 \qquad \text{(iv)}$$

Example 1: Coefficients of consecutive terms in binomial expansion of $(1 + x)^n$ are in the ratio 1 : 7 : 42. Find n.

Here $\quad {}^nC_{r-1} : {}^nC_r : {}^nC_{r-1} = 1 : 7 : 42 \qquad$ (i)

$$\therefore \quad \frac{{}^nC_{r-1}}{{}^nC_r} = \frac{1}{7}$$

This gives on simplification

$$\frac{r}{n - r + 1} = \frac{1}{7}$$

i.e., $\quad n - 8r + 1 = 0 \qquad$ (ii)

Again from (i) $\quad \dfrac{{}^nC_r}{{}^nC_{r+1}} = \dfrac{7}{42}$ which gives on simplification

$$\frac{r + 1}{n - r} = \frac{1}{6}$$

i.e., $\quad n - 7r - 6 = 0 \qquad$ (iii)

Solving (ii) and (iii), the two relations in n and r, we obtain $n = 55$, $r = 7$.

Example 2: In how many ways cricket 11 can be chosen out of 15 selected players if there is no restriction.

Desired number of ways is $^{15}C_{11} = \dfrac{15!}{11!\ 4!} = \dfrac{12 \times 13 \times 14 \times 15}{4 \times 3 \times 2 \times 1} = 1365$ ways

Supposing captain has already been named and he has to be in selected 11. Then we have to chose 10 players out of remaining 14. So it will be

Desired number of ways is $^{14}C_{10} = \dfrac{14!}{10!\ 4!} = \dfrac{11 \times 12 \times 13 \times 14}{4 \times 3 \times 2 \times 1} = 1001$ ways

Example 3: There are 12 persons in a party and each shakes hands with the other. How many hand shakes takes place? Next suppose these 12 persons come to the party separately and each newcomer shakes hands with those who are already present. Then how many hand shakes take place?

Clearly a hand shake takes place between two persons, so we have to find possible number of combinations of 2 out of 12. Hence it is

$$^{12}C_2 = \frac{12 \times 11}{1 \times 2} = 66 \text{ ways}$$

In the second case the first shakes hands with none, second with one, third with two and so on and 12th with 11. So total number of handshakes

$$1 + 2 + 3 + 4 + 5 + 6 + 7 + 8 + 9 + 10 + 11 \quad \text{(an A.P.)}$$

$$= \frac{11}{2}[2 + (11 - 1)]$$

$$= \frac{11}{2} \times 12 = 66 \text{ ways}$$

which is again same as earlier and 6 as expected. Why is it so? This is because

$$^{12}C_2 = \frac{12 \times 11}{1 \times 2} = \frac{11}{2}[2 \times 1 + (11 - 1)]$$

9.2 Binomial Theorem for any Index

Let n be a rational number (positive or negative) and x a real number such that $|x| < 1$. Then according to Binomial theorem

$$(1 + x)^n = 1 + nx + \frac{n(n-1)}{2!}x^2 + \frac{n(n-1)(n-2)}{3!}x^3 + \ldots\ldots + \frac{n(n-1)(n-2)\ldots(n-r+1)}{r!}x^r + \ldots. \quad \text{(i)}$$

This expansion goes upto infinity and is in fact series expansion of $(1 + x)^n$.

If n is positive index then it reduces to binomial expansion for positive index of $(1 + x)$. Replacing x by $-x$ and n by $-n$, we get

$$(1 - x)^{-n} = 1 + nx + \frac{n(n+1)}{2!}x^2 + \frac{n(n+1)(n+2)}{3!}x^3 + \ldots\ldots + \frac{n(n+1)(n+2)\ldots(n+r-1)}{r!}x^r + \ldots. \quad \text{(ii)}$$

Putting $n = -1, -2$ in (i), we get

$$\frac{1}{1+x} = (1 + x)^{-1}$$

$$= 1 - x + x^2 - x^3 + x^4 \ldots\ldots\ldots + (-1)^r x^r + \ldots. \quad \text{(iii)}$$

$$\frac{1}{(1+x)^2} = (1 + x)^{-2}$$

$$= 1 - 2x + 3x^2 - 4x^3 + 5x^4 \ldots\ldots\ldots\ldots + (-1)^r (r + 1) x^r + \ldots.$$

Similarly putting $n = 1, 2$ in (ii), we get

$$\frac{1}{1-x} = (1-x)^{-1}$$

$$= 1 + x + x^2 + x^3 + x^4 \text{..........} + x^r + \text{....} \qquad \text{(iv)}$$

$$\frac{1}{(1-x)^2} = (1-x)^{-2}$$

$$= 1 + 2x + 3x^2 \text{..........} + (r+1)\,x^r + \text{....}$$

(iii) and (iv) are series expansions of $\frac{1}{1+x}$ and $\frac{1}{1-x}$ respectively.

In reverse we could say that sum of the infinite series

$$1 + x + x^2 + x^3 + x^4 \text{..........} \text{ is } \frac{1}{1-x}$$

$$1 - x + x^2 - x^3 + x^4 \text{..........} \text{ is } \frac{1}{1+x} \text{ and so on.}$$

Similarly $$\frac{x}{1+x} = \frac{1}{1+\frac{1}{x}} = \left(1+\frac{1}{x}\right)^{-1}$$

$$= 1 - \frac{1}{x} + \frac{1}{x^2} - \frac{1}{x^3} + \frac{1}{x^4} \text{........} (-1)^r \frac{1}{x^2} \text{........}$$

and $$\frac{x}{1-x} = \frac{1}{1-\frac{1}{x}} = \left(1-\frac{1}{x}\right)^{-1}$$

$$= 1 + \frac{1}{x} + \frac{1}{x^2} + \frac{1}{x^3} + \frac{1}{x^4} + \text{........} + \frac{1}{x^r} \text{........}$$

These results show that sum of series of infinite terms need not always be infinite and can be finite under certain conditions. As stated earlier in the section on series, a series of infinite terms whose sum is finite is called a convergent series.

9.3 Applications of Binomial Theorem

Binomial theorem has been used to solve a variety of problems. We have already seen how it helps in finding sums of certain infinite series. We illustrate here some more applications of general interest.

Example 1: Suppose we are required to decide out of 10000^{1000} and 10001^{999}, which is greater?

Clearly it is not so convenient to obtain their actual values and then compare. Decision off hand is also not easy. In this case help can be taken of Binomial theorem. According to this theorem

$$10001^{999} = (1000 + 1)^{999} = 1000^{999} + {}^{999}C_1\, 1000^{998} \times 1 + {}^{999}C_2\, 1000^{997} \times 1^2 + ...$$

$$= 1000^{999} + 999 \times 1000^{998} + \frac{999 \times 998}{2} 10000^{997} \text{ ..... (1000 terms)}$$

$$< 1000^{999} + 1000^{999} + 1000^{999} \text{ .... (1000 terms)}$$

$$< 1000^{1000}$$

So, $$10000^{1000} > 10001^{999}.$$

A remarkable thing about this problem is that it defies calculations even by ordinary calculators even when used in double precision. Computations involved in this problem are of course possible using softwares such as 'Mathematica'. But use of binomial theorem has helped in solving it even without the use of a hand calculator!

Example 2: Which is greater: $(1 + .000001]^{1,000,000}$ or 2.

This is again problem similar to example 1. However since in this case we have $(1 + x)^n$ where x is less than one, binomial theorem for any index will be more appropriate. Using binomial theorem for any index

$$(1 + .000001)^{1,000,000} = 1 + 1,000,000 \times .000001 + \frac{1000000 \times 99999}{2}(.000001)^2 + ...$$

$$= 1 + 1 + \text{ .... positive terms.}$$

Thus, $$(1 + .000001)^{1,000,000} > 2.$$

10. Algebra as Tool for Solving Problems

Algebra has been extensively used for solving real life problems as well as problems of mathematics. In section 2 of this chapter we saw how some real life problems can be solved by expressing these in form of algebraic equations and then solving these. In the last subsection 9.3 of the previous section we saw how binomial theorem can be helpful in checking correctness of inequalities. In this section, we present two more examples which illustrate how it can be helpful in certain other types of situations also.

Example 1: Show that $n^5 - n$ is divisible by 5 for all positive integer values of n.

Solution: We know that

$$n^5 - n = n(n - 1)(n + 1)(n^2 + 1).$$

Now if n is a positive integer and ends with digits 0, 1, 4, 5, 6 or 9, then one of the factors, on the right hand side of $n^5 - n$ will be divisible by 5 and hence $n^5 - n$ also divisible by 5.

However, if n ends in digits 2, 3, 7 or 8, then $n^2 + 1$ is divisible by 5. Thus, whatever positive integer n is $n^5 - n$ will always have a factor which is divisible by 5 and so $n^5 - n$ will be divisible by 5

Example 2: Show that one plus the sum of the squares of any three consecutive odd integers is divisible by 12.

Solution: Given a positive integer n, $2n - 1$, $2n + 1$ and $2n + 3$ are three consecutive odd integers. So we have to show that

$1 + (2n - 1)^2 + (2n + 1)^2 + (2n + 3)^2$ is divisible by 12 for all integer values of n.

$$\text{Now } 1 + (2n-1)^2 + (2n+1)^2 + (2n+3)^2 = 1 + 4n^2 - 4n + 1 + 4n^2 + 4n + 1 + 4n^2 + 12n + 9$$
$$= 12 + 12n^2 + 12n$$
$$= 12(1 + n + n^2)$$

which is always divisible by 12.

11. Pascal Triangle

Pascal triangle is one of the most well known triangular arrangements of numbers. It is named after Blaise Pascal (1623–1662). It has many interesting properties besides its use in probability distributions of statistics.

To construct a Pascal triangle, write 1 and then beneath it write 1 a little left to itself and another 1 a little to its right. Now write third row of numbers below this second row beginning with writing 1 a little to the left of the first one of second row. Now write next number as the sum of the numbers in the second row to the left and right of 1 in the second row. End this third row by writing 1, a little to the right of second one of the second row. Continue writing successive rows one below the other each row starting with one a little to the left of one of the preceding row and ending again with one written a little to the right of one of the preceding row. The intermediate numbers being sum of numbers to the left and right of the number in the preceding row. Each row will have one more entry than the preceding one. In the figure given below is shown how the Pascal triangle can be formed

1

1   1

1   1 + 1   1

1   2 + 1   2 + 1   1

1   3 + 1   3 + 3   3 + 1   1

1   4 + 1   4 + 6   6 + 4   4 + 1   1

1   5 + 1   5 + 10   10 + 10   10 + 5   5 + 1   1

....

This on simplification gives the Pascal Triangle as

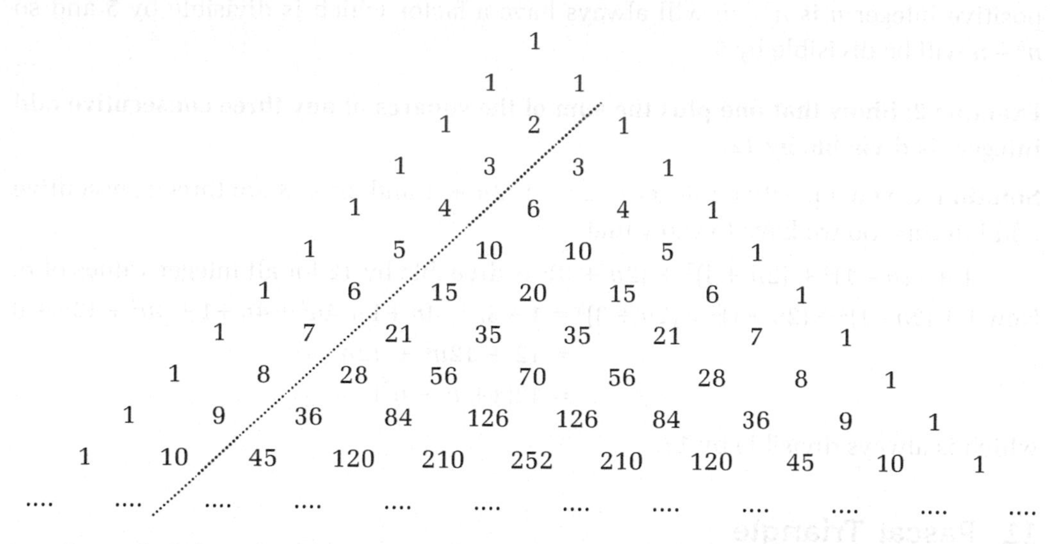

1
1 1
1 2 1
1 3 3 1
1 4 6 4 1
1 5 10 10 5 1
1 6 15 20 15 6 1
1 7 21 35 35 21 7 1
1 8 28 56 70 56 28 8 1
1 9 36 84 126 126 84 36 9 1
1 10 45 120 210 252 210 120 45 10 1
....

In probability distributions Pascal triangle emerges in the following example. Suppose we toss coins and calculate the frequency of heads that we can expect in toss of n coins (n = 1, 2, 3, ...), then we obtain

Number of coins tossed	Number of heads	Number of possible occurance
1	0 1	1 1
2	0 1 2	1 2 1
3	0 1 2 3	1 3 3 1
4	0 1 2 3 4	1 4 6 4 1
5	0 1 2 3 4 5	1 5 10 10 5 1

and so on.

It may be of interest to note that number of possible occurrences of heads (and also tails) for each number of coins tossed are the numbers in the next row of the Pascal triangle! Thus for example when 4 coins are tossed possible occurrences of 0, 1, 2, 3, 4 heads are the respective numbers in the 5th row of Pascal triangle. Thus on generalization when $n(n = 1, 2, 3, 4, ...)$ coins are tossed possible occurrence of 0, 1, 2, 3, ... n heads are the entries in the $n + 1$th row of Pascal triangle.

Pascal triangle is not only helpful in statistics it is helpful in other fields of mathematics as well. It possess several interesting features. We list here some of these:

(i) Number of the entries in n^{th} row is $(n - 1)^{th}$ power of 2. For example in the first row it is $2^{1-1} = 2^0 = 1$, in the second $2^{2-1} = 2$, in the third it is $2^{3-1} = 4$, in 4th it is $2^{4-1} = 8$ and so on.

(ii) If we consider digits in each row as a single number for example 1's in the second, as 11, digits in the third as 121, digits in the fourth as 1331 and so on, we notice all of these are powers of 11^{n-1} upto 6th row. Subsequent rows are also $n - 1$th powers of 11 but with some rearrangement of terms.

(iii) A dotted line has been drawn in Pascal Triangle. Whereas the numbers above the oblique dotted line are natural numbers 1, 2, 3, 4, 5, ... numbers immediately below the dotted line are sums of natural numbers upto point immediately above it across the dotted line. For example 3 is sum of 1 and 2, 6 is the sum of 1, 2 and 3, 10 is sum of 1, 2, 3 and 4 and so on. So the sum of natural numbers upto 8 is 36 and upto 9 is 45 and so on.

(iv) Sum of two consecutive numbers below the dotted line is a perfect square.

For example
$1 + 3 = 4 = 2^2$
$3 + 6 = 9 = 3^2$
$6 + 10 = 16 = 4^2$
$10 + 15 = 25 = 5^2$
$15 + 21 = 36 = 6^2$
$21 + 28 = 49 = 7^2$
$28 + 36 = 64 = 8^2$
$36 + 45 = 81 = 9^2$ and so on.

(v) Numbers in the nth row can be expressed as ${}^{n-1}C_r$, $r = 0, 1, 2, 3, ..., n$. For example

In the first row $n = 1$ so ${}^{n-1}C_0 = {}^0C_0 = 1$

In the second row $n = 2$ so numbers in second row are ${}^{2-1}C_0 = {}^1C_0 = 1$ and ${}^1C_1 = 1$.

In the third row numbers are ${}^2C_0 = 1$, ${}^2C_1 = 2$, ${}^2C_3 = 1$.

In 4th row the numbers are ${}^3C_0 = 1$, ${}^3C_1 = 3$, ${}^3C_2 = 3$, ${}^3C_3 = 1$

In the 5th row the numbers are ${}^4C_0 = 1$, ${}^4C_1 = 4$, ${}^4C_2 = 6$, ${}^4C_3 = 4$, ${}^2C_4 = 1$ and so on.

(vi) If we evaluate the sum of entries in various rows then these are 1, 2, 4, 8, 16, 32, 64, 128, etc. In other words sum of entries in n^{th} row is 2^n. Now, since (as has been shown above in (i)) entries in n^{th} row can be written as ${}^nC_0$, ${}^nC_1$, ${}^nC_2$, ..., ${}^nC_n$. Therefore we prove a well known result of algebra that

$$ {}^nC_0 + {}^nC_1 + {}^nC_2 + \dots\dots + {}^nC_r + \dots\dots + {}^nC_n = 2^n $$

It is possible to generate Fibonacci numbers from Pascal triangle by adding numbers along indicated slanting lines shown in the following figure

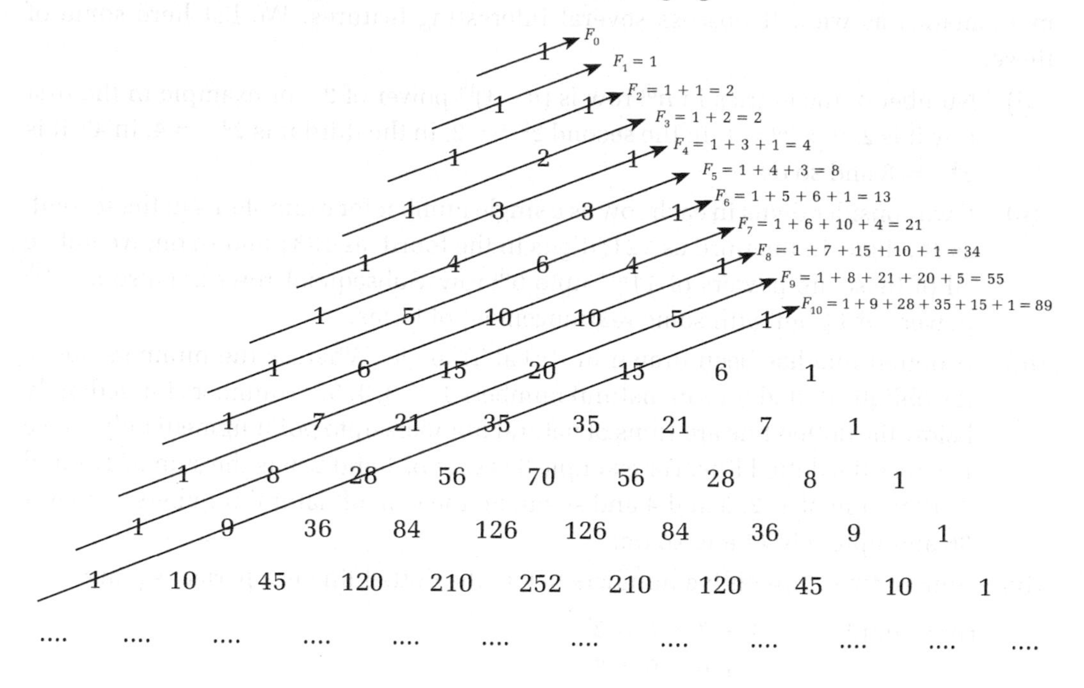

These are $F_0 = 1$, $F_1 = 1$, $F_2 = 2$, $F_3 = 3$, $F_4 = 5$, $F_5 = 8$, $F_6 = 13$, $F_7 = 21$, $F_8 = 34$, $F_9 = 55$, $F_{10} = 89$, ... etc.

12. Pythagorean Triplets

Most of us are familiar with Pythagoras theorem according to which if ABC is a right angled triangle with right angle at C and sides opposite angles A, B, C of the triangle are denoted by a, b and c respectively then

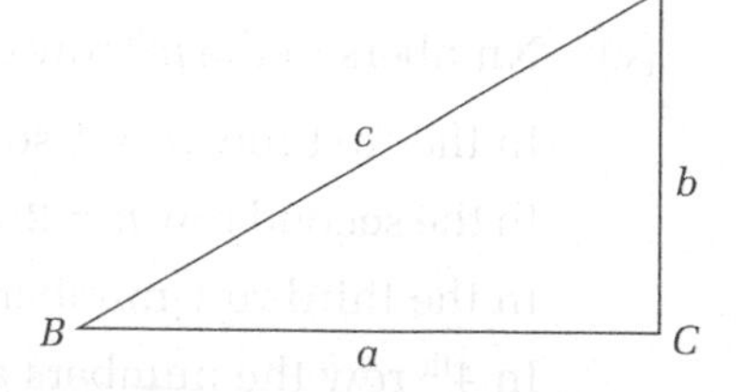

$$ a^2 + b^2 = c^2 \qquad (1) $$

Positive integers a, b, c satisfying (1) are called Pythagorean triplets. These are triplets of integers in which sum of the squares of two integers is equal to the square of the third integer. Few of these are (3, 4, 5), (5, 12, 13), (8, 15, 17), (7, 24, 25). Are there more? The answer is yes. But how to find these? There is a formula for generating Pythagorean triplets. This is as under.

12.1 Formula for Generating Pythagorean Triplets

Let m and n $(m > n)$ be two relatively prime numbers of which one is even and the other odd. Then

$a = m^2 - n^2$, $b = 2mn$ and $c = m^2 + n^2$ will be Pythagorean triplets.

It is easy to verify the correctness of this statement, as

$$a^2 + b^2 = (m^2 - n^2)^2 + 4m^2n^2 = m^4 + n^4 + 2m^2n^2 = (m^2 + n^2)^2 = c^2$$

In fact if m and n are relatively prime, it can be shown that so will be $a = m^2 - n^2$ and $b = 2mn$

Some sets of Pythagorean triplets are:

m	n	a	b	c
2	1	3	4	5
3	2	5	12	13
4	1	15	8	17
4	3	7	24	25
5	2	21	20	29
5	4	9	40	41
6	1	35	12	37
6	5	11	60	61
7	2	45	28	53
7	4	33	56	65
7	6	13	84	85

Some interesting results about Pythagorean triplets are:

(i) Every Pythagorean triplet has one number divisible by 3 and one number divisible by 5.

(ii) Product of Pythagorean triplets is a multiple of 60.

13. Algebra of the Golden Section

If AB is straight line and C a point on it such that

$$\frac{AC}{CB} = \frac{CB}{AB}$$

1–x x
A C B

then the point C is said to divide the straight line AB in golden ratio. Golden section is supposed to have aesthetical beauty. Golden rectangle formed with sides in golden ratio is considered by psychologists to be aesthetically a pleasing rectangle, compared to other rectangles. The design of Parthenon in Athens, Greece is based on the shape of a golden rectangle.

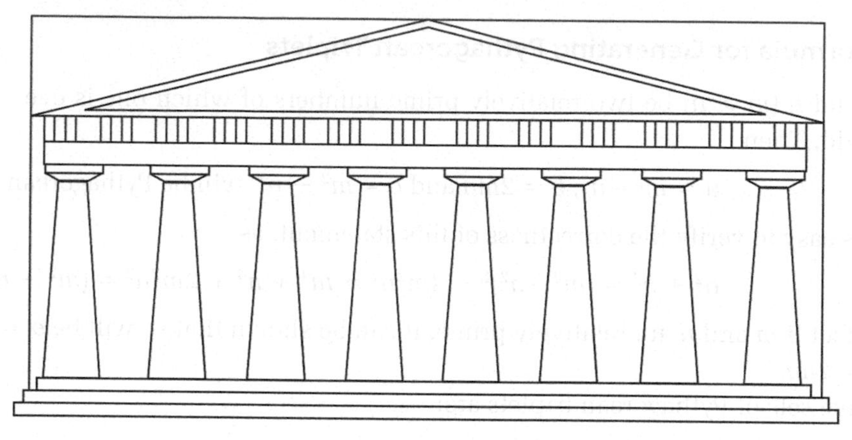

Design of Parthenon in Athens

Besides its aesthetic beauty, golden section has also several other interesting properties.

Writing $x = CB$ and $AB = 1$, we get for golden ratio of AB

$$\frac{1-x}{x} = \frac{x}{1}$$

i.e.
$$x^2 + x - 1 = 0 \qquad \text{(i)}$$

which is a quadratic equation having roots

$$x = \frac{-1 \pm \sqrt{5}}{2}$$

So, one of the root is $\frac{\sqrt{5}-1}{2}$. If we donate this root by $\frac{1}{\phi}$, i.e., if

$$\frac{1}{\phi} = \frac{\sqrt{5}-1}{2}$$

Then
$$\phi = \frac{2}{\sqrt{5}-1} = \frac{2(\sqrt{5}+1)}{(\sqrt{5}-1)\ (\sqrt{5}+1)} = \frac{\sqrt{5}+1}{2}$$

So,
$$\phi \times \frac{1}{\phi} = \frac{\sqrt{5}-1}{2} \times \frac{\sqrt{5}+1}{2} = \frac{5-1}{4} = 1$$

Also
$$\phi - \frac{1}{\phi} = \frac{\sqrt{5}+1}{2} - \frac{\sqrt{5}-1}{2} = 1$$

In fact ϕ is only such a number for which $\phi \times \dfrac{1}{\phi} = 1$ as well as $\phi - \dfrac{1}{\phi} = 1$

(Try if you can find another number with such a property). More surprisingly ϕ is a transcendental number. Its value is

$$\phi = 1.61803398874989.....$$

and $$\frac{1}{\phi} = .61803398874989..... \qquad \text{(ii)}$$

so that $$\phi - \frac{1}{\phi} = 1$$

ϕ can be expressed in recurring decimal form also

$$\phi = \cfrac{1}{1+\cfrac{1}{1+\cfrac{1}{1+\cfrac{1}{1+\cfrac{1}{1+\cdots}}}}} \qquad \text{(iii)}$$

To verify that it is really so, we notice

$$\phi = \frac{1}{1+\phi} \quad \text{or} \quad \phi^2 + \phi - 1 = 0$$

i.e., ϕ is root of the quadratic equation defining golden section. Again ϕ can also be expressed as

$$\phi = \sqrt{1+\sqrt{1+\sqrt{1+........}}} \qquad \text{(iv)}$$

To show that it is really so, we notice

$$\phi = \sqrt{1+\phi}$$

$$\phi^2 = 1 + \phi$$

Hence, ϕ is a root of the quadratic equation of type $x^2 - x - 1 = 0$

Its roots are $x = \dfrac{1 \pm \sqrt{5}}{2}$. One of these is $\dfrac{\sqrt{5}+1}{2}$ which is ϕ.

Another interesting fact to be noted about it is that if we find the powers of ϕ, then we find

$$\phi^2 = \left(\frac{\sqrt{5}+1}{2}\right)^2 = \frac{5+1+2\sqrt{5}}{4} = \frac{6+2\sqrt{5}}{4} = \frac{3+\sqrt{5}}{2} = 1 + \frac{1+\sqrt{5}}{2}$$

So
$$\phi^2 = 1 + \phi = F_0 + F_1\phi$$
$$\phi^3 = \phi \times \phi^2 = \phi(\phi + 1) = \phi^2 + \phi = \phi + 1 + \phi$$
$$= 2\phi + 1 = F_2\phi + F_1$$
$$\phi^4 = \phi \times \phi^3 = \phi(2\phi + 1) = 2\phi^2 + \phi = 2(\phi + 1) + \phi$$
$$= 3\phi + 2 = F_3\phi + F_2$$
$$\phi^5 = \phi \times \phi^4 = \phi(3\phi + 2) = 3\phi^2 + 2\phi = 3(\phi + 1) + 2\phi$$
$$= 5\phi + 3 = F_4\phi + F_3$$
$$\phi^6 = \phi \times \phi^5 = \phi(5\phi + ..) = 5\phi^2 + 3\phi = 5(\phi + 1) + 3\phi$$
$$= 8\phi + 5 = F_5\phi + F_4$$
$$\phi^7 = \phi \times \phi^6 = \phi(8\phi + 5) = 8\phi^2 + 5\phi = 8(\phi + 1) + 5\phi$$
$$= (13\phi + 8) = F_6\phi + F_5$$

and so on.

In general, $\phi^n = F_{n-1}\phi + F_{n-2}$, $n = 2, 3, 4, 5, 6, ...$
where $F_0, F_1, F_2, F_3, ..., F_n$, are Fibonacci numbers defined by $F_0 = F_1 = 1$ and $F_{n+1} = F_n + F_{n-1}$

It is perhaps in view of these fascinating properties and its aesthetic beauty that this ratio is called Golden ratio.

14. Concept of Limit

Concept of limit should not be taken lightly. It is a sophisticated concept and can be easily misunderstood and may sometimes lead to curious situations. We illustrate this with following examples.

14.1 Limiting Lengths

Example 1: Suppose we have a zig-zag type of staircase as shown in the accompanying diagram. Clearly the total length of this zig-zag staircase is $a + b$ (as can be found by summing separately the horizontal and vertical components).

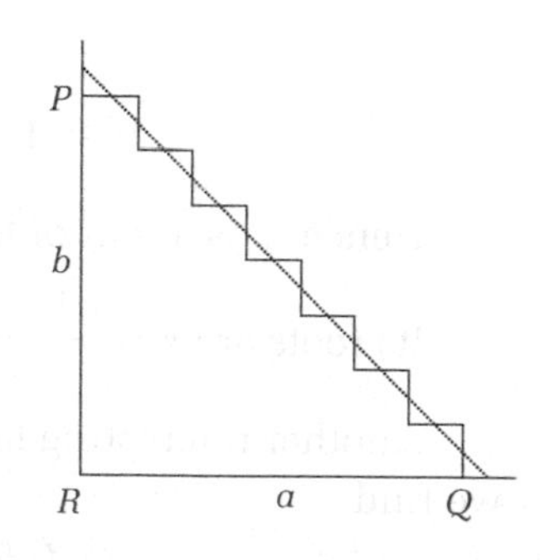

Now suppose we keep on increasing the number of steps indefinitely. The sum continues to remain same, i.e., $a + b$.

Next consider the limiting situation where number of steps become very very large and as a result of which the length and height of each step goes on becoming small and small and gradually approaches zero and the staircase virtually becomes the straight line joining P and Q. But PQ is hypotenuse of the right angled triangle PQR right angled at R.

Hence $$PQ = \sqrt{PR^2 + RQ^2} = \sqrt{b^2 + a^2}$$

Thus in limit the length of staircase as the number of steps increase indefinitely becomes $\sqrt{b^2 + a^2}$ and not $a + b$ as it is when number of steps is finite.

Why limiting length of staircase is different from the finite length? This is because as n, the number of steps approaches infinity, the height and length of each step approach zero. Thus limiting length is $0 \times \infty$ i.e., $0 \times (1/0)$, or $0/0$ which is indeterminate and can assume different values in the limit under different circumstances. (In mathematics $0/0$, ∞/∞, $\infty \times 0$ are not defined. Their limiting values can be found under specified conditions and so can be different in different situations)

Example 2: A similar situation arises when smaller semicircles arc drawn inside a bigger semicircle as shown in the following figure.

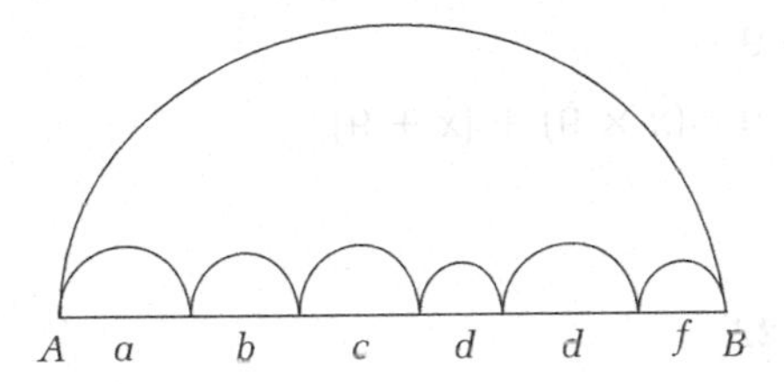

Arc length of the bigger semicircle is

$$\frac{1}{2}\pi AB = \frac{1}{2}\pi\,(a + b + c + d + e + f)$$

Again total arc lengths of small semicircles drawn within the bigger semicircle is

$$\frac{\pi}{2}a + \frac{\pi}{2}b + \frac{\pi}{2}c + \frac{\pi}{2}d + \frac{\pi}{2}e + \frac{\pi}{2}f$$

$$= \frac{\pi}{2}(a + b + c + d + e + f) = \frac{1}{2}\pi AB$$

So the sum of the arc length of smaller semicircles is same as length of the bigger semicircle. This will be so irrespective of the number of small semicircles drawn. Now suppose the number of small semicircles is increased indefinitely, then the lengths of the smaller semicircles will each approach the length of diameter on which these are drawn. Hence the sum of lengths of these semicircles approaches AB as $n \to \infty$.

But length of bigger semicircle is still $\frac{\pi}{2}AB$. Thus $\frac{\pi}{2}AB = AB$, which is not correct.

The reason is again the same. In the limit as $n \to \infty$, the length of each small circle approaches zero. So total length approaches $0 \times \infty$ which is **indeterminate** as in the earlier case.

15. Algebraic Entertainments

It may look strange to regard algebra as a means of entertainment. But it is so. In this section we present some instances where algebra can be used to make sense of certain phenomena observed in behaviour of numbers and some other areas of mathematics, with the object of providing some entertainment and excitement to the reader.

1. An Odd Relation

Any two digit numbers ending in 9 can be expressed as the sum of the product of its digits and the sum of its digits.

Example: $79 = (7 \times 9) + (7 + 9)$

Justification. Let number be x 9.

Then its value is $10x + 9$

Now $10x + 9 = (x \times 9) + (x + 9)$

Hence the result.

2. Mysterious Number 22

Select any three digit numbers whose all digits differ from one another. Write all possible two digit numbers which can be formed with its three digits. Now find their sum and divide this sum by the sum of digits of the original three digit number. The answer will always be 22.

Example: Suppose three digit number is 365.

Sum of its digits is $3 + 6 + 5 = 14$

The sum of possible two digit numbers which can be formed with its digits 3, 6 and 5 are:

$$36 + 35 + 63 + 53 + 65 + 56 = 308$$

Now, $308/14 = 22$

Why is it so? Algebra can explain it.

Let the chosen number be say xyz.

Then sum of its digits is $x + y + z$. (i)

Sum of possible two digit numbers which can be formed with x, y, z is

$$xy + xz + yx + yz + zx + zy$$

and their value is $(10x + y) + (10x + z) + (10y + x) + (10y + z) + (10z + x) + (10z + y)$

$$22x + 22y + 22z = 22\,(x + y + z) \quad \text{(ii)}$$

Dividing (ii) by (i) the answer is 22 whatever digits be x, y and z.

16. Fallacies and Paradoxes

Incorrect use of the rules of algebra may sometimes lead to fallacies and paradoxes. We present here some of these.

1. Misuse of square root

We know that $16 - 36 = 25 - 45$

Adding 81/4 on both sides we can write

$$16 - 36 + \frac{81}{4} = 25 - 45 + \frac{81}{4}$$

or $$(4)^2 - 2 \times 4 \times \frac{9}{2} + \left(\frac{9}{2}\right)^2 = (5)^2 - 2 \times 5 \times \frac{9}{2} + \left(\frac{9}{2}\right)^2$$

i.e., $$\left(4 - \frac{9}{2}\right)^2 = \left(5 - \frac{9}{2}\right)^2 \quad \text{using } a^2 - 2ab + b^2 = (a-b)^2 \qquad \text{(i)}$$

Taking square root on both sides

$$4 - \frac{9}{2} = 5 - \frac{9}{2} \qquad \text{(ii)}$$

or $4 = 5$? Incorrect result.

But where have we gone wrong? It is while taking square root of (i). It may be noted that the square root of $(a - b)^2$ has two values $\pm (a - b)$. Had we taken –ve sign on one side of (ii) we would have got

i.e., $$4 - \frac{9}{2} = -\left(5 - \frac{9}{2}\right)$$

or $-\frac{1}{2} = -\frac{1}{2}$ which is correct.

2. Misuse of Zero

$$a = b$$

or $$ac = bc$$

adding $ab - b^2$ on both sides

$$ab - b^2 + ac = ab - b^2 + be$$

Transferring ac to the right hand side

$$ab - b^2 = ab - b^2 + bc - ac$$

or $$b(a - b) = b(a - b) - c(a - b)$$

or $$b(a - b) = (a - b)(b - c) \qquad \text{(i)}$$

Cancelling $(a - b)$ on both side

$b = b - c$ which is true only when c is zero.

However nowhere in our analysis we had assumed c to be zero.

Where is the fallacy? Fallacy occur red when we divided (i) by $(a-b)$ on both sides. Result is valid only if $a - b \neq 0$. However we had assumed in the very beginning that $a = b$.

3. 1 = 2?

Let $\quad a = b$

Then $\quad a^2 = ab$

or $\quad a^2 - b^2 = ab - b^2$

i.e. $\quad (a-b)(a+b) = b(a-b)$

Cancelling $(a-b)$ on both sides

$$a + b = b$$

or $\quad a + a = a$ since $b = a$

i.e., $\quad 2a = a$ or $2 = 1$?

Here again the fallacy is in canceling $a - b$ on both sides. Since we started with the assumption that $a = b$ so $a - b = 0$ and division by zero is not permitted.

4. 1 = –1?

We know that $\sqrt{a} \times \sqrt{b} = \sqrt{ab}$

Putting $a = b = -1$, we get

L.H.S. is $\quad \sqrt{-1}\,\sqrt{-1} = (\sqrt{-1})^2 = -1$

R.H.S. is $\quad \sqrt{(-1)\times(-1)} = \sqrt{(-1)^2} = \sqrt{1} = 1$

So $\quad 1 = -1$

This is not correct. Where is the flaw?

The flaw is in writing $\sqrt{a} \times \sqrt{b} = \sqrt{ab}$. This is valid only when at least one of a and b is non-negative. Since in the present case both a and b are negative, $\sqrt{a} \times \sqrt{b} = \sqrt{ab}$ is not applicable.

5. All numbers equal

Title is of course preposterous. However consider the following

$\dfrac{x-1}{x-1} = 1$ $\qquad$ Putting $x = 1$, L.H.S. is $\dfrac{0}{0}$, R.H.S. is 1.

$\dfrac{x^2-1}{x-1} = x + 1$ $\qquad$ Putting $x = 1$, L.H.S. is $\dfrac{0}{0}$, 0 R.H.S. is 2.

$\dfrac{x^3-1}{x-1} = x^2 + x + 1$ $\qquad$ Putting $x = 1$, L.H.S. is $\dfrac{0}{0}$, 0 R.H.S. is 3.

$\dfrac{x^4-1}{x-1} = x^3 + x^2 + x + 1$ $\qquad$ Putting $x = 1$, L.H.S. is $\dfrac{0}{0}$, 0 R.H.S. is 4.

$$\frac{x^n - 1}{x - 1} = x^{n-1} + x^{n-2} + \ldots + x^2 + x + 1 \quad \text{Putting } x = 1, \text{ L.H.S. is } \frac{0}{0}, \text{ 0 R.H.S. is } n.$$

Thus, $\frac{0}{0} = 1 = 2 = 3 = 4 = \ldots\ldots = n$. Hence all numbers equal!

This however is not true. This result also shows that an indeterminate form can assume different values since division of zero by zero is not uniquely defined. It is indeterminate and can assume different values under different circumstances. In fact only its limiting value can be found and this limiting value can be different under different circumstances as happened in the above cases. As mentioned earlier not only 0/0 but ∞/∞, $0 \times \infty$, $\infty - \infty$ arc also intermediate forms.

6. Fallacies in Sums of Infinite Series

Let $S = 1 - 1 + 1 - 1 + 1 - 1 + 1 - 1 + 1 - 1 + \ldots\ldots\ldots$ (i)

Then $S = (1 - 1) + (1 - 1) + (1 - 1) + (1 - 1) + (1 - 1) + \ldots\ldots\ldots = 0$

Again $S = 1 - ((1 - 1) + (1 - 1) + (1 - 1) + (1 - 1) + \ldots\ldots\ldots) = 1 - 0 = 1$

Thus $0 = 1$?

What has gone wrong? It is essentially the misuse of the rules of mathematics. In the first case the series is not convergent. This is an oscillatory series. In such a series sum oscillates between two numbers as we keep on changing the number of terms in the series. In the present case sum oscillates between 0 and 1. It is zero if the number of terms is even and 1 if the number of terms is odd.

Again consider the infinite series

$$S = 1 + 2 + 4 + 8 + 16 + 32 + 64 + \ldots\ldots \quad \text{(ii)}$$

Then,
$$S - 1 = 2 + 4 + 8 + 16 + 32 + 64 + \ldots\ldots$$
$$= 2(1 + 2 + 4 + 8 + 16 + 32 + 64 + \ldots\ldots)$$
$$= 2S$$

$\therefore S - 1 = 2S$ or $S = -1$ which is impossible as sum of a series of positive terms can not be negative.

To appreciate the fallacy in this case, consider

$$S = 1 + \frac{1}{2} + \frac{1}{4} + \frac{1}{8} + \frac{1}{16} + \frac{1}{32} + \ldots\ldots \quad \text{(iii)}$$

$$2S = 2 + 1 + \frac{1}{2} + \frac{1}{4} + \frac{1}{8} + \frac{1}{16} + \ldots\ldots$$

$$= 2 + S$$

So, $S = 2$ and this is correct.

Error in the previous case lies in summing series which is not convergent. An infinite series of the type $1 + r + r^2 + r^3 + \ldots\ldots r^n + \ldots\ldots$ is convergent only if $|r| < 1$.

In that case it has a finite sum which is $\frac{1}{1-r}$.

However, if $|r| > 1$, the series is not convergent. It is said to be divergent. The sum S of such a series is not finite. Its value keeps on increasing as more and more terms are added. Whereas (iii) is a convergent series (ii) is a divergent series.

16. When Algebra may not be so Helpful

Although algebra helps in obtaining solutions to many otherwise complex problems, occasionally it is the other way round. A result which can be intuitively guessed may not be conveniently possible to prove using algebra. Here are some examples.

Suppose we want to find four consecutive integers whose product is 120. Let us try to obtain these using algebra.

Let the numbers be x, $x + 1$, $x + 2$, $x + 3$

Then, $x(x + 1)(x + 2)(x + 3) = 120$

or $x(x^2 + 3x + 2)(x + 3\} = 120$

or $x^4 + 6x^3 + 11x^2 + 6x = 120$

It is not easy to solve this quartic equation for finding an integer value of x for which this equation is satisfied.

However, one could have intuitively guessed such four integers to be 2, 3, 4 and 5 as $2.3.4.5 = 120$.

Similarly it may not be easy to prove the following results on prime numbers using algebra.

(i) Every number greater than 2 can be expressed as a sum of two primes.

Examples $3 = 2 + 1$, $4 = 3 + 1$, $5 = 3 + 2$, etc.

By the way this result is known as Goldbach's conjecture and has not been satisfactorily proved as yet by algebra or any other method.

(ii) Every odd number greater than 5 can be expressed as sum of three primes.

Example. $7 = 2 + 2 + 3$, $11 = 7 + 3 + 1$ etc.

Even here algebra is not of any help in proving it.

Mind Teasers

1. Fourteen students in a class study Hindi, eight Sanskrit and four both. How many students are there in the class if (a) each student in the class studies at least one of these two languages, (b) there are 5 students in the class who study none of these two languages.

2. Mohan's brother is 5 years senior to him in age. Sum of the ages of Mohan and his father is 58. Father is 23 years older than Mohan's elder brother. What are the present ages of Mohan, his brother and his father?

3. It has been shown in section 4 that sequences $3n + 1$ and $3n + 3$ end up in loops for any positive integer choice of the value of n. Does it happen (i) if n is a negative integer, (ii) in the case of sequences $3n + 5$ and $3n + 7$?

4. A prince was in love with a princess and wanted to marry her. The king, the father of the princess, was of mathematical bent of mind. He set the following condition for the prince: 'I shall write three two digit numbers, say a, b and c. You then have the choice to name three numbers, say x, y and z. After this I shall tell you the sum of the expression $ax + by + cz$. Based on this information if you could tell me the correct values of numbers a, b and c thought by me, you will be allowed to marry the princess, otherwise not.' Help the prince.

5. A mischievous student tore out 25 consecutive pages from a book. Can their sum be 1990?

6. Find the last digit of the sum of series $1^2 + 2^2 + 3^2 + \ldots\ldots + 99^2$.

7. Some bacteria are placed in a glass. One second later each bacteria splits itself into two bacteria. Next second again each bacteria splits itself into two. After one minute the glass is full of bacteria. When was the glass half full?

8. In the problem of section 5.2 of determining the number of containers which can store entire water of the world, suppose the containers are to be built spherical in shape instead of cubical shape, then how many containers will be needed and what will be the radius (in kilometers) of the largest container?

9. In how many ways can four balls coloured red, black, blue and green be arranged (i) in a row, (ii) in a circle. What if all the balls are of the same colour?

10. Given k persons. (i) How many pairs can be made from them? (ii) How many subsets can be formed? (iii) In how many ways can they be seated on k chairs? (iv) In how many ways can they be seated on m chairs ($m < k$)? (v) How many groups of m persons ($m < k$) can be formed?

11. There are ten girls and four boys in a class. In how many ways can they be seated around a round table so that (i) no two boys are next to each other, (ii) all girls are seated together.

12. Determine the total number of 7 letter codes such that (i) no letter is repeated in the code, and (ii) two specific letters (say a and b) are next to each other in the code.

13. Veena has cans of 8 different colours and she wants to paint 4 squares of 2×2 on a board in such a way that no two neighbouring squares are painted in the same colour. In how many ways can she do it? (two colour schemes are to be considered identical if one can be obtained from the other by rotation).

14. Find missing numbers in the following Pascal triangle

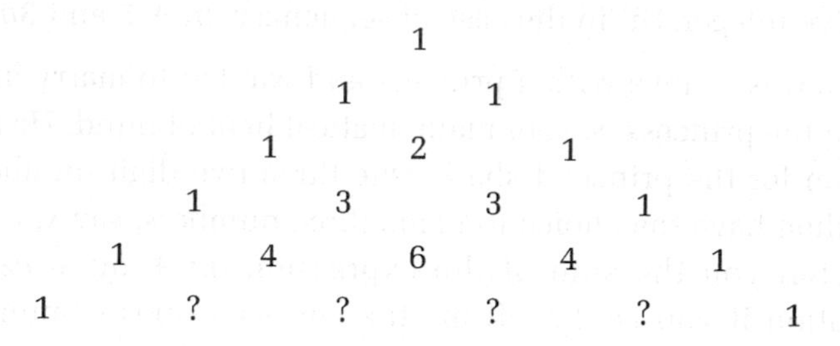

15. Find missing numbers in the following Pascal triangle

1

2 4

3 6 9

4 ? ? 16

5 ? 15 ? 25

6 ? ? 24 ? ?

16. Show that if $a + b + c$ is divisible by 6, then so is $a^3 + b^3 + c^3$.

17. If $a^2 + b^2 = c^2$ where a, b, c are natural numbers, then show that at least one of the numbers a, b, c is divisible by 3.

Chapter 6
Geometry

1. Introduction

Geometry deals with figures drawn on a plane or in space or on some non plane surface. There are two main aspects with which geometry usually deals. One is the drawing of geometrical figures and measuring lengths, areas, volumes and surface areas etc. The other aspect of geometry is establishing general types of results related to specific geometrical figures such as the sum of internal angles of every triangle is 180°. While establishing such general results effort is made to be as much unbiased, objective and impersonal as possible and to rely upon facts and results which are universally valid. We shall begin with plane two dimensional geometry and then proceed to consider three dimensional geometry in space as well as geometries on surfaces other than plane surfaces.

2. Two Dimensional Plane Geometry

Two dimensional plane geometry deals with figures drawn on a plane. This part of the geometry was the first to catch the attention of early human civilization as it related to their day to day problems. The starting point of plane geometry is a point. Even though mathematicians say that it has no dimensions it is usually denoted by a dot on the plane. If one keeps on plotting points consecutively side by side we get a line. If the line is straight, it is called a s**traight line** else a plane curve. If we draw three straight lines in a plane each straight line crossing the other two we get a triangle. Thus a **triangle** is a plane figure enclosed by three straight lines. Similarly a **quadrilateral** is a plane figure enclosed by four straight lines, **pentagon** a figure enclosed by five straight lines and **hexagon** a figure enclosed by six straight lines and so on. Figure enclosed by a closed curve is called a **circle** if the distance of each of its points is same from a fixed point inside this closed curve. This point is called the centre of the circle and fixed distance its radius. Other plane curves which have been extensively studied are **ellipse**, **parabola** and **hyperbola**.

3. Tracing a Plane Geometrical Figure without Raising Pencil

While drawing a geometrical figure one has to frequently raise pencil. We consider in this section certain geometrical figures which can be drawn without raising pencil even once. Some geometrical figures at first sight appear so complex that it does not seem possible that these can be drawn without raising pencil but with a bit of diligence it is seen that these can be drawn. On the contrary there are some simple looking graphs which cannot be drawn without raising pencil.

D C O A B

Fig. 1(a)

D C A B

Fig. 1(b)

Consider a square or a rectangle ABCD whose both diagonals AC and BD are joined. One can check that it is not possible to draw this diagram without raising pencil at least once. Next consider the same square or rectangle but only with one diagonal say AC. Is it possible to draw it without raising pencil? The answer is both yes and no. It depends from where we start.

If we start from corners A or C it is possible. However if we start from corner B or D it will not be possible. The problem of deciding whether a graph can be drawn without having to raise pencil even once is an interesting problem. It has been investigated in detail by renowned mathematician Euler. He came up with the following rules.

(i) Count the number of vertices at which odd number of edges meet. If their number is more than two (Fig. 1(a)) then it is not possible.

(ii) If number of such vertices is two (Fig. 1(b)), then it is possible provided the drawing of the graph is started from one such vertex.

(iii) If graph has no vertex where odd number of edges meet (for example a triangle, a square or a rectangle without a diagonal) then it will always be possible to draw such a graph without raising pencil starting from anywhere.

We can use these rules to decide which of the following figures can be traced without raising pencil and without revisiting any part of the graph.

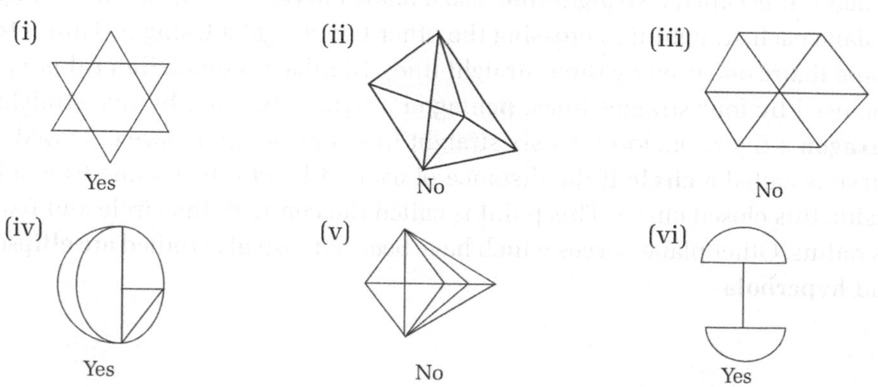

4. Seven Bridges Problem

Tracing graphs without raising pencil even once has an interesting application in what is now popularly known as seven bridges problem. The problem is as under:

In 17th century there was a town by the name of Konigsberg in Prussia (now Germany). Through this town passed river Pregel which divided itself into two branches with two islands in between. Seven bridges in all had been constructed to connect the two banks and the two islands roughly sketched as under:

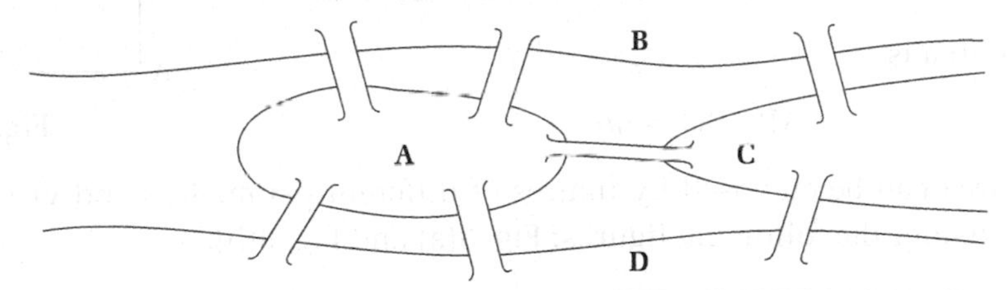

Figure: The layout of the seven bridges of Königsberg

Town folks used to take walk along the two river banks and these two islands using these seven bridges. A question which was often asked was: Is it possible to take a walk in such a way that one can start from one bank and visit both the islands as well as the opposite bank and return back to the starting place without using any of these bridges more than once?

To answer this problem, Consider the land masses A, B, C, D as four vertices and connecting bridges as the edges. We roughly get a graph as shown below:

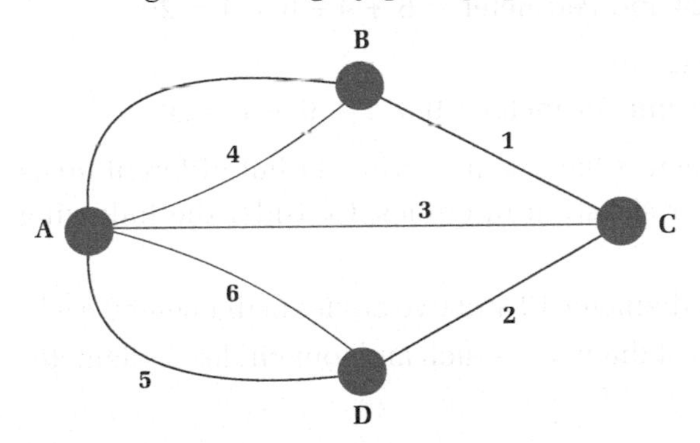

In this graph there are more than two odd vertices, these are B, D and C, as well as A. Hence this problem has no solution(you can try).

Sometimes significant results emerge from apparently simple looking problems. Königsberg problem led Euler to think of a general problem relating to connections between points. This in turn led to a more sophisticated branch of mathematics which is now called **Graph theory**. More over results obtained by Euler do not depend upon specific shapes and sizes of the objects. This concept of shape independent results has given rise to another subject in mathematics known as **Topology**.

5. Areas and Perimeters

Where as the perimeter of a closed figure is the total length of the sides enclosing the figure, its area is a measure of the space enclosed by the perimeter of the figure. For example, in the case of rectangle ABCD whose sides are of length a and b. (Fig. 2). its parameter is:

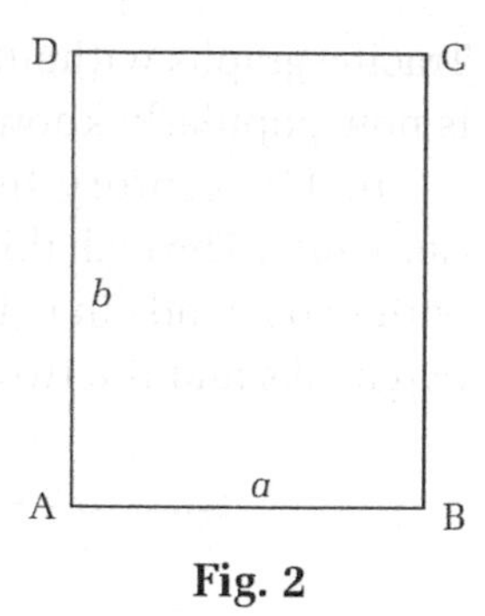

Fig. 2

$$AB + BC + CD + DA = a + b + a + b = 2a + 2b$$

and its area is

$$AB \cdot BC = ab.$$

Same area can be enclosed by figures of different parameters and vice-versa. For example consider the following figures: Fig. 3(a) and Fig. 3(b).

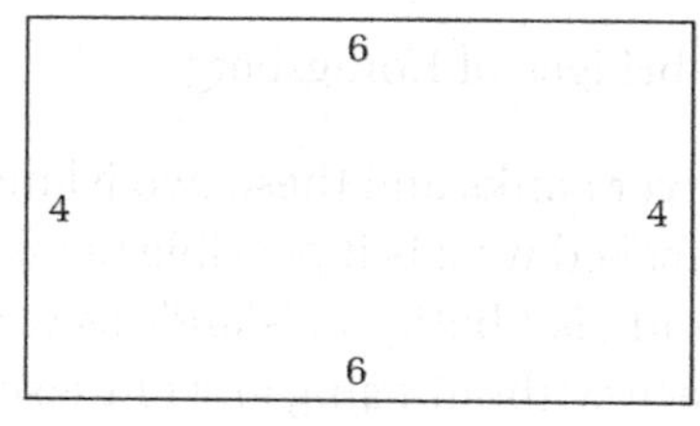

Fig. 3(a)

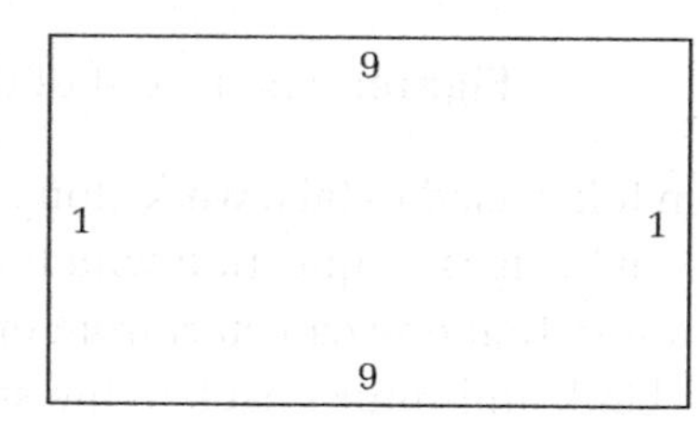

Fig. 3(b)

In the case of the rectangle of Fig. 3(a).
Area = 4.6 = 24 and Perimeter = 6 + 4+ 6 + 4 = 20

In the case of Fig. 3(b)
Area = 1.9 = 9 and Perimeter= 9 + 1 + 9 + 1 = 20

So these two figures have same perimeter but different areas We can have similar situation in circles, Consider the following circles:

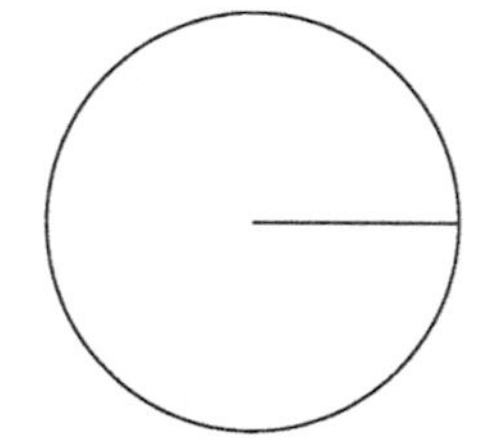
Fig. 4

(i) One circle of diameter 12 and two circles of diameter 6 each.
(ii) Three circles of diameter 4 each and four circles of diameter 3/2 each.

Case:	Number of Circles	Diameter ($2r$):	Perimeter ($2\pi u$)	Area (πr^2)
(i)	1	12	$2 \cdot \pi \cdot 6 = 12\,\pi$	$\pi \cdot 6 \cdot 6 = 36\,\pi$
	2	6	$2 \cdot 2\pi \cdot 3 = 12\,\pi$	$2 \cdot \pi \cdot 3 \cdot 3 = 18\,\pi$
(ii)	3	4	$3 \cdot 2\pi \cdot 2 = 12\,\pi$	$3 \cdot \pi \cdot 2 \cdot 2 = 12\,\pi$
	4	3/2	$4 \cdot 2\pi \cdot \frac{3}{2} = 12\,\pi$	$\cdot 4\pi \cdot \frac{3}{2} \cdot \frac{3}{2} = 9\,\pi$

Thus where as one circle of diameter 12, two circles of diameter 6 each, three circles of diameter 4 each and four circles of diameter 3/2 each, all have the same perimeter 12π their areas are 36π, 18π, 12π, 9π, which are different.

All this seems to go against one's intuition. We expect equal areas to be enclosed by equal perimeters. In fact in the case of original circle even if we divide diameter 12 into hundred equal parts, we shall still get their total circumference to be same (that is 12π); however the total area enclosed by these hundred circles will be $100 \cdot \pi\ (6/100) \cdot (6/100) = 9\pi/25$ which is extremely small as compared to the area 36π enclosed by the original circle.

5.1 Deceptive border

Most of the printed books in order to appear attractive usually leave some border on all the four sides of the printed page. It will be surprising to note that by leaving such borders a lot of precious space of the page is wasted.

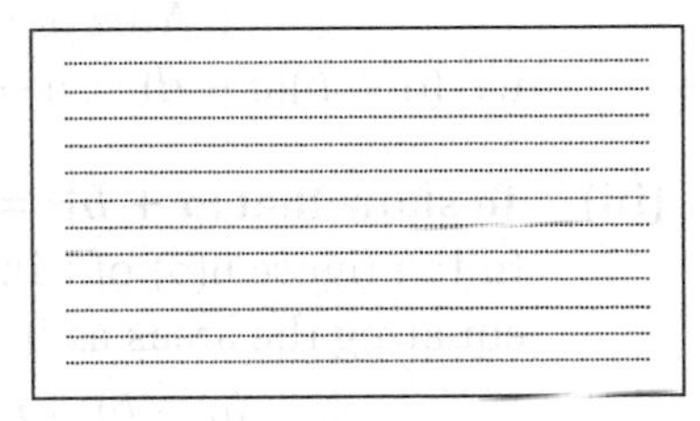

Fig. 5

To get a clear idea of this suppose there is 10 cm by 8 cm page and publisher has left a modest border of 1 cm each on all the four sides. Then the dimensions of the printed part of the page are $10 - 2 = 8$ cm by $8 - 2 = 6$ cm. Thus where as total area of the page is $10 \times 8 = 80$ cm^2, area of its printed portion is $8 \times 6 = 48$ cm^2. Therefore the unutilized portion of the page is $80 - 48 = 32$ cm^2 i.e 40% of the total area of the page. What a huge waste! Is n't it.

6. Use of Geometry in Algebra

Geometry can help in proving certain results of algebra in a simple and straight forward way. We illustrate it with simple examples.

(i) To prove that $a(b + c) = ab + ac$

Draw a rectangle ABCD where one side AB is of length $b + c$ and other AD of the length a.

Let E be a point on AB such that AE $= b$ and EB $= c$. Draw EF parallel to AD. Then from the figure.

Area of rectangle ABCD = area of rectangle AEFD + area of rectangle EBCF

i.e. $a(b + c) = ab + ac$ Proved.

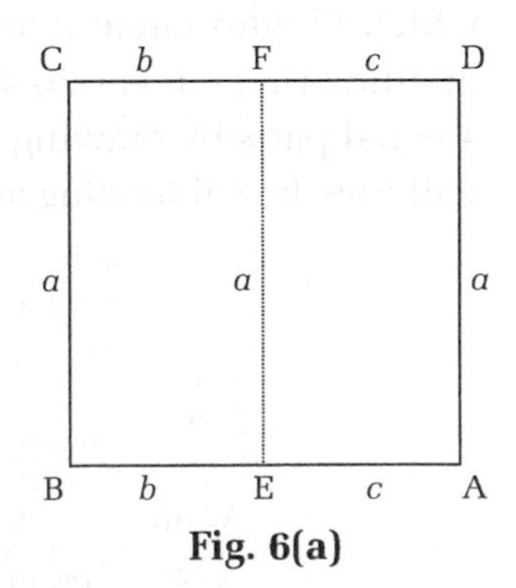

Fig. 6(a)

(ii) To Show that $(a + b)(c + d) = ac + ad + bc + bd$

Draw a rectangle ABCD where side AB is of length $a + b$ and side BC is of length $c + d$.

Let E be the point on AB such that AE $= a$ and EB $= b$. Again let F be a point of BC such that BF $= c$ and FC $= d$. Draw lines EOG and FOH through E and F parallel to BC and AB respectively crossing each other at O. Then from the figure 6(b) Area of rectangle ABCD

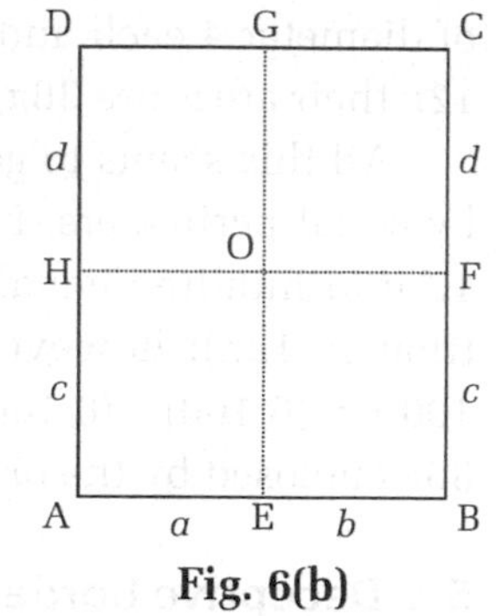

Fig. 6(b)

= Area of rectangle AEOH
+ Area of rectangle EBFO
+ Area of rectangle OFCG
+ Area of rectangle HOGD

i.e. $(a + b)(c + d) = a \cdot c + a \cdot d + b \cdot c + b \cdot d$ Proved

(iii) To show that $(a + b)^2 = a^2 + b^2 + 2ab$

In the figure 6(b) of (ii), let $c = a$, $d = b$, so that ABCD is a square. In this case equating the areas as in (ii):

$$(a + b) \cdot (a + b) = a \cdot a + a \cdot b + b \cdot a + b \cdot b$$
$$= a^2 + ab + b^2 + ab$$
$$= a^2 + b^2 + 2ab \quad \text{Proved}$$

7. Circumference and Area of a Circle

Most of us are familiar with the facts that whereas the area of a circle of radius r is πr^2 its circumference is $2\pi r$. But what is the justification? In this section we briefly try to explain as to how these results can be justified.

In order to appreciate as to how the formula for area of a circle of radius r being πr^2 came into existence, draw a circle of radius r and then sub divided it into 16 equal parts by drawing 8 equally spaced diameters. This may be done by first drawing one diameter which divides circle into two equal parts, then another diameter perpendicular to it dividing ring circle into 4 equal parts and then further subdividing each of these into 4 equal parts by drawing 4 more equally angled diameters in between. Each diameter will now be subtending an angle of $360°/16 = 22\frac{1}{2}°$ at the centre.

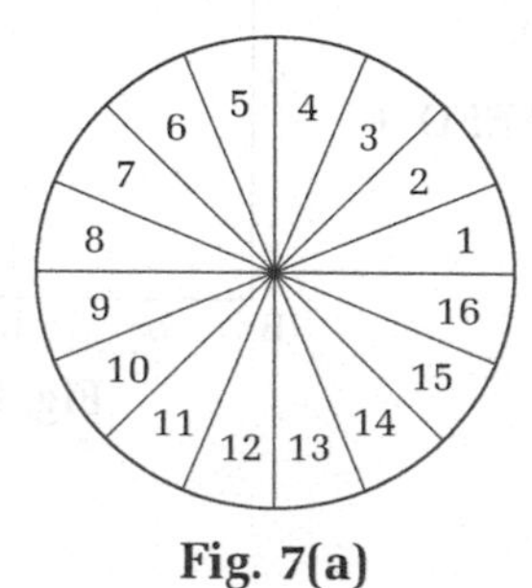

Fig. 7(a)

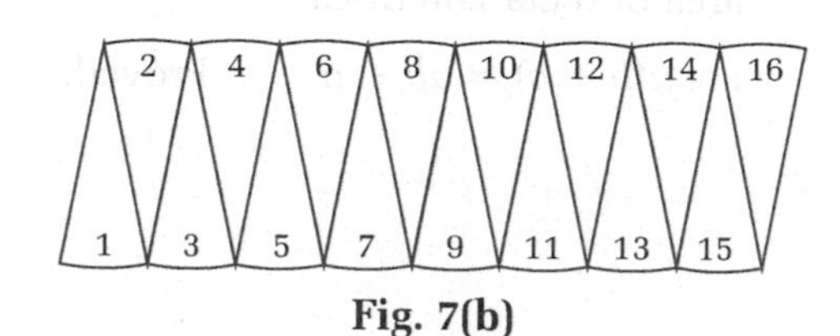

Fig. 7(b)

Now cut these 16 portions and place these side by side as shown in Fig. 7(b). Such a placement suggests that if the circle is subdivided into more and more equal parts, it can be replaced by a parallelogram. Hence the area of the circle will be same as the area of such a parallelogram obtained when the number of sub divisions is very very large.

Now the area of a parallelogram = base x altitude.

In the present case the base is half the circumference (πr) and altitude is r the radius.

Therefore the area of this parallelogram is $= \pi r \cdot r = \pi r^2$

The next question which now has to be answered is as to what is the justification for taking the circumferences of a circle of radius r to be $2\pi r$?

It can be geometrically verified that an arc of a circle which is of the same length as its radius subtends equal angle at the centre irrespective of the length of the radius of the circle. In Fig. 8(a) arc AB = radius OA and arc A′B′ = radius OA′. These subtend same angle AOB at the centre O. This angle is one radian.

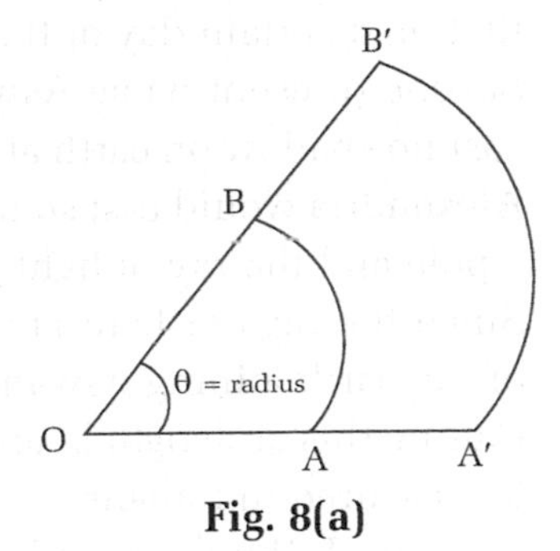

Fig. 8(a)

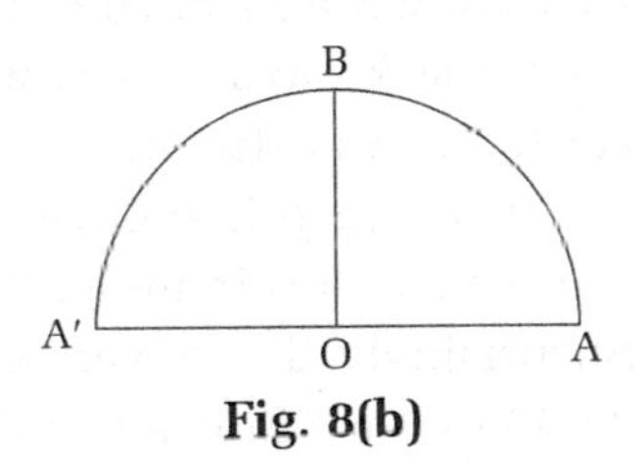

Fig. 8(b)

Thus if an arc of a circle of radius r subtend angle θ radians at the centre then its length is $= r\theta$. In particular in the case of a semi circle, the angle which its arc ABA′ subtends at the centre O is radian measure of 180°.

In radian measure it is taken to be π. Thus the length of arc of a semicircle is πr. So the circumference of the circle is $2\pi r$.

7.1 Value of π

π is a non dimensional measure of the angle in terms of radians. We have already mentioned that π radian is 180°. But what is the value of the non-dimensional measure π. In fact mathematicians have shown that π is a transandental number. In day to day calculations its value is often approximated by 22/7. In decimal form it is taken to be 3.1416 upto 4 decimal places. More accurate values are also available. However being a transcendental number its exact value in decimal form is not possible.

7.2 Circumference of the Earth

Earth is roughly taken to be a sphere. Measuring its radius is not easy. For this one has to go inside it to its centre. However the knowledge of the length of its radius r is necessary to calculate its circumference $2\pi r$.

Surprisingly the length of the radius of the earth had been estimated a long way back and that too with reasonable accuracy (error less than 2%) by Greek mathematician Eratosthenes as early as 230 B.C. To make this measurement, he made use of a result from geometry that alternate interior angles of parallel lines are equal. His argument was roughly as under:

He knew that the sun being very far off from the earth, its rays fall in more or less parallel directions at the same time on two different points on the earth which are not very far off from each other and where the sun is visible at that time. He also knew that on a certain day of the year, the sun was right at the head at noon in the town of Sayene (present name Aswan) in Egypt. As a result a vertical pole fixed there would cast no shadow on earth at that time. However at that very time a vertical pole fixed in Alexendria would cast some shadow. Eratosthenes measured the angle formed by such a pole and the ray of light going past the top of the pole to the far end of the shadow. Since the rays of light are parallel,this is the angle which arc drawn on the surface of the earth joining Aswan with Alexendria would subtend at the centre of the earth. Calling this arclength L, and the radius of earth as r, clearly $L = r\theta$ or $r = L/\theta$, where θ is measured in radians.

Now L the distance between Aswan and Alexendria was known and having measured θ, r the radius of the earth was known. Once r was known circumference of the earth $2\pi r$ was also known.

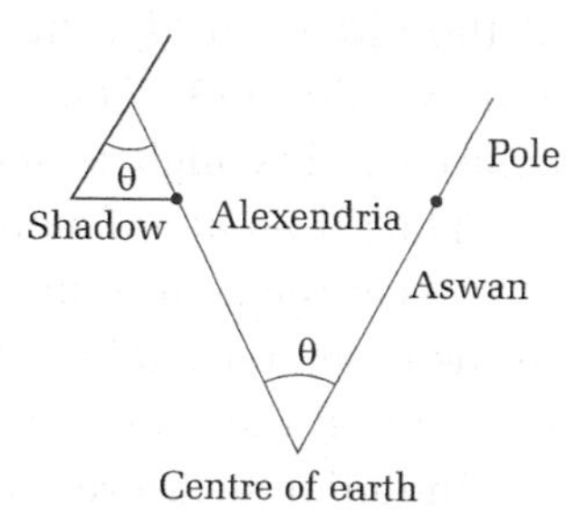

Fig. 9

By using, this approach, Eratosthenes came up with an estimate of the circumference of the Earth as 24,660 miles. Surprisingly this is very close to its modern estimate of 24,900 miles. Eratosthene's estimate of the radius of the earth was 3923 miles where as its recent more accurate estimate based on circumference of 24,900 is 3961 miles, just a difference of 38 miles!

7.3 A Rope Around the Earth

Example: A king ordered the prepration of a long rope which would fit tightly around the earth. Unfortunately by mistake the rope was 1 meter longer than the exact circumference of the earth. As a result it fitted earth loosely. Now the question is: was this loose fit sufficient enough for a mouse to crawl under the rope?

Solution: At first sight it might appear that it would not be possible. But one may be surprised to learn that space would be more than sufficient for not only a mouse to crawl under the rope but also even a bigger animal. Let us see how:

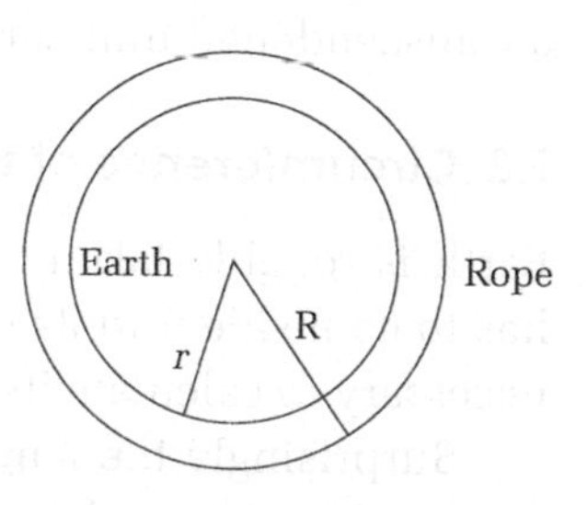

Fig. 10

If R is the radius of the earth then its circumference is $C = 2\pi R$.

Let us suppose the circumference C of the earth is expressed in meters. Then length of the rope is C + 1 meters. It will form a circle around the centre of the earth of radius R′ where

$C + 1 = 2\pi R'$.

Hence $R' - R = (C + 1)/2\pi - C/2\pi = 1/2\pi = 1(2 \times 22/7) = 7/44$ meters = 16 cm.

Thus R′ – R is around 16 cm. So the rope will be fitting loosely around the earth at a height of 16 cm! Hence a mouse can easily crawl under it. Apparently an astonishing result. An increase of just 1 meter in the circumference has increased the radius by around 16 cm.

8. Relating Concentric Circles

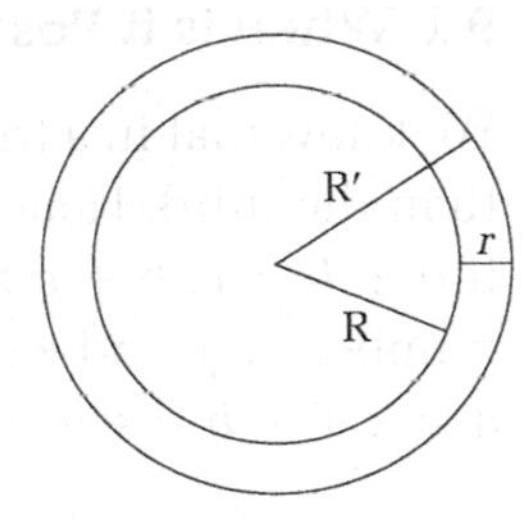

Fig. 10(a)

It may be of interest to note that our end result $R' - R = 1/2\pi$ is independent of Rand C. So whatever the circumference of a circle (be it small or even twice the circumference of the earth), the end result of $R' - R$ will always be $1/2\pi$.

In fact even if the inner circle is made smaller and smaller till its radius R becomes zero, $R' - 0 = 1/2\pi$

Where R′ is radius of the outer circle which now has a unit circumference (as $2\pi R' = 1$).

Similar is the situation with difference in the circumference of two concentric circles. The difference between the circumference of two concentric circles of radius R′ and R will be $2\pi R' - 2\pi R$. If $R' = R + r$ then this difference is $2\pi r$ i.e. the circumference of a circle of radius r.

Thus the difference between two concentric circles and radius R′ and R will always be the circumference of circle of radius equal to the difference between the radii of the two circles. Therefore if we have a circle of radius R and draw another concentric circle of radius R + 10, then the difference between circumferences of two circles will be circumference of a circle of radius 10, whatever be value of R (including zero value).

9. Relationship between the Number of Sides and Measures of Internal Angles of a Regular Polygon

A polygon is an n sided closed figure. Suppose we have a regular polygon of n sides whose all sides are of equal length, then all its internal angles will be equal. The question is whether there is any relationship between the measure of each of these internal angles and n the number of its sides?

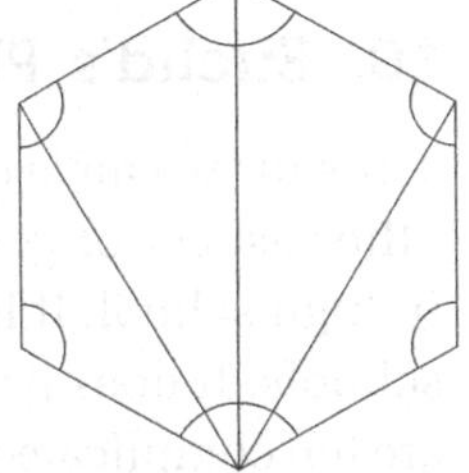
Fig. 11

A polygon of sides n will have n vertices. Let O be one of these; Join it with remaining vertices with the help of straight lines. Then we shall have in all $n - 2$ triangles formed. Now the sum of the internal angles of each of these $(n - 2)$ triangles is 180°. Hence the sum of all internal angles of polygon will be $180(n - 2)^0$. Since polygon is regular all its internal angles are equal and each of these will be $180(n - 2)/n$ degrees.

Examples: In case of an equilateral triangle $n = 3$, therefore each of its internal angle is $180(3 - 2)/3 = 60°$. In case of a square $n = 4$, therefore each of its internal angles is $180(4 - 2)/4 = 90°$,In case of a regular pentagon $n = 5$, Hence each of its internal angles is $180(5 - 2)/5 = 108°$. Similarly in the case of a hexagon $n = 6$ each of its internal angles is $180(6 - 2)/6 = 120°$ and so on.

It may be noted that it is only when n sided polygon is regular, each of its internal angles is $180(n - 2)/n$. In case of a non-regular polygon, different angles will have different measures. However the total sum of all its internal angles will be still $180(n - 2)°$.

9.1 When is it Possible to form of Tringle given Three Sides

We know that in a triangle the sum of the lengths of any of its two sides must be greater than the third. Hence given three sides a, b, c they will be able to form a triangle only if $a + b > c$, $b + c > a$, $c + a > b$. This will be possible if and only if three positive numbers x, y and z can be found such that $a = x + y$, $b = y + z$, $c = z + x$; because if $a = x + y$, $b = y + z$, $c = z + x$, then

$$a + b = x + 2y + z = (x + z) + 2y = c + 2y,$$

thus $a + b > c$ as y is positive,

Similarly $$b + c = y + z + z + x = (y + x) + 2z = a + 2z,$$

thus $b + c > a$ as z is positive,

and $$c + a = z + x + x + y = (z + y) + 2x = b + 2z,$$

thus $c + a > b$ as x is positive.

Conversely if a, b, c form a triangle, then $a + b > c$, $b + c > a$ and $c + a > b$.

Now Let $$a + b = c + 2z, \text{ where } z \text{ is positive.} \quad \text{(i)}$$

$$b + c = a + 2x, \text{ where } x \text{ is positive.} \quad \text{(ii)}$$

and $$c + a = b + 2y, \text{ where } y \text{ is positive.} \quad \text{(iii)}$$

Then adding (i) and (ii) $a + 2b + c = a + c + 2z + 2x$ or $b = x + z$,
adding (i) and (iii) $2a + b + c = b + c + 2z + 2y$ or $a = y + z$,
and adding (ii) and (iii) $b + 2c + a = a + b + 2x + 2y$ or $c = x + y$,

10. Euclid's Plane Geometry

Thus for we have considered practical applications of geometry. Let us now turn to the other aspect of geometry namely Euclid geometry. Euclid's plane geometry is taught in high school. It is often remarked who needs plane Euclid geometry as it is taught in schools. It does not have much of real applications and most of the times proofs given are too complicated and tricky consisting of statements such as 'if-then','since','therefore', 'hence' etc. which are not easy to appreciate.

However that is not really true. It is a misconception which generally makes beginners lose interest in the subject. Besides having several practical applications, Euclid geometry is a wonderful play ground for developing logical and consistent thinking. It may be regarded as a game which has to be played using axiomatic rules created by ancient Greeks which stand to reason and are not arbitrary. Euclid and his predecessors were quite convinced that these rules adequately reflect the laws of the real world. In fact Euclid geometry may be regarded as a game which has to be played under laid down rules. As a game it can be compared with chess. No one can claim to know all the secrets of any of these two great games created by mankind.

Euclid geometry is based on certain axioms and potulates. Starting with these one uses deductive reasoning to establish its various results. Where as axioms are defined as general statements of results which are not subject specific, postulates are truth statements which are specific to that particular subject. Assuming these axioms and postulates whose truth is self evident, the entire edifice of the subject is built in a deductive manner. Axioms are simple common sense statements such as two halves of a quantity are equal to each other or two halves of a quantity taken together make the whole quantity etc. From Euclid's view point, these axioms are akin to the road map where by the construction of geometrical principle is to be directed as well as constructed. Moreover axioms are not numerous.

Postulates by contrast are specific to the subject. For example in geometry we say that a point has no dimensions, through two given points one and only one straight line can be drawn. Similarly two straight lines cannot have more than one point in common and if they have no common point then these straight lines are parallel straight lines. Moreover two parallel straight lines never cross each other however long these are extended on either side. Based on such axioms and postulates, the entire edifice of Euclid geometry has been built. In this and the subsequent sections, we briefly review some of its interesting features.

The entire edifice of Euclidian geometry is built on logical reasoning. Assuming that when we make one complete revolution about a point, we make an angle of 360°, then after half a revolution (when we are exactly in the opposite direction to the original direction from which we started) we make an angle of 180°at a point on it. Thus a straight line is said to subtend an angle of 180° at a point on it.

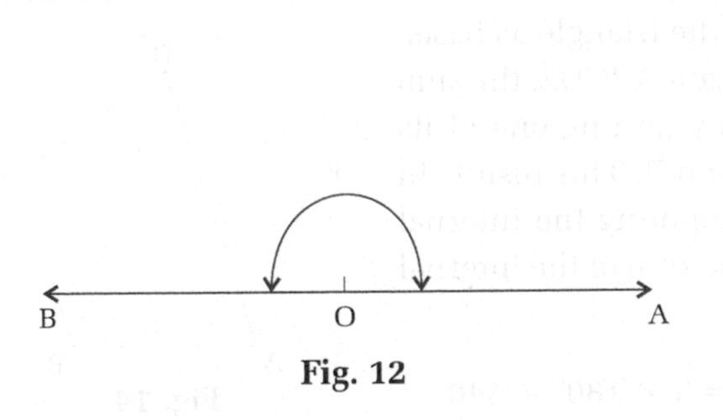

Fig. 12

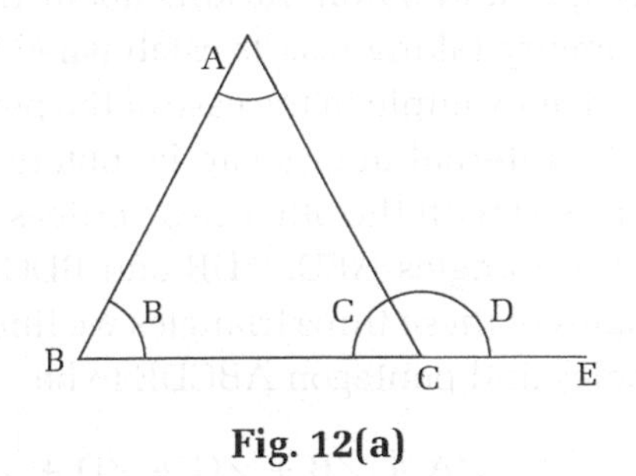

Fig. 12(a)

Again a triangle is a figure enclosed by three straight lines (not parallel to each other). An interesting feature about a triangle is that the sum of three internal angles of every triangle is 180°. It can be verified by measuring internal three angles of any triangle and summing these up. It is also proved logically in high school geometry.

Thus in the figure 12(a) $\angle A + \angle B + \angle C = 180°$ (i)

Let angle ACE be external angle of triangle at point C. Then since BCE is a straight line, therefore

$$\angle C + \angle D = 180° \qquad \text{(ii)}$$

Hence from (i) and (ii) $\angle A + \angle B = \angle D$ (iii)

In other words external angle of a triangle is always equal to the sum of its other two opposite internal angles.

A well known result in high school geometry.

Similar results for figure enclosed by four straight lines (called quadrilateral) can also be established.

Considering quadrilateral ABCD as sum of two triangles ABC and ACD,

we get $\angle A + \angle B + \angle C + \angle D = 180° + 180° = 360°$

(equivalent to one complete revolution about a point).

A quadrilateral becomes a trapezium if one pair and its opposite sides is parallel. It becomes a parallelogram if both pairs of its opposite sides are parallel. A parallelogram whose one internal angle is 90° becomes a rectangle. A rectangle is a square when all its four sides are equal. Several interesting results have been established in Euclid geometry about three sided and four sided figures.

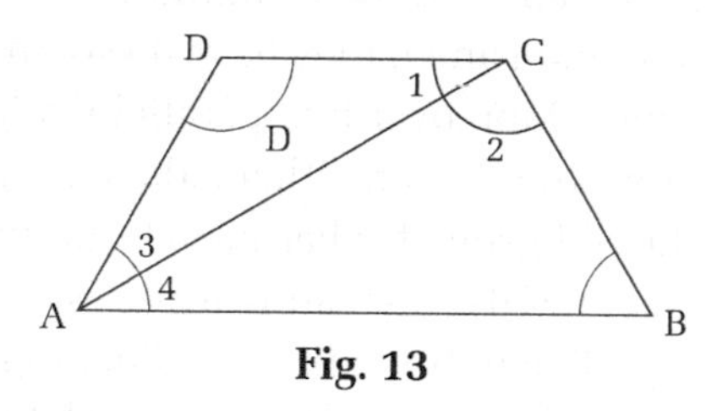

Fig. 13

Figure with sides more than four are also considered. A five sided figure enclosing some area is called a pentagon, a six sided figure a hexagon, an eight sided figure an octagon and so on. Results about these figures have also been established in Euclid geometry taking results established for the triangle as base.

For example in the case of the pentagon ABCDE, the sum of its internal angles can be obtained by joining one of its vertex D with the other two vertices A and B. This results in three triangles AED, ADB and BDC .Summing the internal angles of these three triangles we find the sum of the internal angles and pentagon ABCDE to be

$$\angle A + \angle B + \angle C + \angle D + \angle E = 3 \times 180° = 540°$$

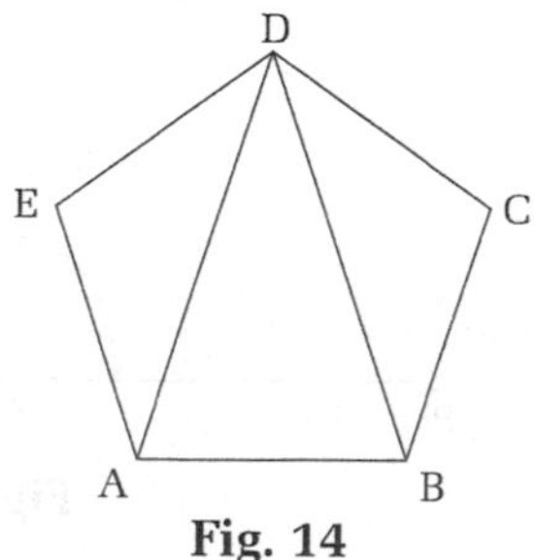

Fig. 14

Similarly sum of the internal angles of not only hexagon and octagon but also of the some other types of figures such as accompanying star type of figure A, B, C, D and E can also be found. It can be easily verified using appropriate triangles that in the case of this star shaped figure

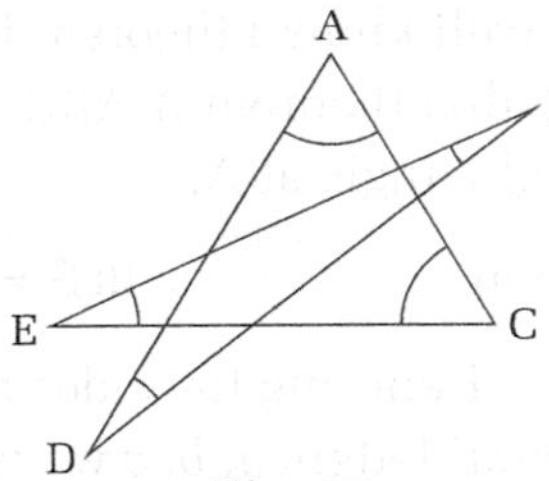

$$\angle A + \angle B + \angle C + \angle D + \angle E = 180°$$

Fig. 15

In Euclid geometry we are not only concerned with the angles but also the sides as well as areas and relationship between the sides and the angles of geometrical figures.

Besides the above type of n sided figures another figure which has been extensively studied in Euclid geometry is circle (which we considered earlier).

10.1 Golden Rectangle

If a point C is chosen on a straight line AB such that AC/CB = CB/AB,then point C is said to divide line AB in golden ratio.

This proportion was known to Egyptians and Greeks. It was perhaps named as 'Golden Ratio' or 'Golden Section' by Leonardo da Vinci who drew geometric diagrams for Fra Luca Pacioli's book in 1509 and was very much fascinated by its properties (discussed in the previous chapter on Algebra).

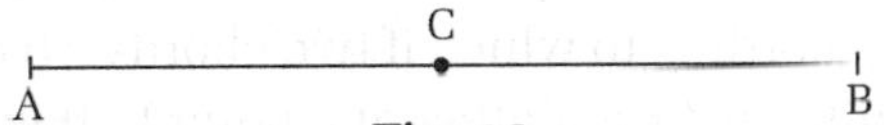

Fig. 16

A rectangle whose length and breadth form golden ratio is called a 'Golden Rectangle'. Such a rectangle has been considered by psychologists to be esthetically very pleasing compared with rectangles having other dimensions. It has often been used in art as well as in architecture in the design of buildings. Parthenon in Athens(Greece) is based on the shape of a golden rectangle.

A question naturally arises as to how to construct a golden rectangle? This can be done as follows. To construct a golden rectangle, begin with a square ABEF. Let M be middle point of one of its sides. Now make a circular arc with M as its centre and ME as its radius. Let D be the point where this arc intersects the side AF extended. Now erect perpendicular to CD at D and let it meet BE extend in C. Rectangle ABCD thus formed is a golden rectangle.

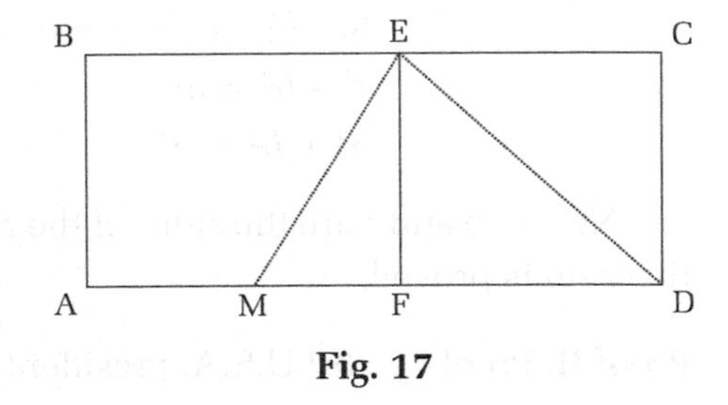

Fig. 17

10.2 Some Alternate Proofs of Pythagoras Theorem

In Euclid geometry we have a number of basic results called Theorems. These results are established by using basic axioms of Euclid geometry, deductive reasoning and certain previously established results in the form of Theorems. Pythagorus Theorem is a well known theorem in Euclid geometry. According to this theorem if ABC is a right angled triangle with right angle at A,

then $$BC^2 = AB^2 + AC^2$$

Denoting the sides opposite the angles A, B, C by small letters a, b, c we write this result as:

$$a^2 = b^2 + c^2$$

Standard proof of this theorem is available in text books of high school geometry.

However in view of its wide spread applications and utility, attempts have been made from time to time to prove it in alternative ways also. In this section we present some interesting alternative proofs.

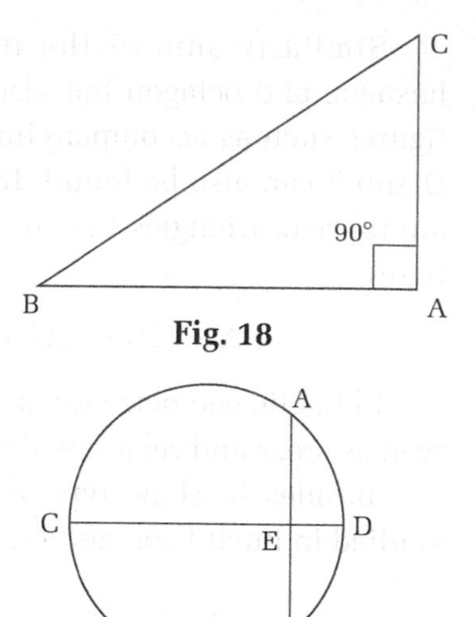

Fig. 18

Fig. 19

Proof 1: This proof makes use of a result of geometry according to which if two chords AB and CD of a circle intersect each other at a point E, (Fig. 19)

then $$AE \cdot EB = CE \cdot ED$$

Now consider a particular case when AB is a diameter and CD is perpendicular to it. (Fig. 20) In this case E will be mid point of DC. So if C is the centre of the circle and CE = a and OE = b,

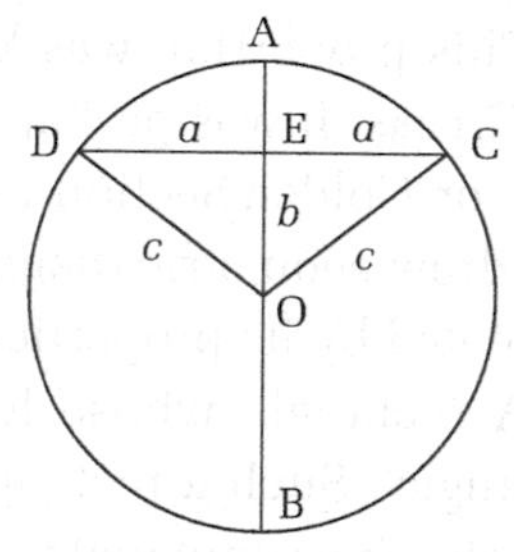

Fig. 20

then
$$AE \cdot EB = CE \cdot ED$$
$$(c - b) \cdot (c + b) = a \cdot a$$
$$c^2 - b^2 = a^2$$
$$a^2 + b^2 = c^2$$

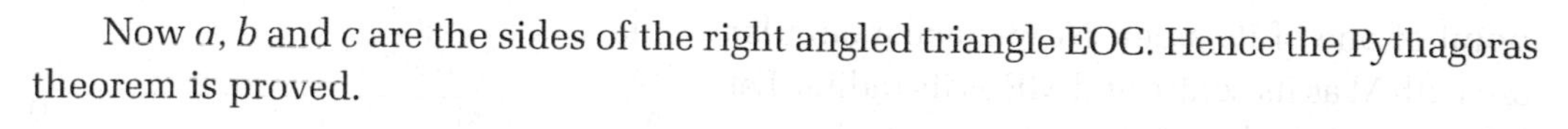

Now a, b and c are the sides of the right angled triangle EOC. Hence the Pythagoras theorem is proved.

Proof II: Proof by 20th U.S.A. president James A. Garfield

Twentieth U.S.A. president James A. Garfield also gave a proof of Pythagoras Theorem using an entirely different approach. President Garfield had not even formally studied mathematics and geometry. His study of geometry was informal self study. Fascinated by Pythagoras theorem and feeling not comfortable with its formal proof, he came up with a proof which is rather simple and straight forward. The proof is as follows:

Consider two congruent triangles ABC and DEC (congruent triangles have corresponding sides and corresponding angles equal) which have a common vertex C and right angles at B and D respectively. The triangles are so placed that DCB is a straight line. i.e. points D, C and B are collinear. (For this angle CAB of upper triangle will correspond to its equivalent angle DCE in the lower triangle. Moreover angle BCA of upper triangle will correspond to its equivalent angle CED in the lower triangle.) (Fig. 21)

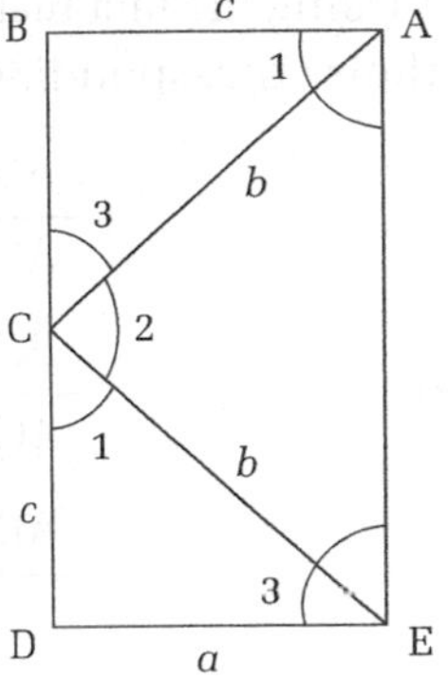

Fig. 21

Since DCB is a straight line, therefore

$$\angle 1 + \angle 2 + \angle 3 = 180°$$

Also $\qquad \angle 1 + \angle 3 = 90°$

(being angles in a right angled triangle other than the right angle.)

Therefore $\qquad \angle 2 = 90°$

Hence AEC is also a right angled triangle with right angle at C.

Now ABDE is a trapezium *as AB is parallel to DE. Hence area of ABDE is

$$\frac{1}{2}(\text{DE} + \text{BA}) \cdot \text{BD} = \frac{1}{2}(a + c) \times (a + c) = \frac{1}{2}(a^2 + c^2 + 2ac) \quad \text{(i)}$$

Also the area of trapezium ABDE is the sum of the areas of triangles CDE, ACE and ACB

i.e. $$\frac{1}{2}ac + \frac{1}{2}b \cdot b + ac = ac + \frac{1}{2}b^2 \qquad \text{(ii)}$$

Equating (i) and (ii) $\frac{1}{2}(a^2 + c^2 + 2ac) = ac + \frac{1}{2}b^2$

or $\qquad a^2 + c^2 = b^2 \qquad$ Proved.

(This proof also appears as one of the five alternative proofs of Pythagoras theorem presented in chapter 37 of the book on Vedic Mathematics by Jagat Guru Swami Sri Bharti Krishna Tirarathji Maharaj)

Proof III: Another Proof is as under.

Consider right angled triangle ABC with right angle at B. Draw BD perpendicular to the hypoteneous AC of the triangle. Then it is easy to verify that triangles ABC and ABD and BDC are similar triangles (in similar triangles ratio of corresponding sides are equal).

* *Trapezium is quadrilateral whose one pair of opposite sides is parallel. Its area is half the sum of parallel sides multiplied with perpendicular distance between parallel sides.*

Using the fact that in case of similar triangles their areas are in the ratio of the square of their corresponding sides, we have

$$\frac{AB^2}{AC^2} = \frac{\text{Area of } \Delta \text{ ADB}}{\text{Area of } \Delta \text{ ABC}}$$

and
$$\frac{BC^2}{AC^2} = \frac{\text{Area of } \Delta \text{ BCD}}{\text{Area of } \Delta \text{ ABC}}$$

$$\frac{AB^2+BC^2}{AC^2} = \frac{\text{Area of } \Delta \text{ ADB} + \text{Area of } \Delta \text{ BCD}}{\text{Area of } \Delta \text{ ABC}}$$

$$= \frac{\text{Area of } \Delta \text{ ABC}}{\text{Area of } \Delta \text{ ABC}} = 1$$

i.e. $AB^2 + BC^2 = AC^2$ Proved.

Proof IV: Algebraic proof:

Although a geometric proof of Pythagoras theorem is usually given, an algebraic proof is also possible. We present here this algebraic proof.

Draw a square ABCD whose each side is of length of $a + b$. (Fig. 22). Let E, F, G and H be points on the sides of this square, such that AE = a, EB = b, BF = a, FC = b, CG = a, GD = b, DH = a and HA = b. Join EFGH. Then EFGH will also be a square. Let each of its sides be c. Moreover EFB, FGC, GHD and HEA are all right angled triangles with right angles at B, C, D and A respectively.

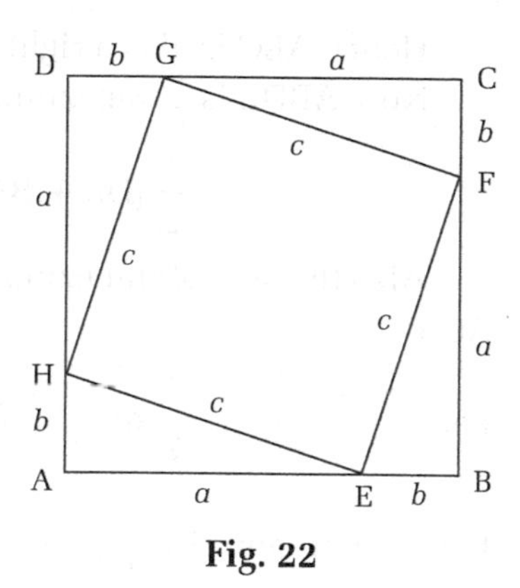

Fig. 22

Now, Area of square ABCD

= Area of square EFGH+ Area of AEBF + Area of AFGC +Area of AGHD + Area of AHEA.

Hence, $(a + b)^2 = c^2 + \frac{1}{2}ab + \frac{1}{2}ab + \frac{1}{2}ab + \frac{1}{2}ab$

or, $a^2 + b^2 + 2ab = c^2 + 2ab$

i.e. $a^2 + b^2 = c^2$ Proved.

Surprisingly even though there are several alternative proofs of Pythagoras theorem some based on algebra and some on geometry, however none is based on trigonometry. Why is it so? The fact is that there can be no proof of Pythagoras theorem based on trigonometry because trigonometry itself is based on Pythagoras theorem. Thus using concepts of trigonometry to prove Pythagoras theorem would amount to circular reasoning.

10.3 Nine Points Circle

One of the joys in geometry is to observe how some unseemly related points are in reality related to each other. In Euclidean geometry, it is shown that a circle can be found to pass through any set of given three non-collinear points (points not all lying in a straight line). It has been shown that in the case of a triangle there exist not three but nine well known points which lie on the same circle. These nine points are:

- Middle points C′, A′, B′ of the sides AB, BC and CA of the triangle ABC.
- Feet of the altitudes D, E, F (points where perpendiculars drawn from vertices of the triangle meet opposite sides).
- Middle points K, L, M of segments joining the centre H of the triangle to the three vertices of the triangle (the centre of a triangle is a point in the triangle where three altitudes drawn from the vertices to opposite sides cross each other. All of them cross at same point).

Fig. 23: Nine points circle

10.4 Brianchon's Theorem

In 1806, at the age of 21, Braianchon a student in polytechnique established a result in Euclid geometry which is known after his name. According to this theorem in a hexagon ABCDEF circumscribed about a conic section such as a circle, three of its diagonals AD, BE and CF cross each other at the same point O. (Fig. 24)

In fact Brianchon's theorem is now regarded as dual of Pascal's theorem with terms, points and lines interchanged. According to this theorem in a hexagon inscribed in a conic section, points of intersection of pairs of its opposite sides are collinear.

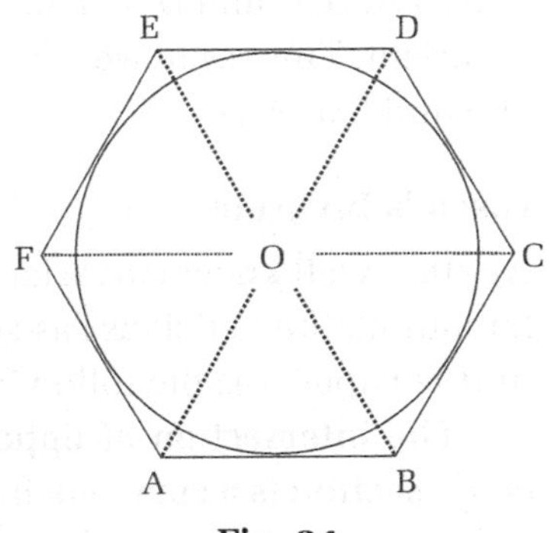

Fig. 24

10.5 Invariance

A result which is not affected by change in data is called invariant. There are several invariant results in geometry. We consider some of these here.

A point of invariance in an equilateral triangle

If any point P is chosen inside an equilateral triangle (which has all sides of equal length and each of its internal angles 60 degree)and perpendiculars PL, PM and PN are drawn on sides BC, AB and AC respectively, then it can be verified (and even proved) that sum of these lengths i.e PL + PM + PN is invariant. In other words it will remain same where ever the point P is chosen. (Fig. 25) As a matter of fact this length is equal to an altitude (perpendicular drawn from a vertex on the opposite side such as AD).

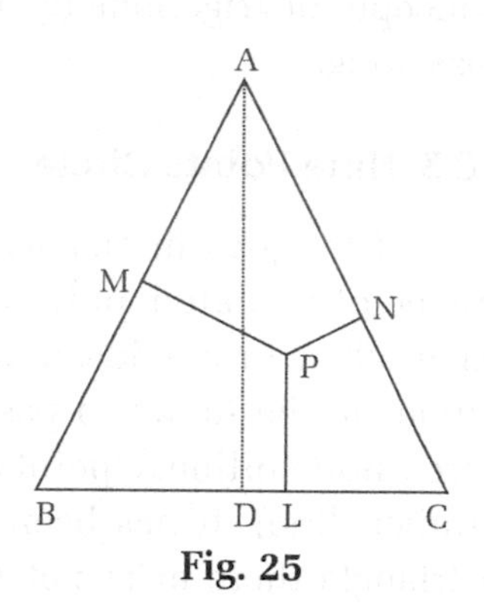

Fig. 25

Pappus Invariant

Draw any two lines on a plane and mark any three points on each of these two lines say A, B, C on the first line and D, E, F on the second line. Now connect the points on the first line with points on the second line ensuring that corresponding points are not connected (A is not connected with D, B not with E and C not with F). Let G, H and I be the points of intersection of these connecting lines. Then points G,H and I will always be collinear (i.e. on the same straight line), whatever be the position of chosen points A, B, C and D, E, F on these two lines. (Fig. 26). This result is known as Pappu's Invariant. This result was discovered by Pappus who lived in Alexendria, Egypt between 300 and 350 A.D.

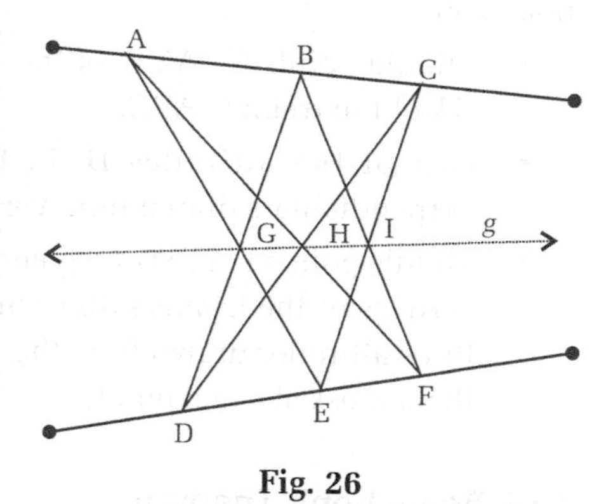

Fig. 26

Pascal's Invariant

Another well known invariant result is attributed to Pascal. In 1640 at the age of sixteen famous mathematicians Pascal published a one page paper. The result that he published in this paper was the following:

The intersection of opposite sides of a hexagon inscribed in a conic section (a conic section is a curve such as circle, ellipse, parabola, hyperbola etc.) are collinear. A formal proof is possible and is available in literature.

10.6 Looks Difficult but Easy

Proving results in geometry is not always easy. However there are sometime situations where proving a certain result apparently seems difficult where as in reality it is not so. We give here examples.

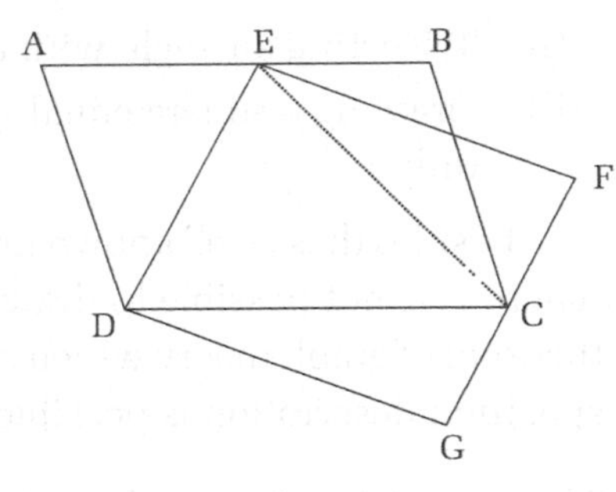

Fig. 27

Example 1: Suppose ABCD is a parallelogram and a point E lies on the side AB and vertex C of the parallelogram lies on a straight line FG such that EFGD is also a parallelogram (Fig. 27). Given the area of parallelogram ABCD say 20 units, what is the area of the parallelogram DEFG?

It apparently looks difficult. However it is not so. Join EC. Then ECD is a triangle which is on the same base DC as the parallelogram ABCD. Moreover both have same altitude (as vertex E is on line AB). Therefore

$$\text{Area of parallelogram ABCD} = 2 \cdot \text{Area of triangle ECD} \quad \text{(i)}$$

Now parallelogram EDGF and triangle EDC are on the same base ED and have same altitude (as the vertex C is on line FG). Hence area of 11gm DEFG is twice the area of triangle ECD. Combining this result with (i) areas of parallelogram ABCD and DEFG are equal.

As another Example

Example 2: Suppose P is a point on the circumference of a circle having centre O. Perpendiculars PE and PF are drawn from P on two perpendicular diameters AOB and COD meeting these in points E and F (Fig. 28). What is the length of straight line EF?

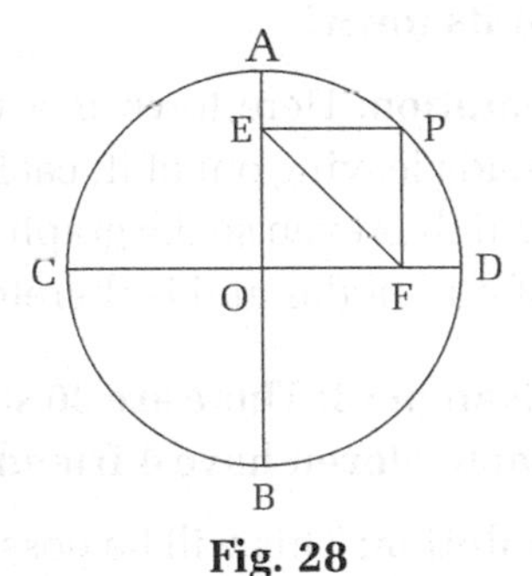

Fig. 28

At first sight it may look difficult. However it can be found in a simple way. Since Pis any point on the circumference of the circle, we can take it to be A also. In that case PF becomes AO and PE becomes zero. In other words F is now the centre O and E the point A and thus EF = AO = radius of the circle. How simple!

11. Some Impossible Constructions in Euclidian Geometry

High school plane geometry has two parts. One is concerned with basic theorems and their proofs and the other with geometric constructions which are usually to be done with a ruler and a compass. There are several of these such as bisecting a given angle, dividing a line in golden ratio, drawing a circle passing through three given points etc. However in Euclidian geometry there are certain constructions which are impossible using a ruler and compass only. Some of these are:

(i) Trisecting an angle with compass and ruler only.

(ii) Drawing a square equal in area to a given circle again using ruler and compass only.

Besides these well known impossible constructions there are several other situations where it is not possible to draw a figure in a specified manner. There is a well known theorem of graph theory which can help in taking a decision whether a graph answering specified description is possible or not.

Theorem: In a connected graph the number of odd vertices must be even

A vertex is said to be odd if number of edges terminating at this vertex is odd. It is said to be even if the number of such edges is even. Number of terminating edges at a vertex is called the degree of the vertex.

Example 1: Can 9 line segments be drawn in a plane each of which intersects exactly 3 others?

Solution: Here taking line as a vertex and its intersection with other segments as edges, we have 9 vertices of odd order(3). Hence such a graph will not be possible. This result has practical applications also.

Example 2: A country has 100 towns. Is it possible that 3 roads lead out from each of its town?

Solution: Here town is a vertex and road an edge. Since each town is to have three roads leaving out of it, each vertex is of odd degree. Since total number of towns is 100 which is even so the graph of connecting roads will be graph having even number (100) of odd nodes and is therefore possible.

Example 3: There are 30 students in a class. Is it possible that 9 of them have 3 friends each, eleven have 4 friends each and 10 have 5 friends each?

Solution: This will be possible if we can draw a graph with 30 vertices of which 9 have degree 3, 11 degree 4 and 10 degree 5. In this case there are $9 + 10 = 19$ odd vertices. Since the number of odd vertices is not even such a graph is not possible. Hence desired format of friendships is not possible in the class.

Example 4: Suppose that cosmic liasons have been established amongst nine planets of the solar system and rockets travel along the following routes:

Earth-Mercury, Pluto-Venus, Earth-Pluto, Pluto-Mercury, Mercury-Venus, Uranus -Neptune, Neptune-Saturn, Saturn-Jupiter, Jupiter-Mars and Mars-Uranus. Is it possible for a traveler to reach Mars from Earth following available connections?

Solution: A diagram of connecting paths is as under. (Fig. 29)

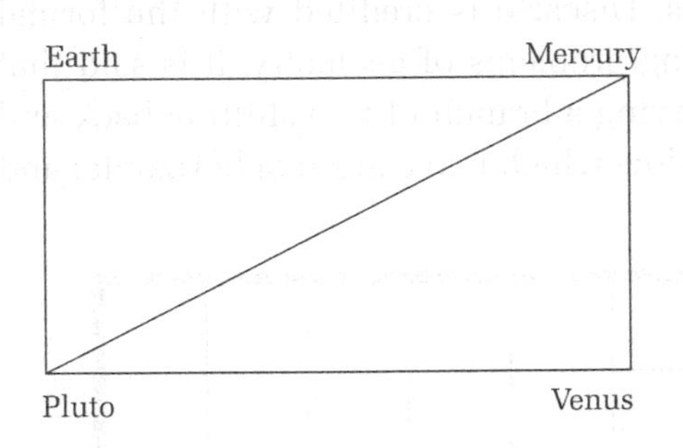

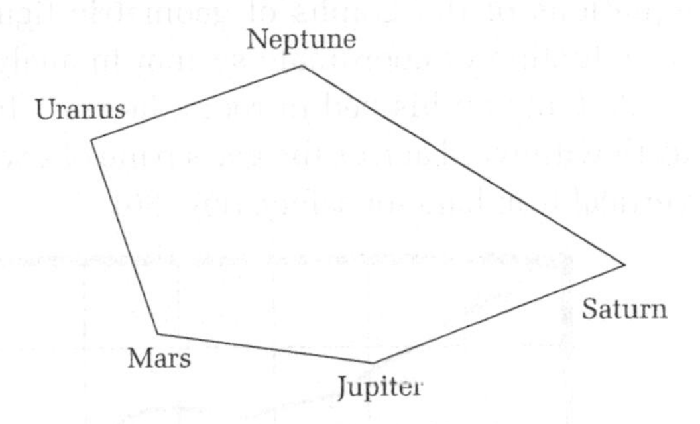

Fig. 29

This clearly shows that it is not possible to travel to mars from earth as there is no connecting route.

Example 5: After coming back from tour of a foreign country, a person told his friends that in the country that he visited there were several lakes each of which was connected with the other through rivers. He also told them that whereas three rivers flowed out of each lake, four rivers flowed into each lake. Show that it is not possible and he is telling lie.

Solution: Each river has two ends (which in the present case are lakes). It flows from one end (one lake) to the other end (other lake). So the total sum of rivers flowing in and flowing out of n lakes must be same. However according to him where as $3n$ rivers flow out, $4n$ rivers flow into each lake and that is not possible.

Example 6: In a certain country each town is connected with every other town with one way road. Show that there is a town in that country from where one can travel to any other town of the country.

Solution: A possible way of proving this is by induction. Let there be n towns .Suppose the result is true for n–1 towns (excluding town say A). That is in case of n–1 towns there is one town say B from which we can travel to any of the remaining n-2 towns. In case road connecting A and B leads from B to A, then we can reach from B to all remaining n–1 towns. Again the result is true for any two towns Hence by induction it is true for all n towns.

12. Coordinate Geometry

Coordinate geometry is essentially the use of algebra in analyzing problems of geometry. Some of the results which are not easy to prove in Euclid Geometry get easily proved when analyzed using coordinate geometry.

Coordinate geometry can be regarded as representation in terms of algebraic equations of the graphs of geometric figures. Discarte is credited with the formal introduction of coordinate system in analyzing problems of geometry. It is said that while lying on his bed in room, he was observing a branch of tree shifting back and forth with wind across the glass pane of a window which had a mesh of horizontal and vertical iron bars for safety. (Fig. 30)

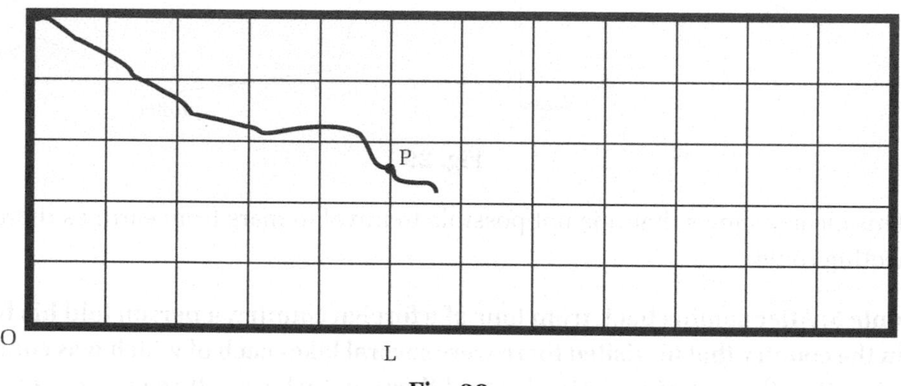

Fig. 30

Descarte realised that he could specify the location of every point of the branch on the window pane by selecting a corner of the window 'O' as origin and then counting the number of meshes in horizontal and vertical direction which needed to be moved to reach the desired position on the branch. If for example the chosen point on the branch is P, then knowing OL, LP position of point Pis uniquely determined. These are usually referred to as x and y coordinates of point P, with OL as x-axis and a line through O parallel to LP as y-axis. (Fig. 31)

Now there can be four points in the plane P_1, P_2, P_3, P_4 for which lengths of x and y ordinates can be same. In order to distinguish these we put as a convention + sign before x coordinate if the point is to the right of Y-axis and –ve if the point is to the left of Y axis. Similarly we put +sign before y coordinate, if the point is above X-axis and –ve sign if the point is below X-axis. Thus we write $P_1(x, y)$, $P_2(-x, y)$, $P_3(-x, -y)$ and $P_4(x, -y)$ to distinguish the four possible locations of a point whose x and y coordinates have identical magnitudes. A point on X-axis has its y coordinate zero and a point on Y-axis has its x coordinates zero. Thus L is $(x, 0)$, $M(-x, 0)$, $N(0, y)$ and $Q(0, -y)$. The origin O has coordinates (0, 0).

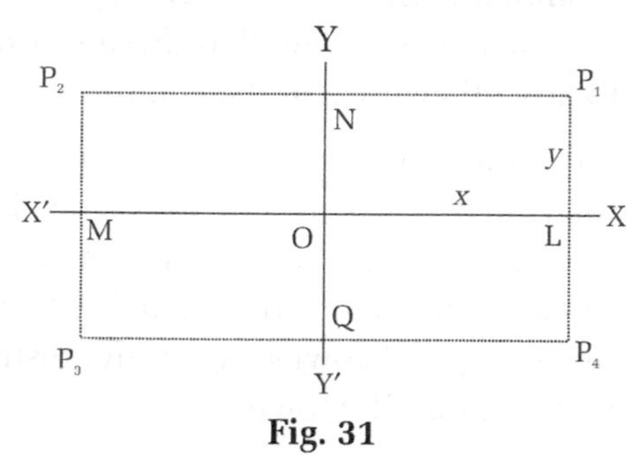

Fig. 31

In fact in two dimensional coordinate geometry effort is made to set up an algebraic relation between the x and y coordinates of the points lying on a given curve. It helps in understanding the behavior and nature of the curve by just looking at such a relationship between x and y without having to even draw or look at the actual graph of the curve. Once algebraic relation is set up help is taken of algebra to analyze its behavior.

For example $2x + 3y = 6$ is a straight line passing through point A(3, 0) on x-axis and point B(0, 2) on y-axis.

Similarly $x - 4y = 14$ is a straight line passing through C(14, 0) and D(0, –7/2). (Fig. 32)

In case we want to find the point P(*x*, *y*) where these two lines cross each other, we may solve the algebraic equations

$$2x + 3y = 6 \quad \text{(i)}$$

and

$$x - 4y = 14 \quad \text{(ii)}$$

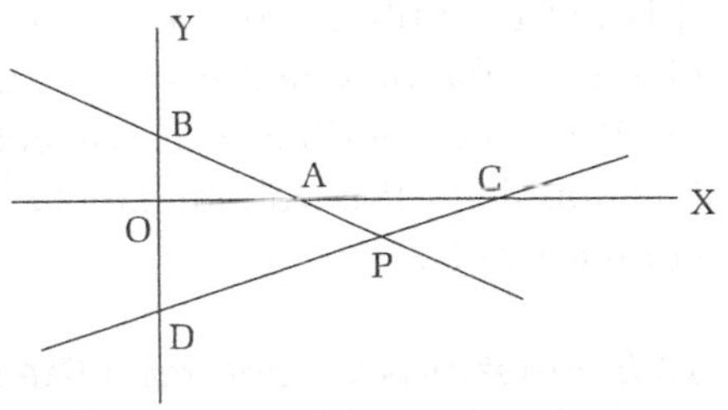

Solving these we find $x = 6$ and $y = -2$. So clearly the two lines cross each other at the point P(6, –2).

Fig. 32

In fact where as a first degree relation between *x* and *y* represents a straight line, a second degree relation between *x* and *y* may represent a pair of straight lines or a conic section. A conic section is a curve obtained when a plane cuts a cone at different angles. Conic sections which are extensively studied using coordinate geometry are circle, ellipse, parabola and hyperbola as these have lot of practical applications. Interested reader can look up any standard book on Coordinate Geometry.

13. Three Dimensional Geometry

Three dimensional geometry (also called solid geometry) deals with figures drawn in three dimensional spaces. Whereas in two dimensional geometry we deal with points, lines and figures drawn in a plane, in three dimensional geometry we consider their more general form in three dimensional space. In three dimensional space we have points, straight lines, curves, planes, surfaces and three dimensional figures such as parallelopiped, cube, cuboids(whose length, breadth and height are not all equal), sphere, spheroid (not all radii equal), cone, cylinder, ellipsoid, paraboliod, hyperboloid etc.

Naturally study of three dimensional figures is more involved as compared to the study of its counterpart two dimensional figures. Help is often taken of three dimensional coordinate geometry to analyze behavior of such figures. Three dimensional coordinate geometry is essentially extension of two dimensional coordinate geometry to three dimensional space. Here the position of a point is represented by P(*x*, *y*, *z*) in place of P(*x*, *y*). (Fig. 33). The third coordinate denoting distance traveled along a straight line through origin O perpendicular to the plane XOY (usually called z-axis).

In this case first degree equation $ax + by + cz = d$ in x, y, z represents a plane. A straight line is represented by the intersection of two planes. Second degree equations represent what are called conicoids whose plane sections are conics. Important conicoids investigated in three dimensional geometry are: sphere, cone, cylinder, ellipsoid and hyperboloid.

13.1 Some Impossible Constructions in 3D

Just as it is not possible to trisect an angle or draw a square equal in area to a given circle in two dimensional space using a ruler and a compass only, in the same way in three dimensional space it is not possible to double a cube. In other words given a cube of side a each so that its volume is a^3, it is not possible to construct a cube which has double of this volume i.e. $2a^3$ using ruler and compass only.

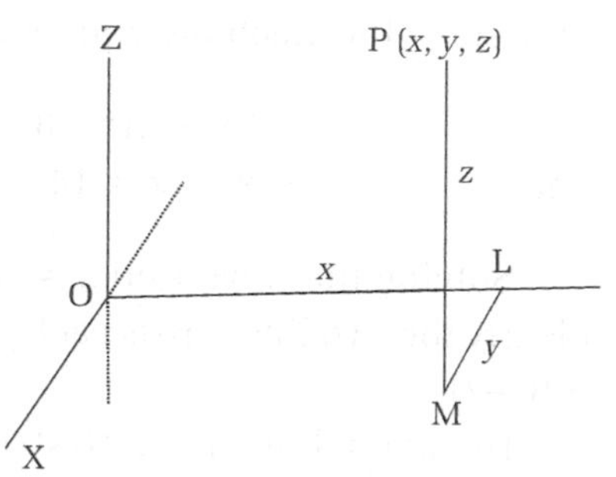

Fig. 33

13.2 Relationship between the Number of Vertices, Number of Edges and Number of Faces of a Polyhedron

A polyhedron is a three dimensional figure bounded by four or more polygon faces in such a way that its faces (F) meet in edges (E) and three or more edges meet in a vertex (V). A cube is an example of a polyhedron.

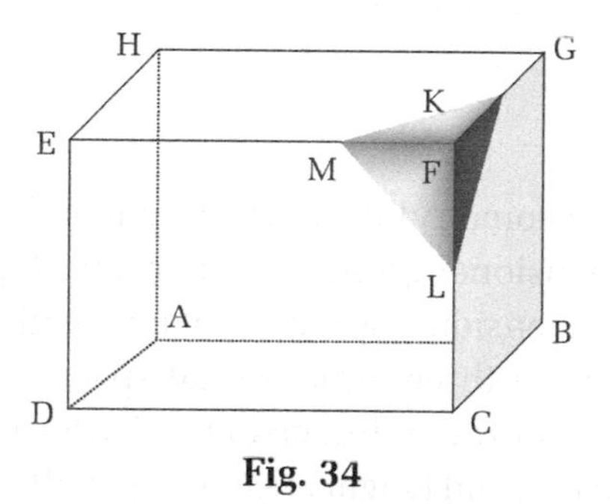

Fig. 34

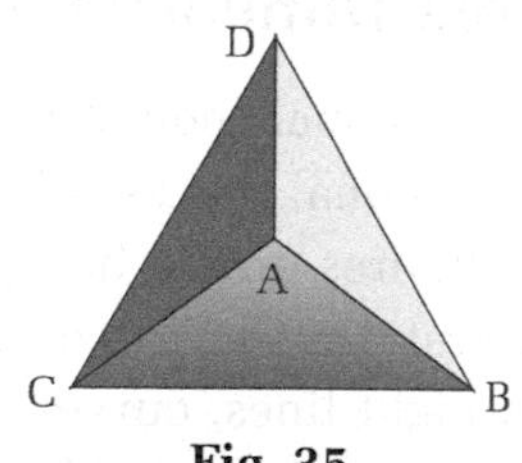

Fig. 35

It has six faces (F = 6), each a square, eight vertices (V = 8)and twelve edges (E = 12) (Fig. 34). Similarly a prism is a polyhedron. It has three faces (F = 3), six edges (E = 6) and four vertices (V = 4). (Fig. 35)

Leonhard Euler in 18th century discovered a remarkable relationship between the number of vertices, number of faces and number of edges of a polyhedron. This relationship is $V + F = E + 2$ or $V + F - E = 2$.

It can be verified that it is true in the case of cube as well as prism. In case we draw a plane (say LKM) cutting all the three edges of vertex F in L, M and K (Fig. 34) and seprate the portion containing this vertex F from the rest of the cube we are left with a polyhedron ABCDEHGKML. It can be seen that in this polyhedron number of faces F = 6 + 1 = 7, number of edges E = 12 + 3 = 15 and number of vertices F = 7 + 3 = 10. So that V + F – E = 10 + 7 – 15 = 2.

13.3 Platonic Solids

The term platonic solids is used to denote polyhedrons in three dimensional spaces with the following properties

(i) All the faces are regular polygons which are congruent (identically similar)

(ii) Same number of faces meet at each vertex.

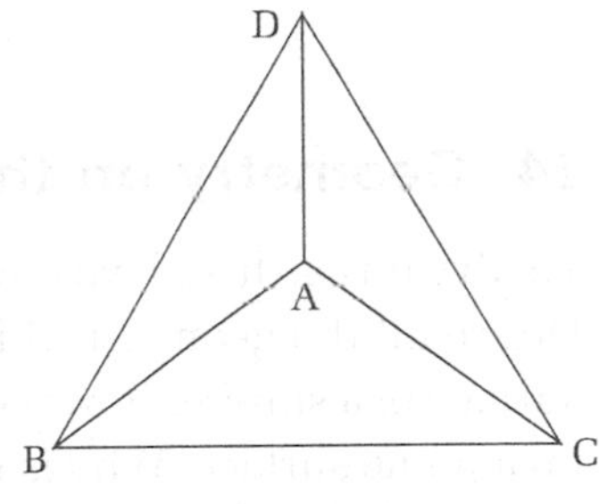

Fig. 35(a)

Cube is an example of a platonic solid. All its faces are identical squares and at each vetex three faces meet. Besides the cube, other four platonic solids are:

(i) **Tetrahedron:** It has four faces each a triangle and at each vertex three faces meet

(ii) **Octahedron:** It has 8 faces each a triangle. At each vertex four triangles meet.

(iii) **Icosahedron:** It has twenty faces each a triangle and at each vertex five triangles meet.

(iv) **Dodecahedron:** It has twelve faces each a pentagon and at each vertex three pentagons meet.

By the way those are the only 5 known platonic solids. (Can you visualize more?) We close this section with a numerical example.

Problem: A cube is inscribed in a sphere. Sphere itself is inscribed in a cone so that the sphere touches the base as well as the slant walls of the cone from inside. Cone is further inscribed in a right circular cylinder. What is the total area of the circular cylinder (including plane top and bottom).

Solution: Accompanying figure (Fig. 36) describes the structure. Let the sides of inner most cube be r each. Since it is inscribed in a sphere, the diameter of this sphere will be the diagonal of this cube and of length

$$\sqrt{r^2 + r^2 + r^2} = \sqrt{3r^2}$$

Fig. 36

So the radius of the sphere will be $\frac{\sqrt{3}}{2} \cdot r$

This sphere touches the base as well as the slant sides of the cone in which it is inscribed.

Hence the diameter d of the base of the cone will have to be $3r$ and its height $\frac{3\sqrt{3}}{2} \cdot r$

Thus the outermost cylinder will have radius $\frac{3}{2} \cdot r$ and height $\frac{3\sqrt{3}}{2} \cdot r$

Therefore its total surface area will be *(twice the area of circular base + area of curved surface).*

$$= 2 \times 2\pi\left(\frac{3}{2}r\right)^2 + 2\pi \times \frac{3}{2}r \times \frac{3\sqrt{3}}{2}r = \frac{9}{2}\pi r^2 (2 + \sqrt{3})$$

14. Geometry on the Surface of a Sphere

We live on earth and we now know that earth is a sphere and we are living on its surface. Hence all the geometrical figures drawn on earth are essentially figures drawn on the surface of a sphere. However for all practical purposes we regard these as figures drawn on a plane surface. Why is it so? This is because earth is a very big sphere whose radius is around 6000 km and the geometrical figures that we consider in our day to day lives such as houses, roads, fields etc. are much smaller in size as compared to the dimensions of the earth. A large spherical surface in a small neighborhood of the observer is a plane for all practical purposes. We observe this when we see the surface of the earth around us. It appears flat rather than spherical. However if long distances are involved such as travel from Delhi to London then the line joining these two cities on the surface of the earth will not be a straight line but an arc of a circle. Thus when geometry of points which are far apart from each other on the surface of earth is to be considered (such a triangle formed by joining Delhi, London and Johnsburg or quadrilateral formed by joining Delhi with London, Johnsburg and Sydney) then such geometrical figures cannot be regarded as plane figures but figures drawn on the surface of a sphere. Thus knowledge of spherical geometry is essential for analyzing geometrical relationships of points on the surface of earth which are far apart. Not only that, it is also essential for analyzing problems related to celestial bodies such as sun, moon and stars which seem to lie on the spherical surface formed by the sky.

A question naturally arises how geometry on the surface of a sphere differs from plane geometry. Euclid's plane geometry is primarily based on the following two postulates:

(i) Any two points lying on a plane can be joined by a straight line which completely lies in the plane and is the shortest distance between these two points.

(ii) From a point not lying on a straight line, several straight lines can be drawn to cross it. However out of these there is one straight line which does not cross the given straight line how so ever long this line be extended on either side. This straight line is said to be the straight line through this point parallel to the given straight line. This line is unique.

Geometry which does not satisfy any one of these two axioms is called a non-Euclidean geometry. In this sense spherical geometry is non-Euclidean as it violates these postulates. For instance if we join two points lying on the surface of a sphere with a straight line, then this straight line does not lie on the surface of the sphere on which the two points lie .It is in fact a chord of the sphere. Line joining these two points which completely lies on the surface of the sphere is a curve and not a straight line. Two points lying on the surface of a sphere can be joined by a number of arcs that lie on the surface of the sphere. Amongst these the arc of shortest length is the great circle arc passing through these two points. (A great circle is a circle on the surface of a sphere whose centre is same as the centre of the sphere). In fact it is this great circle arc which is the shortest distance between two given points on its surface. That is why ships and aeroplanes on international voyages try to follow such great circle arc routes as far as possible instead of trying to move straight in the direction of the destination point.

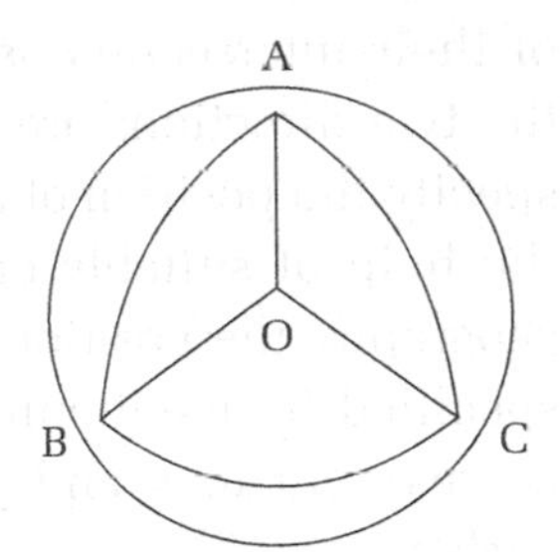

Fig. 37

Again if there are three points A, B and C on the surface of a sphere and great circle arcs are drawn to join these points two by two we get what is called a **spherical triangle** (Fig. 37).

A spherical triangle differs from its counterpart plane triangle in the following ways:

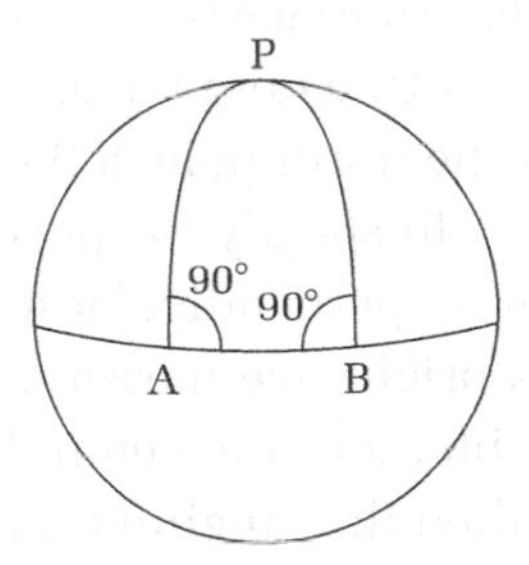

Fig. 38

(i) Its sides are not straight lines but great circle arcs.

(ii) Angle A is now the angle between tangents to the great circle arcs AB and AC drawn at A. Angles B and C are also similarly defined.

(iii) Measures of arc lengths AB, AC and BC which form the sides of this triangle are not distances but the angles which these arcs subtend at the centre O of the sphere. Thus arc AC = angle AOC, arc AB = angle AOB and arc BC = angle BOC.

Thus in a spherical triangle sides as well as angles are expressed in terms of angular measures:

(iv) Another major difference between a plane and a spherical triangle is that where as in a plane triangle ABC, sum of the angles A, B, C, $\angle A + \angle B + \angle C = 180°$ it need not be so in a spherical triangle. This can be appreciated by supposing A, B to be points on the equator of earth and C a pole of the earth (Fig. 38). Then $\angle PAB = 90°$, $\angle PBA = 90°$ (as both great circle arcs PA and PB are perpendicular to the equator) thus in triangle PAB, $\angle A + \angle B + \angle P = 90° + 90° + \angle P = 180° + \angle P$.

Thus spherical geometry differs substantially from plane geometry. In fact where as in plane geometry if a straight line AB is extended on either side the two ends never meet, it is not so in a spherical geometry. In spherical geometry AB is an arc of a circle on the surface of sphere. So if it is extended on either side, the two ends will meet being points on the same circle drawn on the sphere. Spherical geometry has been extensively studied as it has extensive applications in physical geography as well as astronomy.

14.1 Coordinate Systems on Spherical Surfaces

Just as we can fix the coordinates of a point P on a plane surface by fixing two mutually perpendicular straight lines (called axes of coordinates) and choosing the point of their intersection as origin from where distances along the two directions are measured, in the same way we can specify the position of a point P on a spherical surface with the help of suitably chosen great circles. For example in geography the position of a point on the surface of earth is specified by mentioning its longitude and latitude. These correspond to x and y coordinates of a point on plane surface.

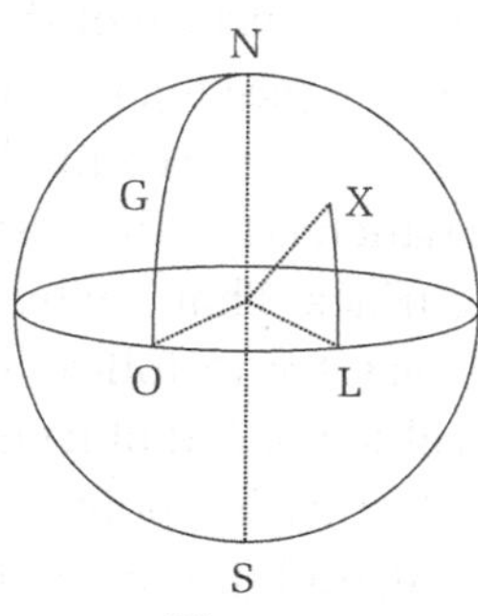

Fig. 39

In the case of earth the great circle of reference which corresponds to x-axis on a plane surface is the equator. Equator of earth as we know is a great circle on the surface of earth whose plane is perpendicular to the axis of rotation NCS of the earth where N is the north pole, S the south pole and C the centre of the earth.

To specify the position of a point X on earth surface we draw XL an arc of a great circle perpendicular to the equator (Fig. 39). Choosing some fixed point O on equator as origin, we measure angle OCL and angle XCL which great circle arcs OL and XL subtend at the centre C of the earth. These angular measures of arcs OL and XL are called the **longitude** and **latitude** of point X. Arc XL when produced will pass through N and S, the north and south poles of earth. Longitude can have any value from 0° to 360° or 0° to 180° East and 0° to 180° west depending upon whether L is to the east or west of O. Similarly the latitude of a place can have any value from 0° to 90° north(N) or 0° to 90° south(S) depending upon whether the point X is in the northern hemisphere or the southern hemisphere. Whereas Asia,Europe and North America are in the northern hemisphere, Australia, New Zealand, South America and South Africa are in the southern hemisphere.

Interestingly north pole has latitude 90° north(N) and south pole latitude 90° south. However they do not have unique values of longitudes since all great circle arcs perpendicular to the equator pass through both these two points. All the points on equator have zero latitude.

Thus far we have not specified as to what is the position of point O on the equator of the earth. Actually any point on equator could have been chosen as origin. However by consensus O is chosen to be the point on the equator where the great circle are joining north pole to observatory in Greenwich (which is in England) meets the equator.

15. A Peculiar Point on Earth's Surface

Problem: Is there a point on earth's surface such that if we first walk one kilometer towards south, then one kilometer towards east and finally one kilometer towards north, we return back to the starting point.

Solution: Obviously north pole is such a point. Is this point unique or are there same more points on the earth with this property? At first sight it might appear that this is the only point with such a property. However a little thinking will reveal that not only north pole but there are infinitely many points in southern hemisphere which have this property. The set of such points lie 1 km north of a small circle near the south pole which has a circumference of one kilometer (Fig. 40). If we walk one kilometer south of such a point X we reach this circle. Walking one kilometer to east along this circle brings us back to the starting point. From this point if we walk one kilometer to north we reach the starting point X.

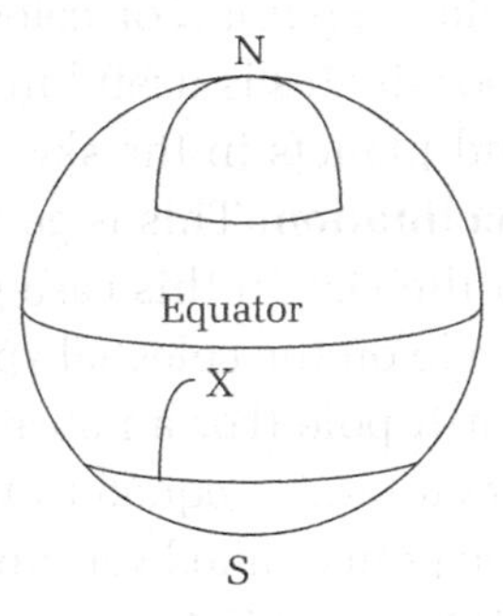

Fig. 40

16. Celestial Sphere

As mentioned earlier visible sky is a hemisphere. Sun, moon, planets and stars appear to be located on its surface. In reality sun, moon, planets and stars are not at the same distance from the observer. However they appear to be so and lying on a spherical surface which we call sky because of their great distances from the earth and the fact that there is a limit to the distance upto what we can see. As a result all these heavenly objects appear to be lying on a spherical surface formed by the limit to which an observer can see in space. The situation can be compared to how distant trees and plants appear. They all appear to be located on the circumference of a circle called horizon where as in reality each of them is at a different distance from the observer.

Therefore in observational astronomy in order to observe and analyse positions and motions of heavenly objects such as the sun, moon, planets and stars, these are assumed to be points located on the surface of a sphere called **celestial sphere** (celestial means heavenly). This sphere has observer as the center.

In order to specify the position of a heavenly object on this celestial sphere, help is taken of a great circle(as in the case of points on surface of earth). However keeping in view a variety of problems which have to be analysed, not one but three great circles have been used as circles of reference leading to three types of coordinate systems (Fig. 41). In one case great circle chosen for reference is horizon and the highest point in the sky (called zenith) corresponds to pole of the earth. System of coordinates corresponding to this is called **Azimuth** and **Altitude** (Azimuth corresponds to longitude and altitude corresponds to latitude of a point on earth). This system is useful in the study

of diurnal (daily) motions of sun, moon and planets in the sky and problems related to their time and place of rising and setting on the horizon. Another system of coordinates which is being used is **Longitude** and **Latitude** (it does not correspond to longitude and latitude of points on earth), In this case great circle of reference is **ecliptic** (which is the path traced out by the sun in the sky in its annual motion around the earth). In this case pole is a point K on celestial sphere which is 90° from points on ecliptic. This system of coordinates is useful in the study of the problems related to the motions of sun, moon and planets in the sky. The third system of coordinates is called **right ascension** and **declination**. This is generally useful in the study of relative positions of distant stars in the sky. In this case great circle of reference is celestial equator (equator is a great circle on the celestial sphere which corresponds to equator of the earth). Its pole is the north pole (For an observer in northern hemisphere).
Great circles equator and ecliptic cross each other at two points called **verinal equinox** and **autumn equinox**. (**Rahoo** and **Ketoo**).

Ecliptic and equator are inclined to each other at an angle of 23½° (called obliquity of the ecliptic). Equinoxes are important points on ecliptic because it is only when sun and moon are both in the vicinity of one of these two points that an eclipse can happen (solar if it is on new moon day and lunar if it is on full moon day).

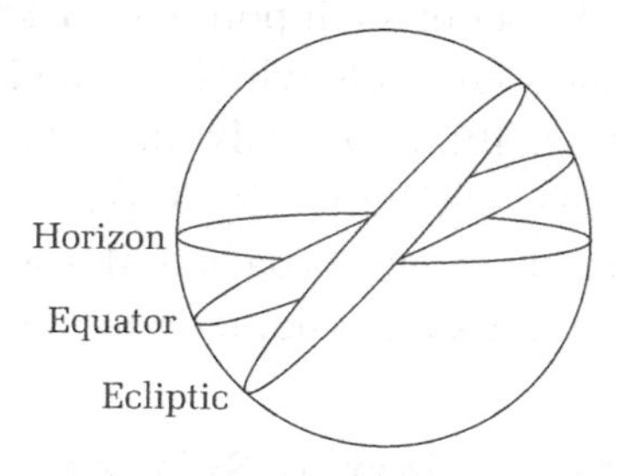

Fig. 41

17. Higher Dimensional Geometries

Mathematicians have not stopped at three dimensional geometry. They have also developed general *n* dimensional geometry (*n* a positive integer). It is useful in problems where more than three coordinates are needed to specify the positions of points. Even though it is difficult to visualize physical situations where more than three coordinates are needed (one coordinate is sufficient to specify a point on a straight line, two to specify a point in a plane and three to specify a point in space). However we do need four coordinates when we talk of space-time continum in which three coordinates specify the position and fourth the time. This is helpful in theory of relativity.

18. Topology

Topology is a more generalized form of geometry in which the concept of unique distance between two fixed points (called metric) is missing. Two figures are said to be topologically equivalent if one can be transformed into the other by one or more operations of distortion, shrinking or stretching. It may appear strange but a tea cup and a doughnut are topologically equivalent. The hole in the doughnut corresponds to the handle of the tea cup.

Tea Cup

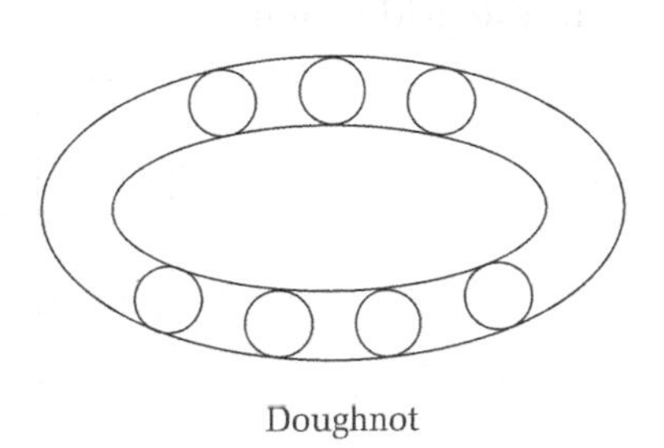

Doughnot

Fig. 42

Topology is sometimes referred to as rubber sheet geometry. If a face of a polyhedron is removed, the remaining figure is topologically equilent to a region of a plane. We can deform the open polyhedron by opening its side faces till it falls flat on its base. Situation is similar to an open rectangular box whose sides are opened up. The transformed figure does not preserve the size or shape of original figure but its boundaries are preserved, edges becoming sides of polygonal region. Interestingly even after this transformation, the value of $V + F - E$ remains unchanged!

Mind Teasers

1. How many triangles and squares are there in the following figure?

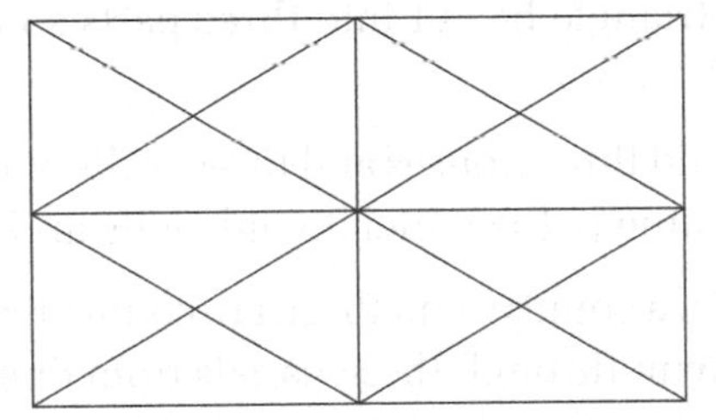

2. Can we draw a closed path with straight segments such that each of these intersects exactly one of the other segments?
3. Is it possible to draw a line consisting of 4 straight segments so that it passes through all the points shown in the following figure.

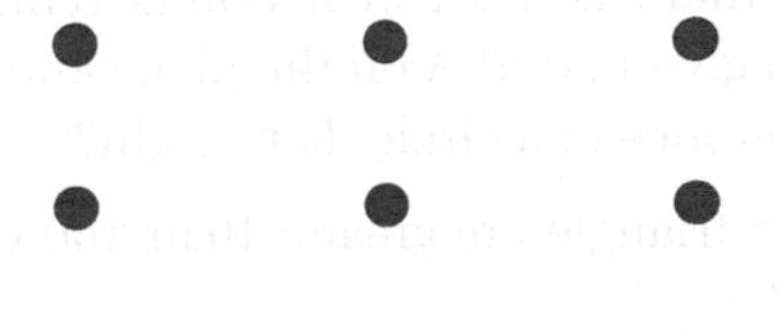

4. Is it possible to form the following grid using (i) 5 line segments of length 8 each. (ii) 8 line segments of length 5 each, given that length of each segment in the grid is one.

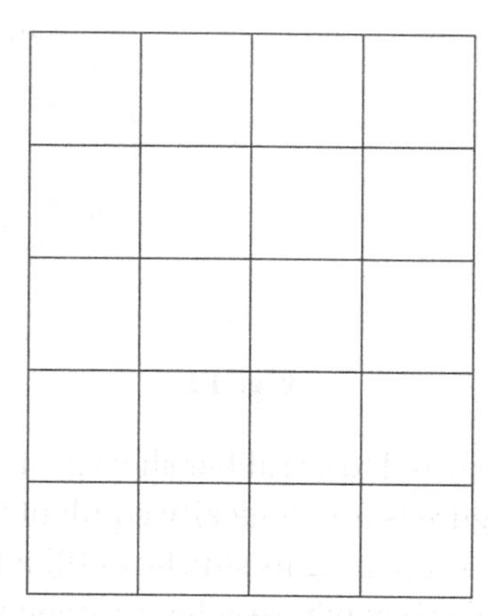

5. In a country any two of its cities are connected either by rail or road. Show that it is possible to choose any type of transport and reach each of its towns from any other town with at most two transfers on the way.
6. Is it possible to cut a 3×9 rectangle into 8 squares? If yes do it.
7. What is the greatest number of regions into which three straight lines of finite length can divide a plane?
8. Can an arbitrary triangle be cut into three parts so that these can be arranged to form a square?
9. Is it possible to build three houses and three wells and then connect houses with each other using nine paths which avoid wells and do not cross each other?
10. There are 7 lakes in a country and these are connected to each other by 10 canals so that one can swim through these canals from one lake to another lake. How many islands are there?
11. Suppose in the seven bridges problem it is desired to come back to the same point from where one started. Is it possible? If not what additional requirements will be needed.
12. Peter claims that there is a set of n points lying in a plane which are all equidistant from a given point A on the plane and claims that all these points lie on the circumference of a circle. Is he right?
13. If all the sides of a triangle are greater than 100 cm, can its area be still less than 1 cm square?

14. Show that the sum of the perpendicular distances from a point P inside an equilateral triangle on its sides does not depend up the position of point (i.e. in the following figure PL + PM + PN = constant).

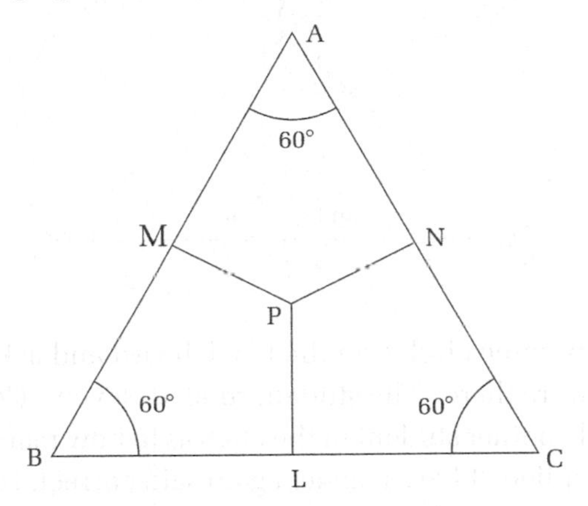

15. Show that the area of a right angled triangle which has sides of integral lengths is always divisible by 60. Moreover this area can not be a perfect square.

16. Diagonals of a parallelogram are perpendicular to each other. Is this parallelogram a rectangle?

17. Of all the parallelograms which have a given parameter, which encloses the greatest area?

18. Of all the rectangles having a given parameter which encloses the greatest area? Justify it analytically also.

19. Is it possible for the accompanying five point star figure to satisfy the inequalities AB > BC, CD > DE, EF > EG, GH > HI and IK > KA?

Also show that sum of angles at vertices B, D, F, H and K is 180°

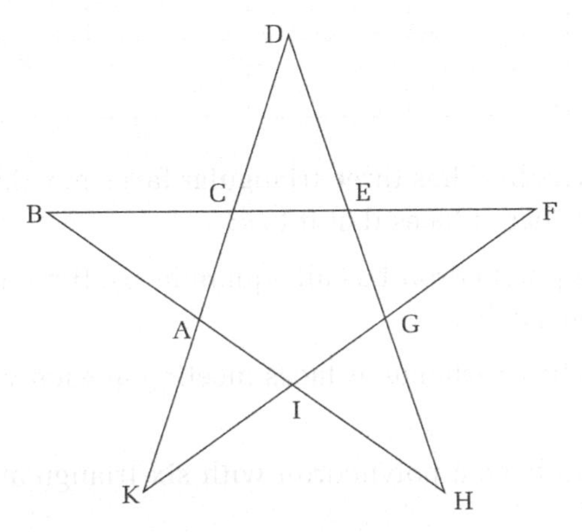

20. Two circles intersect each other at points A and B. AC is a diameter of the first circle and AD a diameter of the second. Are the points C, Band D collinear?

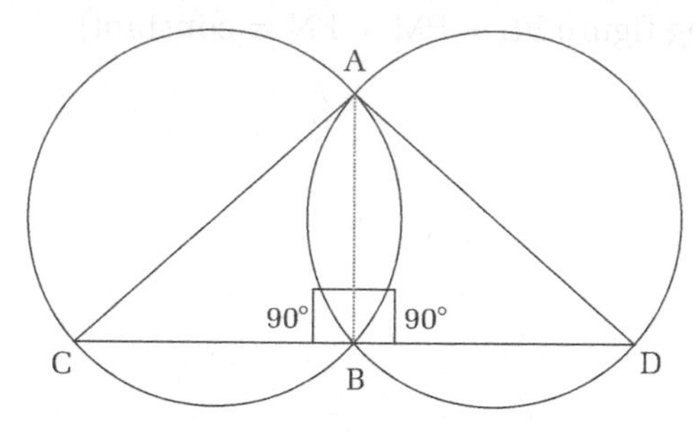

21. A teacher drew some circles on the black board and asked a student as to how many circles were there? The student replied 'seven'. 'Correct' said the teacher. Then he asked another student in the class as to how many circles he had drawn. The student replied, 'Five'. Teacher again said correct. How this is possible and actually how many circles the teacher had drawn?

22. A piece of wire is 120 cm long. Can it be used to form a cube? If yes, what will be the length of each edge? Also, what is the minimum number of cuts that will have to be made in the wire?

23. Imagine a wooden cube of sides 3 inches each with two parallel lines drawn in each direction 1 inch apart. What is the minimum number of cuts that will cut apart all 27 cubes of sizes $1'' \times 1'' \times 1''$?

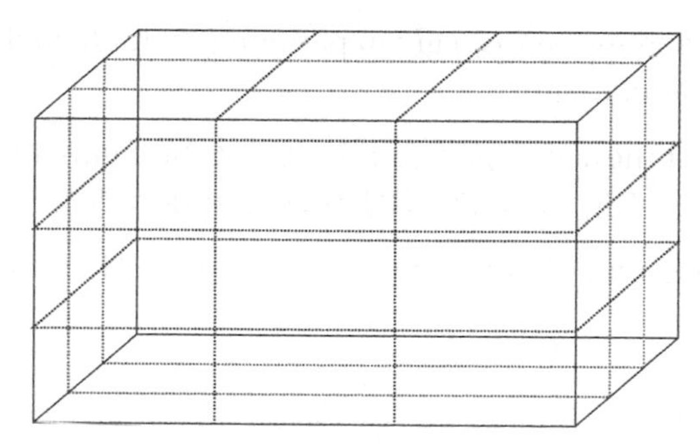

24. A regular polyhedron has three triangular faces meeting at each vertex. What is the total number of faces that it has?

25. Show that if a polyhedron has all square faces, three meeting at each vertex, then it must be a cube.

26. A polyhedron has 5 triangular faces meeting at each vertex. How many faces does it have?

27. Is it possible to have a polyhedron with six triangular faces meeting at each vertex?

28. We saw in section 15 that there are certain points on the surface of the earth such that if we travel from there first 1 km south, then 1 km east and then 1 km north, we reach back to the same point. Are there also points on earth where if we reverse this route (first travel 1 km north, then 1 km west and then 1 km south) we reach the same point?

29. Are there some places on the earth where the sun is exactly vertical at noon at some day of the year? If so, where are these located? Are there also places on the earth where the sun is exactly vertical everyday at noon throughout the year?

30. Due to daily rotation of the earth about its axis(called its diurnal motion), sun, moon and stars appear to be moving on the sky from east to west. Is there any star in the sky which does not partake this motion? If yes, then what is it called?

Chapter 7
Trignometry

1. Introduction

Trigonometry is a branch of mathematics which occupies curious 'no man's land' in prominence in the education of most of the students. It has very wide applications in different fields particularly geography and astronomy. It provides the user many tools with which he can do many useful things such as determining areas of geometrical figures like triangles and quadrilaterals, heights of buildings, towers, hill peaks without having to go to their tops, find the width of a river without crossing it, and even determine heights and sizes of heavenly bodies such as sun and moon without ever being there. In fact when a teacher in a class mentions that mathematics will be of great help to them in their higher studies, he is in particular referring to Trigonometry in case of students opting for engineering studies.

Trigonometry originated in the west with the work of Greek Mathematician Hipparchus in second century B.C. In India, Hindus too were familiar with it even in the ancient past in connection with their studies of heavenly objects. Hipparchus not only formulated basic fundamentals of trigonometry but even prepared tables of trigonometric functions for practical use. He even used these to solve practical problems such as determining the precise directions in which two parties digging a tunnel in a mountain from opposite sides of the hill should dig so that they eventually meet. He also made use of trigonometry to estimate the distances and sizes of sun and moon. In fact these computations done in distant past were quite accurate. Their current values obtained by using more accurate and sophisticated techniques are not far from these values. Needless to say that these estimates of sizes and distances of sun and moon were quite at variance with the commonly held beliefs of those days which regarded sky as a sort of canopy on whose surface heavenly objects such as sun, moon and stars are studded. In spite of his calculations it took centuries for the western civilization to come to terms with the fact that this universe is very very big and sun, moon and stars are not at the same distance from the earth.

2. Significance of Triangle

In fact triangles are the be all and end all of trigonometry. In order to appreciate the basic structure of trigonometry, it is necessary to appreciate the basic structure of a triangle.

A triangle has three sides and three angles. For convenience we denote the angles by letters A, B and C and sides opposite these angles by a, b and c (see figure 1). We know that in every triangle the sum of its internal angles is 180°. Thus

$$A + B + C = 180^\circ \text{ or } \pi.$$

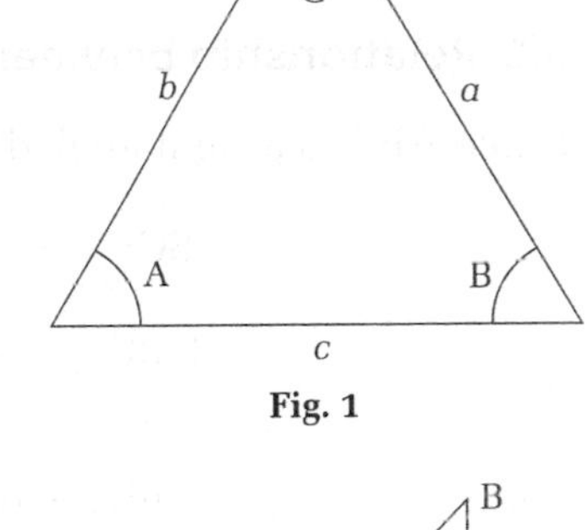

Fig. 1

Hence none of the angles of a triangle can be more than 180°. If all the three angles A, B and C are less than 90° each, we call it an **acute angled triangle**, if one angle is 90° and other two acute, we call it a **right angled triangle**. If one angle is obtuse (more than 90°), we call it **obtuse angled triangle.** (Can two angles of a triangle be obtuse?) If two angles are equal, we call it an **isosceles triangle**. If all the three angles are 60° each, we call it an **equilateral triangle**.

3. Trigonometric Ratios

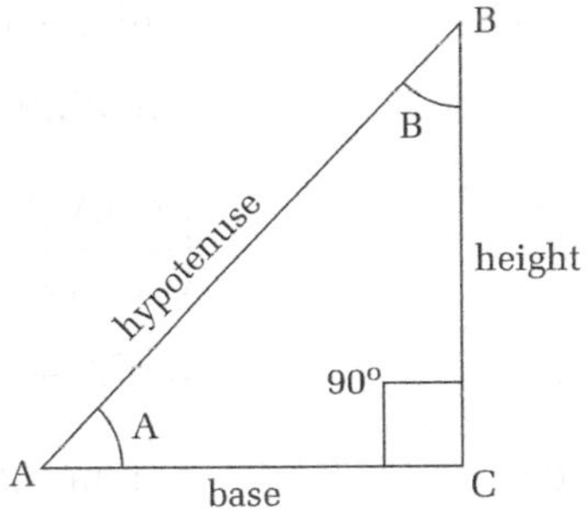

Trigonometry is based on the ratios of sides of a right angled triangle. In a right angled triangle with right angle at C (Fig. 2), we call side AB opposite this right angle as hypotenuse, vertical side BC as height and horizontal side AB as 'base'. We define ratios

Fig. 2

$$\frac{BC}{AB}\left(\text{i.e. } \frac{\text{height}}{\text{hypotenuse}}\right)$$ as **sine of angle A** (denoted by sinA)

$$\frac{AC}{AB}\left(\text{i.e. } \frac{\text{base}}{\text{hypotenuse}}\right)$$ as **cosine of angle A** (denoted by cosA)

$$\frac{BC}{AC}\left(\text{i.e. } \frac{\text{height}}{\text{base}}\right)$$ as **tangent of angle A** (denoted by tanA)

$$\text{Thus, } \sin A = \frac{BC}{AB},\ \cos A = \frac{AC}{AB},\ \tan A = \frac{BC}{AC}$$

Reciprocals of these ratios are defined as

$$\text{cosecA} = \frac{AB}{BC}, \ \text{secA} = \frac{AB}{AC}, \ \text{cotA} = \frac{AC}{BC}$$

respectively and called **cosecant, secant and cotangent of angle A.**

In fact it is obvious that $\text{cosecA} = \frac{1}{\sin A}$, $\text{secA} = \frac{1}{\cos A}$, $\text{cotA} = \frac{1}{\tan A}$

and $\frac{\sin A}{\cos A} = \tan A$

We can also define these trigonometric ratios for angle B.

3.1 Relationship between Trigonometric Ratios

Since ABC is a right angled triangle with right angle at C, therefore

$$BC^2 + AC^2 = AB^2$$

or $$\left(\frac{BC}{AB}\right)^2 + \left(\frac{AC}{AB}\right)^2 = 1$$

i.e. $$\sin^2 A + \cos^2 A = 1 \qquad \text{(i)}$$

dividing by $\cos^2 A$

or $$\left(\frac{\sin A}{\cos A}\right)^2 + 1 = \left(\frac{1}{\cos A}\right)^2$$

i.e. $$\tan^2 A + 1 = \sec^2 A$$

or $$1 + \tan^2 A = \sec^2 A \qquad \text{(ii)}$$

Similarly dividing (1) by $\sin^2 A$, we have

$$1 + \cot^2 A = \text{cosec}^2 A \qquad \text{(iii)}$$

Several other similar relations can be established. These may be found in any standard book on trigonometry.

3.2 Tables of Trigonometric Functions

Values of trigonometric functions, sine, cosine, tangent, etc. for different values of angle varying from 0 to 90° (0 to $\pi/2$) have been computed and are available in the form of tables called trigonometric tables for ready reference. Some of the commonly

Angle	Sine	Cosine	Tangent
0°	0	1	0
30°	1/2	$\sqrt{3}/2$	$1/\sqrt{3}$
45°	$1/\sqrt{2}$	$1/\sqrt{2}$	1
90°	1	0	∞

used values are shown in the accompanying table. Values of sec, cosec and cotangent are the reciprocals of corresponding values of cosine, sine and tangent. Values of these ratios for angles beyond 90° are usually calculated from these ones using appropriate trigonometric formula such as $\sin(90+\theta) = \cos\theta$, $\cos(90+\theta) = -\sin\theta$, $\tan(90+\theta) = -\cot\theta$ for values of θ in the second quadrant (90° to 180°).

For values of θ in the third quadrant (180° to 270°)

$$\sin(180+\theta) = -\sin\theta,\ \cos(180+\theta) = -\cos\theta \text{ and } \tan(180+\theta) = \tan\theta$$

For values of θ between 270° and 360° (fourth quadrant), we use the relations

$$\sin(270+\theta) = -\cos\theta,\ \cos(270+\theta) = \sin\theta \text{ and } \tan(270+\theta) = -\cot\theta$$

(These days calculators and PCs have inbuilt programmes to directly compute these trignometric ratios for any given value of angle)

4. Trigonometric Fallacies

First part of every text book on trigonometry is primarily concerned with establishing relationships between trigonometric functions which prove useful in its subsequent applications. Likc algebra, trigonometry also abounds in fallacies which generally arise because of some oversight or wrong application of a rule. We present here one such example for illustration and caution.

We know that $\sin^2\theta + \cos^2\theta = 1$

Therefore $\cos^2\theta = 1 - \sin^2\theta$

Hence $\cos\theta = \sqrt{1-\sin^2\theta}$ (i)

Thus $1 + \cos\theta = 1 + \sqrt{1-\sin^2\theta}$ (ii)

Squaring $(1+\cos\theta)^2 = \left(1+\sqrt{1-\sin^2\theta}\right)^2$ (iii)

This result has to be true for all choices of the value of angle θ. Suppose we choose $\theta = 180°$.

Then $\sin\theta = \sin 180° = 0$ and $\cos\theta = \cos 180° = -1$. Substituting in (iii)

$$(1-1)^2 = \left(1+\sqrt{1-0}\right)^2$$

i.e. $0 = 4$?

What has gone wrong and where? The fact is that while taking square root in (i), we have ignored negative sign which is also possible. Had we written (i) as

$$\cos\theta = -\sqrt{1-\sin^2\theta}$$

then (ii) would have been

$$1 + \cos\theta = 1 - \sqrt{1 - \sin^2\theta}$$

and (iii) $$(1 + \cos\theta)^2 = \left(1 - \sqrt{1 - \sin^2\theta}\right)^2$$

Now writing $\cos\theta = -1$ and $\sin\theta = 0$, we get

$$(1 - 1)^2 = (1 - 1)^2$$

which is true.

5. Some Practical Applications

In this section we present some simple practical applications of trigonometry.

5.1 Determining the Sides and Angles of a Triangle

Sometimes in a triangle we know values of some of its sides and angles while values of other elements of the triangle are not known and we want to determine their values. In such cases, trigonometrical relations which can be established between the sides and angles of a triangle are often helpful. For example suppose we know the lengths of the three sides a, b and c of a triangle and we want to determine values of angles A, B and C without having to actually draw the triangle. Then we can find angle A using the relation

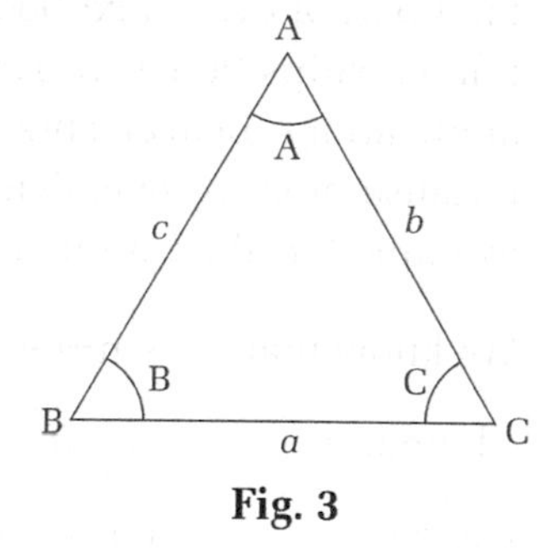

Fig. 3

$$a^2 = b^2 + c^2 - 2bc \cos A \qquad (1)$$

i.e. $$\cos A = \sqrt{\frac{b^2 + c^2 - a^2}{2bc}} \qquad (2)$$

Of course care will have to be taken to be sure whether to use positive or negative sign while writing square root in (2) keeping in mind that if sign taken is +ve then A is an acute angle and if sign taken is –ve, A will be an obtuse angle.

Values of angles B and C can be similarly obtained. In fact once A and B are known, C is just 180° – (A + B).

(1) can also be used to determine side opposite angle A if that angle and lengths of sides b and c including it are known. In fact given three elements (sides and angles are called elements of the triangle), it is generally possible to determine the remaining three using appropriate trigonometric formulae.

5.2 Determining the Area of a Triangle

A number of trigonometric formulae are available which can be used to determine the area of a triangle given some of its elements. One of the most effective formula in determining the area of a triangle is when lengths of its sides are known.

$$\text{Area} = \sqrt{s\,(s-a)\,(s-b)\,(s-c)} \qquad \text{(i)}$$

where $s = (a + b + c)/2$ is called the semiperimeter of the triangle. For example if the sides of a triangle are of lengths 3, 4 and 5 meters (i.e., $a = 3$, $b = 4$, $c = 5$) then its perimeter is $3 + 4 + 5 = 12$ and semi perimeter $s = 6$

Hence, its area

$$A = \sqrt{6\,(6-3)\,(6-4)\,(6-5)}$$

$$= \sqrt{6 \times 3 \times 2 \times 1}$$

$= 6$ square meters

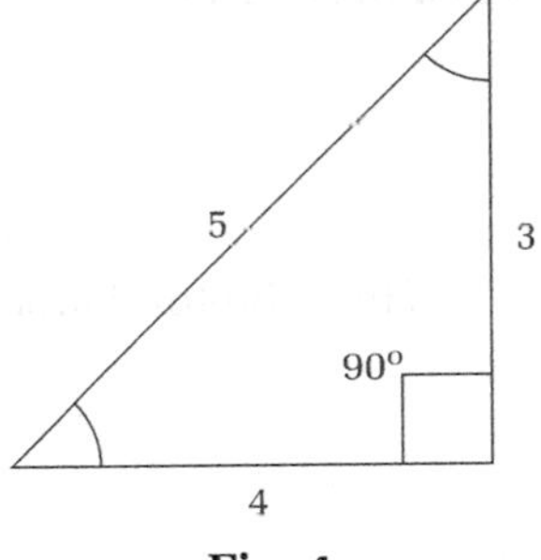

Fig. 4

In fact we know that triangle formed by these three sides is a right angled triangle (as $c^2 = a^2 + b^2$).

so its area is $= \dfrac{1}{2} \times \text{base} \times \text{height}$

$$= \frac{1}{2} \times 4 \times 3$$

$= 6$ square meters.

which is same as obtained earlier using (i)

The above formula of finding area of a triangle can also be used to determine areas of certain other geometrical figures such as quadrilaterals and polygons by first suitably subdividing these into triangles. For example to find the area of a quadrilateral ABCD, shown in Fig. 5 we may join opposite vertices A and C to obtain two triangles ABC and BCD. As the sides of quadrilateral are known, we can use these to find the length of diagonal AC if angle B or D is known. Once the length of AC is known, areas of triangles ABC and ACD can be calculated using (i). Adding areas of these two triangles, we get the area of quadrilateral.

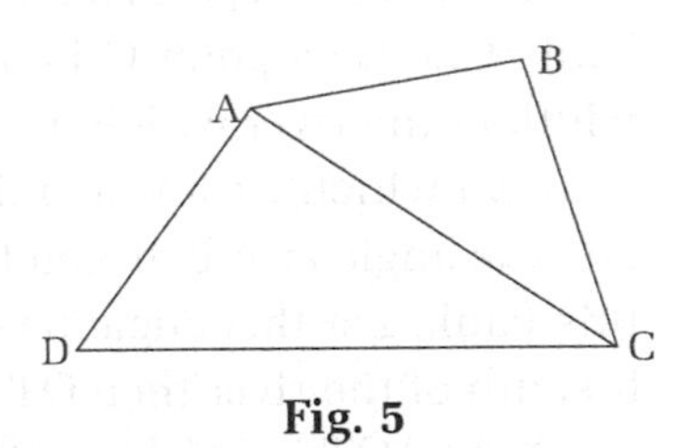

Fig. 5

Example: Find the area of the quadrilateral ABCD given AB = 3, BC = 4, CD = 3.5, DA = 2.5 and diagonal AC = 5 metres.

Solution: In this case

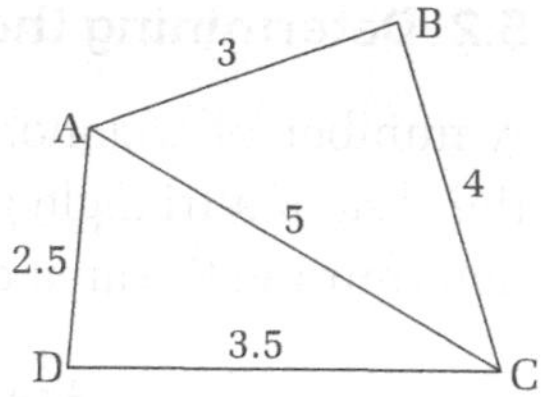

Area of Δ ABC $= \sqrt{6(6-3)(6-4)(6-3)}$

$= \sqrt{6 \times 3 \times 2 \times 1}$

$= 6$ square meters

Fig. 6

Area of Δ ACD $= \sqrt{5.5(5.5-5)(5.5-3.5)(5.5-2.5)}$

$= \sqrt{5.5 \times .5 \times 2 \times 3}$

$= 4.06$ square meters (upto second decimal place)

Thus the total area of quadrilateral

$= \text{Area of } \Delta \text{ ABC} + \text{Area of } \Delta \text{ ACD}$

$= 6 + 4.06$

$= 10.06$ square meters

5.3 Determining the Width of a River without Crossing it

In earlier times people were often interested in determining the width of river without (or before) crossing it so that adequate precautions could be taken while crossing it. Trigonometry helps in determining width of a river without having to cross it.

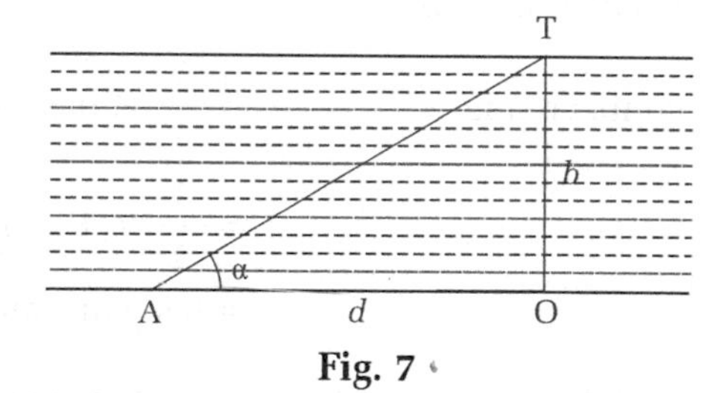

Fig. 7

Suppose we are on one bank of a river and there is a tree or a pole standing on the opposite bank. Choose a point O immediately opposite this tree on this bank. Then OT is the width of the river which has to be determined. Choose some other point, say A on the bank on which we are standing at a suitable distance from O and measure angle TAO, i.e., the angle at A between the direction of tree on the opposite bank and point O on this bank. Let this angle be α. Also measure distance AO. Let this be d. Let b be the breadth of the river then OT $= b$. We have to find b knowing values of d and α.

Now AOT is a right angled triangle with right angle at O. Hence

$$\frac{\text{TO}}{\text{AO}} = \tan \alpha$$

or

$$\frac{b}{d} = \tan \alpha$$

i.e.

$$b = d \tan \alpha$$

Thus the desired width of the river is known as d and α are known.

5.4 Determining the Height and Distance of a Distant Object

Suppose there is a tower or a hill or a tall building say AB located at some distance from the observer O and observer wants to determine the distance of its foot from his position O and also its height AB without having to go to A (may be there is some river between him and the tower). The situation is depicted in the accompanying figure 8.

What the observer needs to do is to first measure α, the angle of elevation of the top of AB as measured from his present position O, then travel some distance (say d), from O towards A to the point O_1 and again measure the angle of elevation, (say β) of the summit from O_1.

Then in the figure

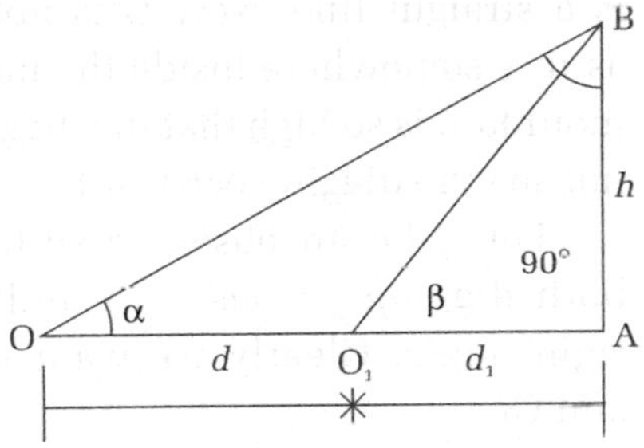

$$\tan\alpha = \frac{AB}{OA} = \frac{h}{d+d_1} \quad \text{(i)}$$

and

$$\tan\beta = \frac{AB}{O_1A} = \frac{h}{d_1} \quad \text{(ii)}$$

Fig. 8

where d_1 is distance of A from O_1 and h the height of the tower (both currently not known)

from (ii), $$O_1A = d_1 = h\cot\beta \quad \text{(iii)}$$

also from (i), $$OA = d + d_1 = h\cot\alpha \quad \text{(iv)}$$

dividing (iv) by (iii)

$$\frac{d+d_1}{d_1} = \frac{\cot\alpha}{\cot\beta}$$

i.e. $$d_1 = \frac{d\cot\beta}{\cot\alpha - \cot\beta} \quad \text{(v)}$$

substituting in (iii),

$$h\cot\beta = \frac{d\cot\beta}{\cot\alpha - \cot\beta}$$

i.e., $$h = \frac{d}{(\cot\alpha - \cot\beta)} \quad \text{(vi)}$$

This is the height AB. Now using (vi) in (iv) distance of AB from O is

$$\frac{d\cot\alpha}{\cot\alpha - \cot\beta} \quad \text{(vii)}$$

5.5 Digging a Tunnel in a Mountain

Example: A tunnel is to be dug in a mountain and to expedite its completion it is desired to start digging from opposite sides along suitable directions so that both the digging parties eventually meet at some point in the mountain.

Solution: Suppose digging is to start from the point A on one side and point B on the other side. Clearly for the two digging parties to meet somewhere inside the mountain (say at D), they should start digging along directions AD and BD so that ADB is a straight line. Now D is not initially known as it is somewhere inside the mountain. Also the mountain is so high that the digging party at A can not see the digging party at B.

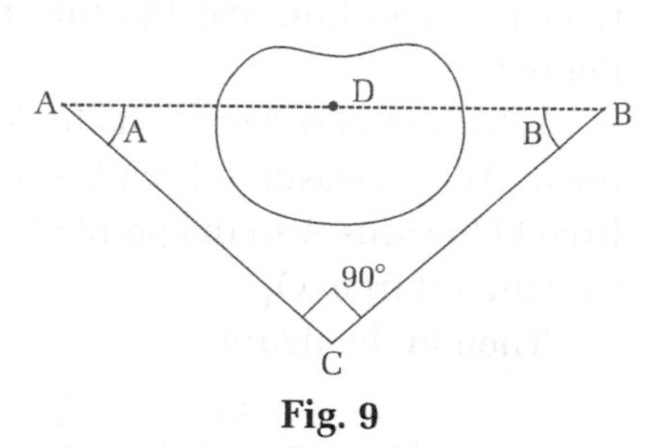

Fig. 9

Let C be an observer suitably stationed outside the hill so that he can see both digging parties at A and B and has so positioned himself that ∠ACB is a right angle. Clearly we want to determine angles A and B knowing lengths AC and CB.

Now $\frac{BC}{AC}$ = tan A. Therefore as lengths BC and AC are known, angle A is known.

Again $\frac{AC}{BC}$ = tan B, so angle B is also known.

(In fact since ABC is a right angled triangle, $\angle B = 90^\circ - \angle A$).

So the party at A should start digging in direction making angle A with AC and part at B should start digging in direction making angle B with BC.

Sometimes it may not be possible to find a point C outside the mountain where $\angle ACB = 90^\circ$, but a point C is available from where both A and B are visible (if not, they can be suitably positioned for this to happen). In such a scenario C is not a right angle (Fig. 10). In such a case in triangle ABC, sides AC, BC and their included angle C is known.

So, $$AB^2 = AC^2 + BC^2 - 2AC.BC.\cos C \qquad \text{(i)}$$

Hence, AB is known.

Knowing AB angles A and B can be computed using

$$BC^2 = AC^2 + AB^2 - 2AC.AB.\cos A \qquad \text{(ii)}$$

and $$AC^2 = AB^2 + BC^2 - 2AB.BC.\cos B \qquad \text{(iii)}$$

Other alternative approaches are also possible.

(For computing an unknown angle we usually prefer computing cosA in place of sinA as for cosA, there is only one possible value of A between 0 and 180° for which cosA has a given value. However in case of sinA, two values are possible as there are two angles A, 180 – A for which sinA has same value).

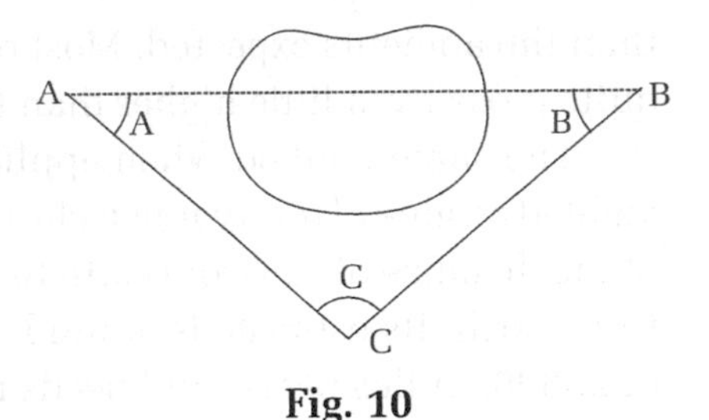

Fig. 10

6. Applications in Astronomy

There are several interesting applications of trigonometry in astronomy. Many of these were known to humanity in the distant past also. We present here some of these.

6.1 Distances of Heavenly Objects

Trigonometry has been used in the past to estimate the distances of heavenly objects such as moon, sun and stars.

Let us first consider moon. In the figure 11, suppose M is the moon and C the centre of earth. Also N is a point on the surface of earth from where moon is just visible above horizon and O is a point on the earth where moon is vertically above at that very instant. Join MN, CN and COM. Then MNC is a right angled triangle with right angle at N. Let $\angle MCN = \alpha$.

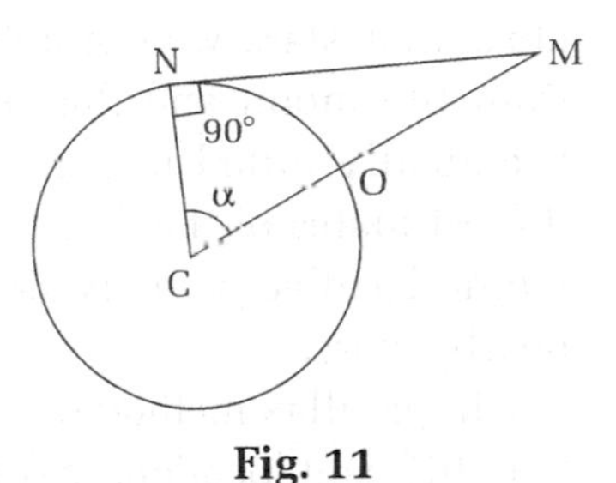

Fig. 11

Then, $$\frac{CN}{CM} = \cos\alpha$$

or $$CM = CN \sec\alpha \qquad \text{(i)}$$

This gives distance CM of the moon from the centre C of the earth since CN, the radius of earth is known. Also α is known in view of the fact that since ON is an arc of a circle of radius CN subtending angle α at the centre C so,

$$ON = CN\,\alpha \qquad \text{(ii)}$$

where α is in radians. Now ON being distance between two points N and O on the surface of earth is known. Also CN being radius of the earth is known. Thus from (ii) α is known. Knowing CN and α, (i) gives the distance of moon from, the centre of earth. Thus distance of moon from the centre of earth is secα times radius of earth which is around 4000 miles.

Actual observations revealed that value of α was quite close to 90°. Substituting this observed value of α in (i) and using value of radius of earth for CN, (i) yielded a value of the distance of moon from the earth around 245,000 miles! A much much larger value than the ancients expected. Most of the persons in ancient times imagined moon to be quite close by, a little higher than the highest mountain!

The same method when applied to determine the distance of the sun from earth yielded a value of α very much close to 90° (though not exactly 90°) leading to an estimate of the distance of sun from earth to be around 370 times more than the distance of moon from earth. Its estimate is around 93 million miles! In fact light traveling at the speed of 286000 miles per second needs more than 8 minutes to reach us from the sun. Thus if sun was to suddenly stop radiating light at this very moment, it will be only after 8 minutes that people on earth will come to know of it.

When astronomers tried the same method to find the distances of stars they were appalled to notice that in the case of practically all the stars, the value of α was so close to 90°, that the difference of observed value from 90° could be easily attributed to observational errors. Thus it became abundantly clear that stars were much much farther from us than the moon and the sun. In fact this method practically failed to provide any reliable clue to the estimates of the distances of even nearby stars. Later on a slight variation of this method (called *parallax* method) enabled humans to get estimates of distances of a few nearby stars.

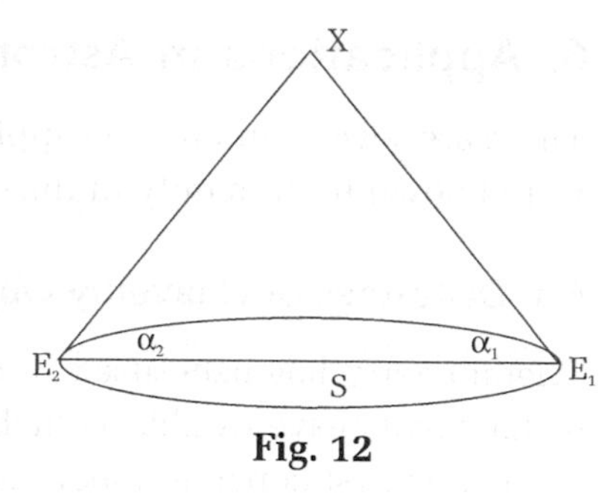

Fig. 12

In parallax method use is made of observations of the same heavenly object X from two different locations, say E_1 and E_2. Knowing distance E_1E_2 and angles α_1 and α_2, XS can be calculated using trigonometry (here S is middle point of E_1E_2). Larger the base E_1E_2, more accurate will be measures of α_1 and α_2. In order to achieve this observations of the same star are taken six months apart so that E_1E_2 is now a diameter of the orbit of earth around sun and therefore about 2×93 million miles. In this case middle point S is centre of sun (For more details of parallax method reader may see some book on spherical astronomy). This technique yielded reliable estimates of distances of only a few nearby stars. In majority of stars even this large base is not sufficient. In fact the nearest neighbouring star to our solar system is the star Proxima Centauri. It is more than 4 light years away. In other words, light emanating from it and traveling at the speed of 286000 miles per second takes more than 4 years to reach us! (By the way this is the nearest star! So one can visualise how vast this universe is).

6.2 Dimensions of Sun and Moon

Both sun and moon appear to us to be small sized circular discs moving in the sky from east to west. Sun during the day and moon at night. But are these really that small as they appear or is it because of the large distances involved. We know that the same object starts appearing smaller and smaller as it is moved farther away from the observer.

In the previous sub-section we have seen that even the ancients knew that sun and moon are far away from us, sun being more farther away than moon. Therefore it is natural to expect that sun and moon are not as small as they appear. They appear small as these are being observed from a great distance. A question then naturally arises. Is it possible to know their exact sizes? The answer is yes and in this trigonometry again proves helpful. Let us see how.

Let O be the observer and M, the centre of the moon. Let its circular disc be of radius r and d the distance of the centre of the moon from the observer O.

Imagine an enveloping cone drawn with observer O as its vertex such that it just envelopes the moon. In that case in section shown in figure, its points of contact A and A′ with the moon M will be such that OA and OA′ are tangents to the circular disc of the moon with M as centre. Hence ∠MAO and ∠MA′O will be right angles.

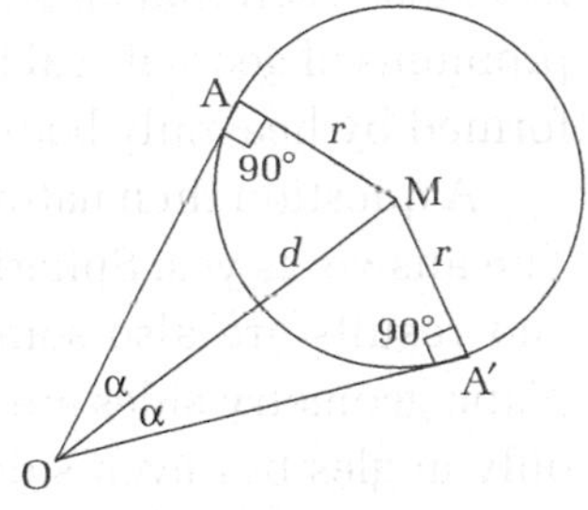

Fig. 13

Let ∠AOM = ∠A′OM = α (say), where α is semivertical angle of the enveloping cone.

Now in the right angled triangle AOM, with right angle at A,

Then, $$\frac{AM}{OM} = \sin\alpha$$

∴ $$AM = OM \sin\alpha$$

or $$r = d\sin\alpha \qquad (1)$$

where d is the distance of moon from observer.

Thus from (1) radius of the disc of moon is $d\sin\alpha$. Since d is known and α is known from observations, radius r of the disc of the moon is known. Calculations show that the diameter of moon is about 3500 km. Same method can be used to find the radius and diameter of the sun. Diameter of sum comes out to be about 1300 thousant km!

Thus sun and moon are much much bigger than they are normally thought to be. Moon is much smaller than earth. Its radius is around one third of the radius of earth. Sun is much much bigger than the earth. Its radius is more than 100 times of the radius of earth! That is why whereas because of mutual pull of gravitation it is the moon which revolves around the earth, in the case of earth and sun, it is the earth which is revolving around the sun. To get an idea of the relative sizes of earth and sun, if sun is compared to a football, then the earth is just the size of a sand grain!

A question naturally arises can we use this technique to determine sizes of stars also? The answer is no. This is because whereas sun and moon appear in the sky as circular discs semivertical angles of whose enveloping cones can be measured, it is not so in the case of stars which just appear as specs of light in the sky virtually having no dimensions.

7. Spherical Trigonometry

In this chapter we have thus far discussed plane trigonometry and its applications. Plane trigonometry deals with geometrical figures such as triangles drawn on a plane surface. In the previous chapter we have seen that besides plane geometry we have also spherical geometry which deals with geometrical figures drawn on the surface of a sphere. We have also seen that spherical geometry has practical relevance as it helps in analyzing problems of geometrical figures drawn on the surface of earth, or of geometrical figures formed by heavenly bodies on the spherical surface of the sky.

A question then naturally arises: is there also a subject like spherical trigonometry? The answer is yes. Spherical trigonometry is similar to plane trigonometry. Definitions and results are also somewhat parallel. The main difference being that whereas in plane geometry sides are lengths, in spherical geometry as seen in the last chapter, not only angles but even sides are also expressed in terms of angles which they subtend at the centre of the sphere. Moreover whereas in plane geometry sum of the internal angles of a triangle is always 180° or π, it need not be so in a spherical triangle. In plane trigonometry we have what is known as sine formula which establishes relationship between the sides and angles of a plane triangle. The formula is

$$\frac{a}{\sin A} = \frac{b}{\sin B} = \frac{c}{\sin C}$$

Its counterpart in spherical trigonometry is

$$\frac{\sin a}{\sin A} = \frac{\sin b}{\sin B} = \frac{\sin c}{\sin C}$$

In this case we have sines of sides also as the measures of sides these are now also in angular measure. The subject has lot of practical applications in spherical astronomy in analyzing relative positions and motions of heavenly objects in the sky as well as in computing the times and places of rising and setting of sun, moon and stars in the sky. This subject however does not form part of the curriculum of school trigonometry. Only students intending to go in for higher studies in astronomy study it at a later stage.

Mind Teasers

1. What are the minimum and maximum possible values of $a\cos^2\theta + b\sin^2\theta$?
2. Two straight roads intersect each other at an angle of 60°. A bus on one of these roads is 2 km away from the intersection while a car on the second road is 3 km away from the intersection. How far apart are they from each other?
3. In a triangle we know that $\frac{a}{\sin A} = \frac{b}{\sin B} = \frac{c}{\sin C}$. If $\frac{a}{\cos A} = \frac{b}{\cos B} = \frac{c}{\cos C}$ also, then what can be said about the nature of the triangle? If length of side a is 2, what is the area of the triangle?

4. Two angles of a triangle are 30° and 45° and the length of the included side is $\sqrt{3} + 1$. What is the area of this triangle?

5. Show that the area of a cyclic quadrilateral ABCD (whose all vertices lie on a circle) and whose sides are of length a, b, c and d respectively is $\sqrt{(s-a)(s-b)(s-c)(s-d)}$, where $2s = a + b + c + d$.

6. On level ground the angular elevation of the top of a tower is 30°. On moving 20 meters towards the tower, the angle of elevation becomes 60°. What is the height of the tower?

7. A flag staff of 5 m height stands on the top of a building 25 meters high. The flag staff and the building subtend equal angles at an observer stationed at a height of 30 meters above the ground on opposite side of the street. What is the width of the street?

8. A tree is broken by the wind. Its upper end bends touching the ground at a distance of 10 meters from the foot of the tree and making an angle of 45° with the ground. What was the original height of the tree?

9. A fighter plane flying at a height of 300 meters above the ground passes vertically above another plane at the instant when angles of elevation of the two planes from same point on ground are 60° and 45° respectively. What is the height of the lower plane?

10. A person standing on the bank of a river observes that a tree standing on the opposite bank subtends an angle of 60° with the bank on which he is standing. When he retraces back 40 meters it subtends an angle of 30°. What is the breadth of the river?

Chapter 8
Probability and Statistics

1. Concept of Probability

All games of chance begin with the concept of probability. The word probability is not very different from the word 'probably'. Like the word 'probably', the word probability is also used to refer to the likelihood of an event happening. When we say that there is 40% chance that it will rain today, it means that there are relatively less chances of its raining today. However if we say there are 80% chance of raining today then it means that there is much greater possibility of its raining today. However in reality it may happen that it rained when there were 40% chances of raining and it did not rain when there were 80% chances of raining. Similarly when an unbiased coin is tossed we say that there is fifty-fifty chance of its falling on head or tail. However, it does not mean that if an unbiased coin is tossed twice we shall get one time head and one time tail. What it really means is that if an unbiased coin is tossed very large number of times say (10,000 times), then each of the ratios (i) number of times we get heads/10,000 and (ii) number of times we get tails/10,000 will be very close to ½ (but may not be exactly ½).

In fact the mathematical study of the subject of probability is said to have begun when some gamblers in France approached a mathematician to advise them as to how to gamble for profit. Mathematicians have assigned a measure to the likelihood of happening of an outcome as a result of an experiment or a trial. This is called probability. The probability of the happening of an absolutely certain event is taken to be one while the probability of happening of an impossible event is taken to be zero. For example the probability of getting either a head or tail in the throw of an unbiased coin is one whereas the probability of getting neither a head nor a tail in the throw of an unbiased coin is zero. Other values of probabilities of happening of an event are always between 0 and 1. Formally we define probability as under:

Let a trial result in n exhaustive, mutually exclusive, equally likely outcomes of which m are favourable to the happening of an event A, then the probability of happening of event A, denoted by P(A) is

$$P(A) = \frac{m}{n}$$

For instance in the throw of an unbiased coin there are only two possible mutually exclusive equally likely outcomes (a head or a tail). So here $n = 2$. Since out of these there is only one possibility of getting a head so $m = 1$. Hence probability of getting head in a throw of an unbiased coin is ½. Similarly probability of getting tail is also ½.

In the same way if an unbiased dice is thrown then the probability of getting a six is 1/6. The probability of getting an even number on the top is 3/6 = 1/2 as there are 3 even numbers (2, 4, 6) out of 6 possible outcomes.

It must be noted that in defining the measure of probability, the use of the adjectives: exhaustive, mutually exclusive and equally likely is very important. Even if one of these conditions is not satisfied, the result will be incorrect. For example, suppose there is a room in which an electric fan is hanging from the ceiling. Now there are only two possibilities: the fan may either fall or it may not fall (here $m = 1$ and $n = 2$). So the probability of the ceiling fan falling is ½. However is it really so? Had it been so we must have been hearing of one or the other fan falling every day. But the falling of a ceiling fan is a rare event. So where have we gone wrong? This is because we have applied the measure of probability in an incorrect manner. All the three conditions : exhaustive, mutually exclusive and equally likely do not hold here. No doubt the possible outcomes of a fan either falling or not falling are exhaustive and mutually exclusive, but these are not equally likely. Possibility of the falling of a fan is not same as the possibility of its not falling as the fan is tightly screwed to the ceiling.

A question naturally arises how to compute the measure of probability in situations like this where all the three requirements of the outcomes being exhaustive, mutually exclusive, equally likely are not satisfied. In such cases we may adopt the following strategy. Suppose a trial is repeated a very large number of times (say n) out of which event A happens m times, then we define $P(A) = m/n$, when $n \to \infty$ (i.e., when n becomes very large).

Thus suppose we want to determine the probability of getting a head in a throw of a biased coin. Then we should throw this coin a very large number of times (say n) and count the number of heads that we get (say m), then the ratio m/n is the estimate of the probability P(A) of getting a head in throw of this coin. Larger the number of trials (n), more accurate will be the estimate of P(A). In fact this test can be used to test whether a given coin is unbiased or not. If P(A) approaches 1/2 as n increases indefinitely, then it is unbiased otherwise not.

1.1 Sex of the other Child

Example: A person goes to the house of a friend. He knows that besides his friend and his wife, their family has two children. However he does not know their sex. He knocks at the door. A child opens the door and the child is a boy. What is the probability that the second child is also a boy?

Solution: There are four possibilities. Both the children can be boys (B, B), elder boy younger girl (B, G), elder girl younger boy (G, B) and both girls (G, G). Since the child who opens the door is a boy, the fourth possibility of both being girls is ruled out. So we have only three possibilities (B, B), (B, G), (G, B) of which only first is favourable to the second child also being boy. Thus probability of second child being also a boy is 1/3.

Now suppose we slightly modify the problem and say that the child who opened the door was elder of the two and a boy. In this case what is the probability that second is also a boy. In this total number of cases are just two (B, B), (B, G). So the probability of second also being a boy is ½.

1.2 Thick Coin Problem

Example: Suppose there is a thick coin which has base diameter D and thickness B (in a normal coin B is very small as compared to base diameter D). Now suppose this thick coin is tossed. What is the probability of its landing on its edge?

Solution: When tossed, the coin can land either on one of its flat faces or edge (if edge is thick). Clearly the probability of its landing on edge will be proportional to the area of the edge surface vis-à-vis total surface area of the coin.

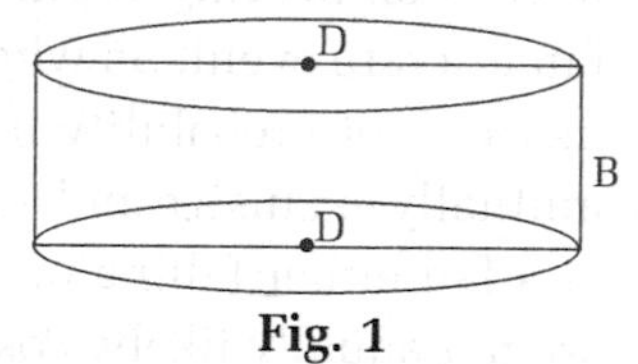

Fig. 1

Now total surface area of the coin = area of its two plane faces of diameter D each and surface area of edge of thickness B

$$= 2\pi\,(D/2)^2 + 2\pi\,(D/2) \times B$$

$$= \frac{\pi D^2}{2} + \pi DB$$

Also, area of edge surface $= \pi DB$

Hence, probability of its falling on edge

$$= \frac{\pi DB}{\frac{\pi}{2}D^2 + \pi DB}$$

$$= \frac{2B}{D + 2B}$$

$$= \frac{1}{1 + \frac{1}{2}\left(\frac{D}{B}\right)}$$

$$= \frac{1}{1 + \frac{R}{B}}, \text{ where R is radius of plane face.}$$

Clearly as B → 0, R/B → ∞ and hence $\dfrac{1}{1+\dfrac{R}{B}} \rightarrow 0$

In other words if thickness B is small probability of its falling on edge is small and almost negligible. That is why in case of normal coins which have edges of very small thickness it is assumed that when tossed the coin will fall on either of its faces with equal probability ½.

1.3 Should the Bet be Accepted

Example: A deck of cards is lying on table with cards face down. You ask your friend to subdivide it arbitrarily into three groups. Then you say: I bet even money that one of the top cards of these three decks is a face card (i.e., either a king or a queen or a jack): should your friend accept the bet?

Solution: Your friend might feel like accepting the bet. His argument being that out of 52 cards there are only 12 face cards (4 kings, four queens and 4 jacks). So probability of a card being face card is just 12 / 52, i.e., 0.25 which is quite small. However, it is not really so. In fact there are three cards out of which just one has to be face card. Now the number of ways in which 3 cards can be selected out of 52 is ${}^{52}C_3$. Out of these there are 40 non face cards. Now a non face card can be selected in ${}^{40}C_3$ ways. Thus probability of all the three being non face cards is

$${}^{40}C_3/{}^{52}C_3 = 0.44706$$

Hence the probability that at least one top card being a face card is 1 – 0.44706 = 0.55294, i.e., more than 55%! So your friend is advised not to accept the bet. Looks somewhat strange. Isn't it?

2. Repeated Trials

If a coin is tossed 10 times we say that trial (of tossing coin) has been repeated ten times. Such a repetitive performance of the same experiment is known as repeated trials. While calculating probability in such like cases we generally make a basic assumption that outcome of each trial is independent of the others. In other words if we have got (say head) in sixth trial, then probability of getting a head or tail in the next trial is not affected by this outcome and is still ½. However most of us are not willing to accept this. If a gambler has been loosing several times in succession he thinks that his luck will favour him next time but sadly this may not happen. A coin or a dice is neutral (if it is unbiased). It does not remember past history. That is why most of the gamblers ultimately ruin themselves as casino rules are designed in favour of the owner.

Not only this many couples who aspire for a male child often think that if they have already got two daughters, then there is greater chance of the next child being male.

But that is not so. That is why some families end up with having six to seven daughters each time hoping that the next child would be a male.

2.1 Aspiring for a Male Child

Example: Once a couple which desired a male child had already got seven daughters. So they decided to give up saying that God did not want them to have a male child. However one of their friends urged them not to give up hope and try once more saying that since they had already got seven daughters, therefore the probability of their getting eighth daughter was very small. According to him it was just $(\frac{1}{2})^8$ or just about .004 as there was equal chance of getting a male or a female child at each birth. But is it really so?

This is what is called misuse of probability (as happens when it is argued that probability of a ceiling fan falling is ½). It must be understood that probability is applied to an event which is yet to occur and has not occurred. Once an event has occurred its outcome is hundred percent known. So when the couple has already seven daughters their definite births can not be brought under the realm of probability. The probability of their getting next child also daughter is still ½ as it was on each birth. The above argument of probability of getting all eight daughters as 1/256 would have been true if it had been made before the birth of the first child. Of course it is rare that a family having 8 children will have all eight daughters, but as mentioned in the beginning unless the probability of happening of an event is exactly zero it can still happen, howsoever small the probability of its happening might be.

In fact after birth of seventh daughter, it can be said that possibility of eighth child also being a daughter is small but not that its probability is small. Possibility is different from probability. (In mathematics we now also have theory of possibility).

2.2 A Historical Anecdote

Example: A historical anecdote from 17th century tells about Italian gamblers who used to bet on total number of dots rolled in throw of three dice. They believed that chance of rolling a total of 9 on throw of three dice was same as the chance of rolling a total of 10. Their argument was like this:

There are six ways to get a total of 9 in a throw of three dice, namely (1, 2, 6), (1, 3, 5), (1, 4, 4), (2, 3, 4), (2, 2, 5) and (3, 3, 3). Similarly there are six ways of obtaining a total of 10 namely (1, 4, 5), (1, 3, 6), (2, 2, 6), (2, 3, 5), (2, 4, 4) and (3, 3, 4). However, their experience showed that 10 turned up a bit more frequently compared to 9. Gamblers approached Galileo (1564–1642) and asked for its reason. Galileo could explain and satisfy them. Can you also do? Try.

Solution: Actually numbers such as (1, 2, 6) can appear on any of the three dice and so it means these can appear in 3 ways. So is the case with (1, 3, 5), (1, 4, 4), (2, 3, 4) and (2, 2, 5). However it is not so in case of (3, 3, 3). It can appear only in one way. Thus there are in fact $3 \times 5 + 1 = 16$ ways (and not six) in which 9 can appear. However in the case of ten, there are no identical numbers at all three places. So ten can appear in $3 \times 6 = 18$ ways. So there is greater probability of sum of appearing numbers being 10 than 9.

3. Expected Returns

Example: Suppose there is a casino which asks for a deposit of Rs. 100 from the customer with the assurance that if in the throw of a dice, he calls correct number, then he will receive Rs. 400. However, if he calls an incorrect number, then he will forfeit his Rs. 100. Is this going to be profitable to the casino owner in the long run?

We can not of course predict the outcome when a dice is thrown. However we know that out of the six possible outcomes only one is favourable to the customer. Thus the probability of his winning is 1/6 and the probability of his losing is $1 - 1/6 = 5/6$

Now suppose 600 customers play this game in the casino. Then out of these 600, one hundred are expected to win and 500 are expected to loose. Thus the casino owner will have to pay Rs. $100 \times 400 = 40000$ while the amount the losing customers will forfeit to the casino owner is Rs. $500 \times 100 = 50000$. Thus the expected overall profit of the casino owner is still Rs. $50000 - 40000 =$ Rs. 10000. In fact no gambler den can survive if it has to pay out more than it receives from the customers. In view of this the bet amount has to be judiciously calibrated. It must be such that casino owner after paying the winners and expenses of running the casino is left with reasonable profit. For instance in this very case if the casino owner promises to pay Rs. 600 to the winner in place of 400 then he will be unable to run the casino as on average he would have to pay to the winning customers more than he receives from the loosing customers.

Lotteries also work on the similar use of probability theory. Suppose lottery tickets costing Rs. 10 each are being sold with the assurance that there will be one first prize of Rs. 10 lacs, two second prizes of Rs. 5 lacs, three third prizes of Rs. 2 lacs, 4 fourth prizes of Rs. 1 lacs and one hundred consolation prizes of Rs. 10000 each, making total prize amount Rs. 40 lacs. Now suppose there is a sale of 1 million tickets. Then the company receives in all 10 million rupees. Out of this it will have to pay just Rs. 40 lacs as prize money. Assuming further expenses of Rs. 20 lacs, towards salary of employees and commission to agents, etc., there is a total outflow of Rs. 60 lacs. So the company is still left with an overall profit of Rs. 40 lacs. Of course a few lucky grow rich. However a vast majority lose their bet amounts. This is how lotteries and casinos work.

In fact not only lotteries even insurance companies work more or less on similar premise. They collect small amounts of premium from a large group of persons and in case of death of an insured person pay the assured amount to the family of the deceased. It is assumed that the number of such pay outs will be small. In fact insurance companies employ specialists to decide the amount of premium to be charged. The amount of premium to be charged depends upon the mortality rate and number of customers expected to purchase the policy. Persons who specialize in this subject (called Actuaries) are very highly paid as their decisions affect the earnings of the insurance companies.

4. Separating Coins into Two Groups

Example: There are twelve coins lying on a table in a dark room. Of these 5 coins are known to have heads up and the remaining seven tails up. Without switching on the light it is desired to separate these coins into two groups (some coins can be flipped upside down if desired but without switching on the light) so that when the light is put on both the groups have equal number of heads showing up: Is that possible?

Solution: At first sight it might appear to be a difficult if not an impossible task and we might be tempted to say that the probability of such a grouping is well nigh zero. However a little thinking will show that it is not really so and there is indeed a clever way of doing it with hundred percent probability. It is like this.

Separate the coins into two groups of 7 and 5 coins. Now flip upside down the coins of the smaller group having 5 coins. Both groups will now have same number of heads up. Strange! However it will be indeed so. The argument is simple. When we have separated the coins into two groups of 7 and 5 coins, let the bigger group of 7 coins have n heads up. Then it has $7 - n$ tails up. Since we know that there were in all 5 heads up. The smaller group will have $5 - n$ heads up and hence $5 - (5 - n) = n$ tails up. Now if these five coins are flipped these will have n heads up. So simple!

5. Putting Letters in Right Envelopes

Example: Suppose we write 10 letters and corresponding 10 addresses on 10 envelopes. Now letters are randomly put in the addressed envelopes. What is the probability that (i) just one letter is put in a wrong address envelope, (ii) two letters are put in wrong envelopes?

Solution: One may start scratching one's head as to how to get this probability. However a slight thinking will tell you that in the first case it is just zero because if one letter is put in a wrong addressed letter, then there will be another also in a wrong addressed envelope. A single letter can not be put in a wrongly addressed envelope.

Next the number of ways in which any two letters can go in wrong envelopes is

$$^{10}C_2 = \frac{9 \times 10}{2} = 45$$

Also total number of ways in which 10 letters can be put in 10 envelopes is 10!. So probability of 2 being placed in incorrect envelopes is

$$\frac{45}{10!} = \frac{1}{80640} \text{ (quite small).}$$

6. Do Men have more Sisters than Women

Example: Suppose we ask a large number of students (both boys and girls) in a school to write down the number of brothers and sisters (excluding himself/herself) which he/she has. We next separate the notes written by the boys and girls. Should we expect boys to have more (or fewer) sisters than girls or there would be no difference.

Solution: The answer is that boys are expected to have more sisters than girls, the prime reason being that a girl while writing her sisters has to exclude herself from being counted whereas it is not so in the case of boys. On the contrary girls are expected to have more brothers than boys because in this case during counting of brothers a girl has not to exclude herself but a boy has to.

7. Is it a Fair Game

Consider the following game of chance.

Example: There are three identical circular coins. One of these coins is painted black on both the sides. Another is painted black on one side and white on the other side. The third is painted white on both sides. These coins are placed on a table so that on the upper side one face is black and two white and covered so that it is not known which side of each coin is on the upper side. A person is asked to pay Rs. 10 and take out a coin from under the cover. If the top side of the coin is black, the game is aborted and coin put back under the cover and coins reshuffled and the person is asked to draw again. However, if the upper face of the coin is white then the other face of this coin is examined. If the other face is black the player wins and gets double the amount, i.e., Rs. 20. However if the other face is also white then he looses Rs. 10 which the organizer keeps. Is it a fair game that gives equal chance of winning to the player?

Solution: At first sight it seems so because the other face can be either black or white. But in reality it is not so. It may be noted that stage two of the game is played only if the face in stage I is white, i.e., it is either coin II or coin III and not coin I. Now if we see the other faces of these two coins II and III, one is white and one is black. So we are to calculate probability of second face being black given that first face is white.

Face I

Face II

Coin I Coin II Coin III

$$\text{So it is } \frac{2}{3} \times \frac{1}{2} = \frac{1}{3} \text{ and not } \frac{1}{2}.$$

This is what is known as **conditional probability**, which does not apply to the case of probability of getting a head in successive throws of the same coin but applies to the probability of getting a male child given that earlier two are already female.

Probability of happening of an event has to be calculated carefully using correct reasoning. An incorrect reasoning often leads to an incorrect answer. In many cases the correct answers are not what we ordinarily expect these to be.

7.1 Conditional Probability

In the problem of three coins one painted black on both sides, another painted white on one side and black on the other and the third painted white on both sides, we got the probability of winning 1/3 and not 1/2 because the outcome of second attempt was dependent upon the outcome of the first attempt. We were to check the colour of second face if the first face was white otherwise not. In probability theory this is called 'conditional probability'. The law of conditional probability states that the probability of the happening of an event A given that an event B has already happened denoted by P(A/B) is given by

$$P(A/B) = P(A \cap B)/P(B)$$

where $P(A \cap B)$ denotes probability of simultaneous happening of events A and B.

In the example of three coins let A denote the event of getting upper face white on the selected coin and B the event of getting first upper face white and then second black. Then in such a case

$$P(A) = 2/3,$$

$$P(A \cap B) = \tfrac{1}{2},$$

therefore $$P(A/B) = \frac{1}{2} \times \frac{2}{3} = \frac{1}{3}$$

Same phenomenon is at work in the following example.

8. Complaint of a Daughter

Example: A mother has two daughters who live in the suburbs of a town where she lives. One lives towards the north of the city and the other towards south of the city. Both are on the same bus route that passes close to the mother's house. Every Sunday mother goes to the bus stop between 2 and 3 p.m. and catches the first bus that arrives hoping that this way she would be fair to both her daughters. The bus service is every half an hour on both sides. Whereas the buses going north arrive at her bus stand between 2.00 – 2.10 and 2.30 – 2.40 p.m., buses going south arrive between 2.15 – 2.25 and 2.45 – 2.55 p.m. However, after a few months, her daughter living towards north complained that she was partial to her sister living south as she had visited her more times than she had visited her. Mother was confused. She could not understand why it was so. Can you help?

Solution: The reason is that she would be sure of catching a bus going north only if she arrived at the bus stand before 2.10 or between 2.25 and 2.40 p.m. which is 25 minutes. However if she arrived anywhere between 2.10 and 2.25 p.m. or 2.40 to 2.55 (i.e., 30 minutes), she would be getting a south going bus. Thus her chances of getting a south going bus are greater than chances of her getting a north going bus.

9. Switch or Not to Switch

Example: You are shown back sides of two cards. On the front of one of these cards is written a number which is double of the number on the other card. However these are not told to you but you know that the number on the face of one card is double of the number on the face of the other card. You are allowed to randomly select one of these cards and its other side shown to you. You may either end the game winning the number of rupees written on the face of the card chosen or switch the card and win the number of rupees written on face of that card. What should be your choice: to switch or not to switch?

The following argument shows that switching may be better. Suppose the number written on the face of the selected card is say n. Then the number on the face of the other card is either $2n$ or $n/2$. If you stick to your original card your expected amount of winning is n. However if you switch, then your expected amount of winning is

$$\frac{1}{2}(2n) + \frac{1}{2}(.5n) = 1.25n \text{ which is more than } n.$$

This reasoning applies to any positive number n. According to this if after switching you are again given a choice to switch it would be better to switch again and this argument continues indefinitely without end.

This is what is known as **'dilemma of probability problems'**. It may be noticed that what we are calculating in the second case are expected returns and not actual returns. Actual returns are often vastly different from expected returns. Same dilemma appears in a well known T.V. serial popularly known as Monty Hall Problem.

10. Monty Hall Problem

Example: Monty Hall problem also known as 'Let's make a deal' is a popular T.V. game show. In this show a person is randomly selected from amongst the audience and asked to come on the stage. He is presented three closed doors and is told that whereas behind one of these closed doors is a big prize (say a car) behind the other two are duds (usually a donkey and a goat). He is asked to select one door. Once the choice has been made the host of the show opens one of the other two doors behind which is a donkey. After this the host asks the person whether he would like to stick to his earlier choice or change it with the remaining closed door.

At this some amongst the audience shout urging him to stay with his earlier choice while others ask him to switch over. The question again is will it really make any difference whether he sticks to his earlier choice or changes it. At first sight it appears that since now one of the two closed doors is concealing the car so chances are fifty-fifty and his shifting choice would not make any difference. However some others argue that it will make a difference and switching over is a better option. Their argument is as under:

Suppose there had been 1000 doors instead of just 3 doors and the person had chosen the 1000th door. The probability of its concealing prize money is very small (just 1/1000) and prize money is more likely to be behind one of the other 999 doors. Probability of one of these concealing prize money is larger (999/1000). Now suppose the host opens one by one 998 of the remaining 999 doors and none of these contains the prize money. Then the left out 999th door will be concealing the prize money. In fact probability of one of the left out doors concealing prize money keeps on increasing as doors not containing prize money are opened one by one. In our game there are just three doors in place of 1000. However the argument still applies.

However a naïve approach of argument can be that once a door has been opened and it did not contain prize money, the probability of prize money being behind either of two remaining doors is now equally likely and so there is no point in switching choice. However even in this case a little analysis can convince us that perhaps shifting would be preferable. Let us see how.

Let us denote car by C, donkey by D and a goat by G. There are six possible ways of concealing these behind three doors. These are:

	Doors		
Ways	1	2	3
1	G	D	C
2	D	G	C
3	G	C	D
4	D	C	G
5	C	D	G
6	C	G	D

Suppose the contestant has chosen door 2. Then out of six possible ways only two are favourable to its concealing car. So the odds are 2 against 4 or 1 against 2 in favour of not switching. So it seems preferable to switch. However one could also argue that the host will not open a door which conceals car. Once it has been opened it will be either goat or donkey. Now only two doors are left : one concealing car and the other a dud (donkey or goat). So the chance of the car being behind either of the two doors is equal.

In her 'Ask Marilyn' column in Parade magazine of U.S.A., Marylin reported to have received thousands of letters subsequent to her publishing the suggestion that in general, it is wise to switch. Of these 90% reprimanded her for being in error. Their main argument was the assumption of equal probability of 1/2 amongst remaining two doors as discussed above, in place of it being 1/3 each at initial stage.

This problem had generated many arguments in academic circles both for and against shifting choice. It had been a topic of discussion in U.S.A. in New York Times and some other popular publications also in early 1990's. John Tierney wrote in New York Times (Sunday, July 21, 1991) that the 'conclusion that it was preferable to shift is perhaps an illusion.'

It must be kept in mind that probabilistic conclusions are influenced to a large extent by how the situation under investigation is analysed. Moreover these just indicate the long run trend if the experiment is repeated several times and do not apply to a specific case. For instance when we say that probability of getting a head in the throw of a coin is 1/2 it does not mean that in two throws of the coin we shall get once head and once tail. Even ten successive throws of a coin may not result in a head. Therefore as mentioned earlier value of obtained probability is just an indicator of the possibility of its occurrence and not a guarantee of its occurrence which many start assuming.

11. How Much Money a Gambler Wins

Example: A gambler is offered to play a game. The rules of the game are the following. There are 100 cards which are put face down. On 55 of these 'Win' is written and on the rest 45 'Loss' is written. Gambler starts with a stake of Rs. 1000. He has to bet one half of the money that he has at that time on a card and turn it. He wins or looses as written on the face of the card. Game ends when all the hundred cards have been turned. How much money do you expect the gambler to have at the end.

Solution: It appears that since he will win 10 times more than he looses so he would end up with something more than 1000. But unfortunately that is not so.

Suppose he wins on the first card. Then he has Rs. 1500. Next suppose he looses on the second card. So he will be left with Rs. 750. Had he lost first and won second time he would still have Rs. 750. So every pair of times he wins and looses, he looses 1/4 of his money. There are going to be 45 such pairs of wins and losses and 10 straight wins.

Thus he will end up with $1000\left(\frac{3}{4}\right)^{45}\left(\frac{3}{2}\right)^{10}$ and this is just 1.38 when rounded off. Really surprising!

Situation is somewhat similar to the case when a person is first given a raise of 10% in salary and next year his salary reduced by 10%. He does not revert back to his original salary. His salary will now be 99% of his original salary.

12. Buffon's Needle Problem

Suppose we have a square sheet of paper of size 10 cm by 10 cm and divide it into two equal parts along one of its diagonals. Imagine that there are detectors fixed all over its surface looking for a shower of cosmic rays from the sky. A particle can fall anywhere on earth and thus anywhere on this surface area. Since the area is very small compared to the surface of earth, the probability p of a cosmic particle falling at any point on this area is very small and the probability of its falling on any of these two equal triangular areas will be $p/2$ each. The probability of its falling on 1 cm by 1 cm area taken anywhere on this plate will be $p/200$.

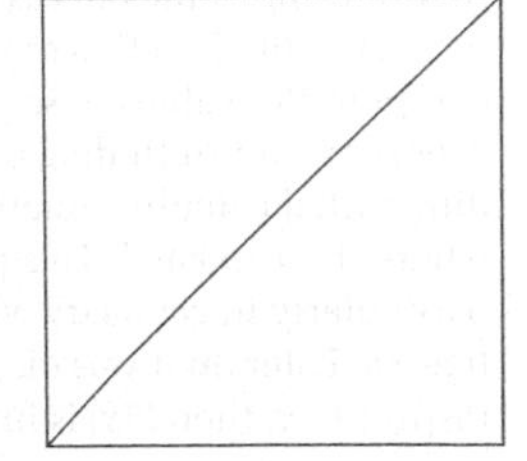

George Louis Conte de Buffon (1707–1788), a great 18th century French mathematician, had posed a similar problem which is now known after his name. The problem is as under:

Suppose there is a large sheet of ruled paper having equidistant parallel lines. The gap between two parallel lines is l which is equal to the length of a needle (or a match stick). *Such a* needle (or match stick) is tossed in air in a random fashion with no particular regard as to where it will fall on the ruled paper. It will naturally fall on the paper in an unpredictable manner. It may or may not cross a ruled line. His question was: what is the probability of it intersecting one of the ruled lines when it falls.

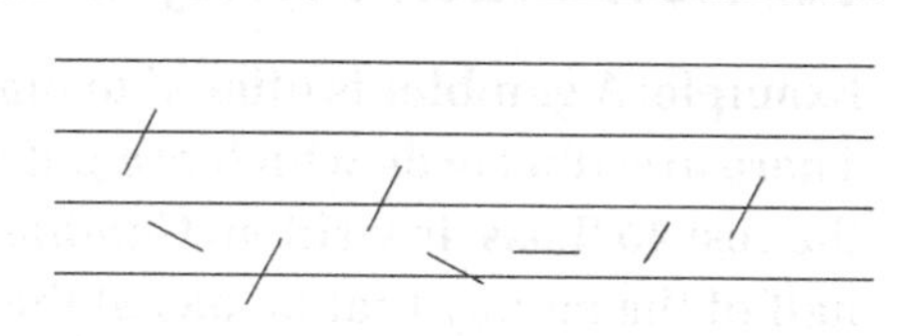

Solution: To answer this question, suppose x denotes the distance of the mid point of the needle from the nearest ruled line. Clearly x can have any value from 0 to $l/2$ and all values are equally likely. Further let θ be the angle which needle makes with the ruled line. Then θ can have any value from 0 to $\pi/2$. Again all values are equally likely. Now draw a rectangle with sides $l/2$ and $\pi/2$.

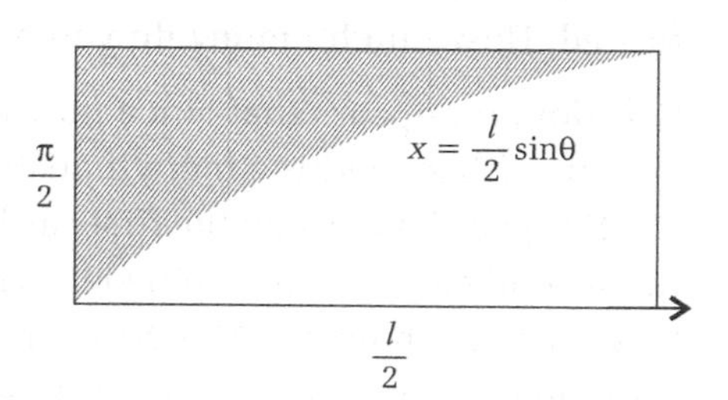

Let us now consider the situation under which match stick intersects the line. It can be shown analytically that it will do so as long as

$$x < \frac{l}{2}\sin\theta \qquad \text{(i)}$$

In the figure the curve $x = \frac{l}{2}\sin\theta$ has been plotted.

Shaded region of the figure consists of points which satisfy (1). It can be shown that this shaded area is $l/2$. (This area can be calculated using techniques of calculus)*.

---

* Calculus is a subject which is usually introduced in 10 + 2 after high school. Newton is credited with having formalised the subject. It essentially arose from the necessity of calculating limiting value of an expression $\frac{f_1(x)}{f_2(x)}$ when both numerator and denominator approach zero as x approaches some real value, say a. For example $\lim_{x \to a} \frac{x^2 - a^2}{x - a}$. Such situations arise in a variety of ways in real life. For example we define velocity at a point as the limiting value of distance travelled divided by time taken in travelling it as both tend to zero. Newton called this limiting values in cases where it existed as the derivative of the function appearing in the numerator with respect to the function or variable (such as time) appearing in denominator. Techniques have been formalised to compute the values of such derivatives where they exist. These have a lot of practical applications in physics, geometry and in finding maximum and minimum values of unknown functions. This subject is now known as differential calculus. Another branch of this subject arose while finding limiting values of sums of infinite series where these existed. This part is now called integral calculus. Integral calculus has also very wide applications, particularly in geometry where it is used for finding lengths, areas and volumes and surface areas of geometrical figures. Later on it was discovered that two branches of calculus are inter-related. If $f(x)$ is derivative of $F(x)$ with respect to x, then $F(x)$ is integral of $f(x)$ with respect to x.

Hence probability p of the needle crossing a ruled line is

$$p = \frac{\text{Shaded area}}{\text{Area of rectangle}} = \frac{l/2}{\frac{\pi}{2} \times \frac{l}{2}} = \frac{2}{\pi}$$

Corollary. What happens if $l > d$, the distance between consecutive parallel lines? (Answer. Needle can now cross even two consecutive lines).

13. Statistics

In order to appreciate the underlying structure of a huge population, knowledge of some of its parameters can be quite helpful. For instance if we want to know the financial health of a group of persons one method is to know exactly the economic status of each individual of the group. This may be alright if the size of the group is not large. However if we are interested in the economic status of persons living in a country, it may not be practically feasible. In such situations we generally try to draw our conclusions based on the analysis of behaviour of specific parameters in samples taken from the population. This is basically the main object of statistics which is now a very well developed subject.

Gaunt is credited with having invented the subject of statistics. He lived in 16th century England. He believed that many social ills of the society that existed at that time could not be eradicated unless society had some idea about the basic characteristics of its population. Gaunt lived in England in times when its population was being periodically decimated by outbreaks of epidemics of plague and other infectious diseases. He believed that knowledge of certain statistical parameters of the population would make it possible to improve it. Gaunt was surprised to discover that the rates at which people in England died those days from certain natural causes such as accidents, suicides, etc. did not change much over time. This prompted him to look for the values of certain types of averages of the population in order to analyse the behaviour of the population. Statistics is now being used to analyse data in different fields of human activity spanning from medical profession to opinion polls. With statistics we try to draw conclusions about the nature of population under investigation using information available from a sample data taken from it.

14. Statistical Parameters

Some of the statistical averages which are commonly used to analyse the behaviour of a population are : mean, mode, median and standard deviation. Whereas **mean** is just the average of the sums of the values of a variate whose behaviour is under investigation, **mode** is the value of that variate which has largest frequency of occurrence. **Median** is the variate which falls exactly in the middle if the values of all variates are arranged in ascending or descending order. **Standard deviation** is the square root of the sum of the squares of deviations of the values of variates from their mean value divided by N the total number of variates.

Suppose we are interested in the statistical analysis of the behaviour of a variate X which takes values $x_1, x_2, x_3, \ldots\ldots, x_n$ in a population with frequencies $f_1, f_2, f_3, \ldots\ldots, f_n$ respectively (f_1 being frequency of x_1 implies x_1 occurs f_1 times). We may write this data in the form of a table as:

Variable, x :	x_1	x_2	x_3	....	x_n
Frequency, f :	f_1	f_2	f_3	....	f_n

Clearly the total size of the population under investigation is:

$$N = f_1 + f_2 + f_3 + \ldots\ldots f_n$$
$$= \Sigma f_i, \quad i = 1, 2, 3, \ldots\ldots, n.$$

The mean $\bar{x}$ of this population is

$$\bar{x} = \frac{\sum_{i=1}^{n} f_i x_i}{N}$$

Its mode is that x_i for which f_i is largest amongst $f_1, f_2, f_3, \ldots\ldots, f_n$. Its median is that f_i which will fall in the middle of all variates $x_1, x_2, x_3, \ldots\ldots, x_n$, if these are arranged in ascending or descending order of their values. Its S.D. (denoted by σ) is

$$\sigma = \sqrt{\frac{1}{N} \sum_i f_i (x_i - \bar{x})^2}$$

Its square is called variance. In fact whereas the mean gives us an idea of the average value of the variate, S.D. tells us how other values of the variate are placed vis-à-vis the mean value. A large value of S.D. indicates that many values of the variate are far away from the mean value (these may be on either side of mean), small value of S.D. implies that most of the values of variates are clustered around the mean value.

There is an interesting remark about the mean value by Australian labour minister. In 1973 he remarked, 'We look forward to the day when everyone will receive more than the average wage!' Is that in any case possible?

We now illustrate the significance of these parameters through examples.

Example 1

Suppose in a class of 40 students, the heights of the students are as under

Height (in inches)	60	62	63	64	65	66	67	68	69	70
Number of students	4	4	3	4	6	8	3	6	1	1

Now if we regard the height of student as variate X and number of students who have same height as its frequency, then the mean (average) height $\bar{X}$ of the students in this class is

$$\bar{X} = \frac{(\Sigma f_i x_i)}{\Sigma f_i}$$

$$= \frac{4\times 60 + 4\times 62 + 3\times 63 + 4\times 64 + 6\times 65 + 8\times 66 + 3\times 67 + 6\times 68 + 1\times 69 + 1\times 70}{4+4+3+4+6+8+3+6+1+1}$$

$$= \frac{2599}{40} \approx 65 \text{ inches, i.e., 5 feet 5 inches (approx.)}$$

Mode is 66, i.e., 5 feet 6 inches. In other words the largest number of students who have identical heights is 8 and their identical height is 5 feet 6 inches.

Median is average height of 20th and 21st student when students are arranged in ascending order of the height. Both these fall in the group of 6 students whose height is 65 inches. Thus median value of this data is 65 inches or 5 feet 5 inches.

Standard deviation of the data is

$$\text{S.D.} = \left[\frac{1}{40}[4(60-65)^2 + 4(62-65)^2 + 3(63-65)^2 + 4(64-65)^2 + 6(65-65)^2 + 8(66-65)^2 + 3(67-65)^2 + 6(68-65)^2 + (69-65)^2 + (70-65)^2]\right]^{1/2}$$

$$= \left[\frac{1}{40}[4\times 25 + 4\times 9 + 3\times 4 + 4\times 1 + 6\times 0 + 8\times 1 + 3\times 4 + 6\times 9 + 16 + 25]\right]^{1/2}$$

$$= \left[\frac{1}{40}[100 + 36 + 12 + 4 + 8 + 12 + 54 + 16 + 25]\right]^{1/2}$$

$$= \left[\frac{267}{40}\right]^{1/2}$$

$$= (6.675)^{1/2} \approx 2.53 \text{ (approx.)}$$

(Note in numerical computations working upto large number of decimal places is not recommended as the motive is to just get an idea of the nature and pattern of data). The above results show that on average height of students in the class is 65 inches, the largest number of students in the class have height 66, and heights are not scattered far from the mean as S.D. is small. Maximum deviation from mean height is just 6 inches. Value of median 65 indicates that the height of half of the students in the class is 65 inches or less and height of other half of the class is 65 inches or more.

Next example illustrates how knowledge of averages can be helpful in taking appropriate decisions about the populations.

Example 2: The following statistics characterize distribution of marks scored in English in two sections of the same class

	Mean	Median	Standard deviation
Section A	**78**	**65**	**10**
Section B	**72**	**73**	**6**

Which class needs more attention of the teacher? Which class has a few gifted students?

Solution: In section A mean is 78 which is more than mean 72 of section B. Moreover S.D. of section A is also more than standard deviation of section B. So the spread of marks in section A is greater than spread of marks in section B. Hence whereas section A is expected to have a few gifted students whose score is very high, it is also expected to have some students whose score is low. Comparatively there is less deviation from mean 72 in marks of students of section B. Moreover whereas median of class A is 65 it is 73 for class B. So whereas fifty percent of students of section A have marks less than 65, fifty percent of students of section B have marks more than 73. So obviously section A needs more attention of the teacher compared to section B as not only 50% of the students of this class have marks less than 65 but some are also having very low marks (as S.D. is 10). No doubt it also has some very gifted students.

14.1 Coefficient of Correlation

Another statistical parameter which gives a good insight into the nature of the relative behaviour of two different parameters of the same population is the coefficient of correlation. We define coefficient of correlation ρ as under

Let X and Y be two variates whose means are $\bar{X}$ and $\bar{Y}$ and S.D. σ_x and σ_y respectively. Let their **covariance** be defined as

$$\text{cov}\,[XY] = E\,[XY] - E\,[X]\,E\,[Y]$$

where E [X] and E [Y] are expected values of X and Y and E [XY] is expected value of product of random variables X and Y.

We define **coefficient of correlation** ρ between X and Y as

$$\rho = \frac{\text{cov}\,[XY]}{\sigma_x\,\sigma_y}$$

ρ is a measure of the linear relationship between random variables X and Y. Its value is always between –1 and 1$(-1 \le \rho \le 1)$. $\rho = 1$ means perfect correlation, $\rho = -1$ perfect opposite correlation. Again $\rho = 0$ means these are totally uncorrelated. Nearer the value of ρ is to 1, greater is correlation between X and Y.

Example: Suppose the heights and weights of 10 students in a class are:

Student	**1**	**2**	**3**	**4**	**5**	**6**	**7**	**8**	**9**	**10**
Weight (in kg) X	**55**	**60**	**62**	**56**	**61**	**58**	**59**	**57**	**55**	**57**
Height (in cm) Y	**152**	**157**	**158**	**154**	**160**	**158**	**162**	**160**	**155**	**158**

Is their any correlation between the heights and weights of the students? If yes, does weight increase with increase in height or with decrease in height?

Solution: In this case E [X] = 58 kg, E [Y] = 157.4 cm and E [XY] = 9133.3

Also, $\sigma_x = 2.3$, $\sigma_y = 2.9$ and cov [XY] = E [XY] – E [X] E [Y] = 4.1

Hence, $$\rho = \frac{\text{cov [XY]}}{\sigma_x \sigma_y} = \frac{4.1}{2.3 \times 2.9} = .6$$

Value of ρ as .6 indicates that there is not very high correlation between height and weight of students. Of course weight in general creases with height.

15. Juggling with Statistics

It is sometimes humorously remarked that statisticians can prove anything by making you look at the data, the way they would like you to see. Here is an example. There was a news in the newspaper which compared the sales of different brands of cars in the years 2009 and 2010. The comparative figures for June 2009 and June 2010 were as under:

S. No.	Company	Sales in June 2010	Sales in June 2009	Percentage change
1.	Tata Motors	27811	17039	50
2.	Hyundai	27366	23016	18.9
3.	Maruti Suzuki	72812	61773	17.89
4.	General Motors	9539	4492	100
5.	Ford India	7261	1982	267
6.	Honda Siel	4595	4067	13

So if we look at the percentages it appears that Ford India outperformed all other companies. But is it really so? Whereas Ford India increased its sales from 1982 units in 2009 to 7269 in 2010, there was an increase of 5287 units in the sales over the year. However Maruti increased its sales from 61773 units to 72812 units which is an increase of 11039 units, almost twice the increase in number of units of sales of Ford. So who performed better? If base is large even a big increase looks small when expressed in percentage form. However if base is small even a small increase looks big.

16. Different Voting Schemes

There are several voting schemes which are used in elections. Three of the most commonly used schemes are:

(i) **Scheme I:** Person getting the highest number of votes gets elected.

(ii) **Scheme II:** Each voter is asked to rank each candidate as per his order of preference. Weighted average rank of each candidate is calculated. One who gets the highest rank is declared winner.

(iii) **Scheme III:** Each voter is asked to rank each candidate as 1, 2, 3, 4, etc. as per his preference. The candidate with highest first preference votes is declared winner if the number of such first preference votes is more than 50%. If no candidate gets more than 50% first preference votes, the candidate with minimum number of first preference votes is eliminated and his/her votes transferred to other candidates as per their second preferences indicated in their ballots. This process of elimination continues till a person with more than 50% first preference votes emerges and he is declared the winner.

Theory of voting is now well developed. However according to Arrow, no voting scheme is completely flawless. These can be manipulated to favour one candidate over the others. The following example illustrates how different voting schemes yield different results even when applied to the same situation.

Example: Suppose in an election there are 33 voters and 3 candidates say A, B and C. Further suppose that each voter is asked to cast his vote indicating his order of preference for the three candidates. The results after counting are tabulated as under.

Rank Order	ABC	ACB	BAC	BCA	CAB	CBA
Number of Votes	10	4	2	7	3	7

Let us analyse this data according to the three voting schemes one by one to decide who the winner is under each of these schemes.

Scheme I: Under this scheme A gets a total of $10 + 4 = 14$ votes, B gets $2 + 7 = 9$ votes and C gets $3 + 7 = 10$ votes. So A is the winner.

Scheme II: Suppose we assign 3 points to first, 2 to second and one to the third preference. Then

Score of A is $10 \times 3 + 4 \times 3 + 2 \times 2 + 7 \times 1 + 3 \times 2 + 7 \times 1$
$= 30 + 12 + 4 + 7 + 6 + 7 = 66$

Score of B is $10 \times 2 + 4 \times 1 + 2 \times 3 + 7 \times 3 + 3 \times 1 + 7 \times 2$
$= 20 + 4 + 6 + 21 + 3 + 14 = 68$

Score of C is $10 \times 1 + 4 \times 2 + 2 \times 1 + 7 \times 2 + 3 \times 3 + 7 \times 3$
$= 10 + 8 + 2 + 14 + 9 + 21 = 64$

Hence A's average rank is $\frac{66}{33} = 2$, B's average rank is $\frac{68}{33} = 2.06$

and C's average rank is $\frac{64}{33} = 1.94$. So, B is the winner.

Scheme III: A has 14 first preference votes, B 9 and C 10. However, none gets more than 50% (i.e., 17 or more votes). So we consider second preference votes. B who gets minimum first preference votes (9) gets eliminated and second preference votes cast to him get transferred to A and C as first preference votes. So, now A has 14 + 2 = 16 votes and C 10 + 7 = 17 votes. So C is the winner.

Thus each scheme yields different person as the winner.

17. A Child has no Time to go to School

There always exist possibility of misuse of statistics to draw incorrect conclusions, sometimes with a purpose. We illustrate this with an example. The following data shows that a child has no time to go to school.

A child spends 8 hours per day (or 1/3 of a day) sleeping. He spends 3 hours a day eating meals etc. A child has 60 days of summer vacation and 21 days of winter vacation. Moreover on each Saturday and Sunday, the school is off. There are also 15 public holidays in an year. Thus we have accounted for

$$\frac{1}{3} \times 365 + \frac{3}{24} \times 365 + 60 + 21 + 2 \times 52 + 15$$

$$= 121 + 45 + 60 + 21 + 104 + 15$$

$$= 366 \text{ days}$$

This is more than an year. So where is the time for the child to go to school or play?

Apparently it seems correct. However, there is a flaw in the argument. Have you noticed?

The flaw is that during 60 days of summer vacation, 21 days of winter break, 104 Saturdays and Sundays in an year and 15 public holidays, he sleeps as well as eats. So for these 200 days, the hours of his sleep and eating meals are being counted twice.

If we correct this error, we have

$$\frac{1}{3} \times 165 + \frac{3}{24} \times 165 + 60 + 21 + 2 \times 52 + 15$$

$$= 55 + 22 + 21 + 104 + 15$$

$$= 217 \text{ days}$$

leaving the student with (365 – 217) × 24 = 148 × 24 = 3552 hours in an year to go to school and play. A reasonable amount of time.

18. Effect of View Point

An important principle of statistics is that results get influenced by the point of view of decision maker, sampling methods used as well as statistical techniques used in analysis of collected data. We illustrate it with examples.

Example: Suppose there are 20 families. Of these 19 families have one child each whereas the 20th family has two children. So there are in all 21 children in 20 families.

Hence average number of children per family $= \frac{21}{20}$

Since 19 families have three members each and one family four members, so average family size $= \frac{19 \times 3 + 1 \times 4}{20} = \frac{61}{20}$

Now $\frac{61}{20}$ is much greater than $\frac{21}{20}$ which means that on average most of the children come from large size families. Is it so?

Corollary: Is it possible to justify that most of the children come from average size families?

Example: Has the Cost of Living Really Gone Up

Last year commodity A was selling at Rs. 500 per quintal and commodity B at Rs. 1000 per quintal. This year the prices have reversed. Commodity A is selling at Rs. 1000 per quintal and commodity B at Rs. 500 per quintal. Has the cost of living gone up or down or is there no change?

Surprisingly any of these three alternatives can be justified. It depends upon what year we take as the base.

If previous year is taken as the base and we calculate how many percentage points the prices have changed we notice that price of commodity A has doubled. Hence there is 200% increase in cost of A. Cost of B has halved. So there is 50% decrease in cost of B. So there is an average increase of (200 – 50)/2, i.e., 75%. So cost of living has gone up. However if we take current year as the base year, A was selling at 50% of its current price last year and commodity B was selling at 200% higher price than present year. So cost of living was (200 – 50)/2 = 75% higher than current year. So cost of living has decreased this year!

Again one could argue that since the price of one commodity has doubled and the price of other halved so there is no overall change in cost of living.

19. Paradoxes

Many paradoxes arise in probability and statistics from incorrect understanding of the concepts of probability or ambiguity in the choice of the way the provided data is analysed. We present here some of these.

1. Paradox in Chance

Example: Suppose three coins are tossed simultaneously. What is the probability of coins turning up alike (i.e., all three heads or all three tails)?

One way of argument could be that as there are two equally likely possible outcomes in the throw of each coin so there are $2 \times 2 \times 2 = 8$ possible outcomes of which two are favourable to the event. Hence probability of all three coins turning up alike is $2/8 = 1/4$.

Against this a person argues as under. For three coins to turn alike at least two must have turned alike. Once this has happened it is even chance that third will show a head or a tail. So the probability of getting all three alike is 1/2 and not 1/4.

Where is the flaw? The flaw is in the assumption that two coins have already turned alike. So it is taken for granted that this event has happened with hundred percent probability. Actually all the three coins are thrown simultaneously so this argument has a flaw and correct answer is 1/4.

2. Conditional Probability Paradox

Some paradoxes arise from incorrect understanding of conditional probability. Here is an example.

Example: Suppose a person undergoes a test for a specific disease. The test is known to be 95% accurate. Test results are positive indicating that there is a great chance that the person has contracted the disease. Is the chance of having contracted disease 95% or less or more, given that in a population of 20,000 only one percent are known to have that disease?

Solution: Since only one percent of 20,000 have disease so there are 200 persons who have this disease. Now if all these 200 persons had been subjected to the test, then the test being only 95% accurate, it would have indicated that amongst these 200 persons only 190 have the disease.

So probability of having disease if the test results are positive

$$= \frac{\text{Number of persons with disease who test positive}}{\text{Total number of persons who test positive for the disease}}$$

$$= \frac{190}{190 + 990} = \frac{190}{1180} \approx 0.1617$$

(The remaining 19800 if subjected to the test 95% will test negative. In other words $19800 \times \frac{95}{100} = 18810$ will test negative and the remaining $19800 - 18810 = 990$ will test positive for the disease even though they do not have disease).

However the probability that a person will test positive for disease given that he has disease is $\frac{190}{200}$ or 95%

Thus if one submits oneself to the test without any prior information of whether one has disease or not and tests positive, his chance of actually having disease is only 16.1%. However if one knows that one has disease and submits for the test then there are 95% chances that he will test positive for the disease.

3. Division of prize money paradox

Example: Two players play a game alternatively in which each has 50% chance of winning. It has been agreed that the first player to win six rounds would be the winner and take the entire money at stake. Due to certain unavoidable circumstances, the game got interrupted and had to be stopped when the player A had won 5 games and B 3 games. The question now is how should the money at stake be divided fairly amongst the two

(This problem was first posed by de Mere to Pascal and was published in 15th century).

Obviously there are two ways in which decision can be made. One way is to divide the money in the proportion of the number of games that each player has won. In that case A's share would be 5/8 and B's 3/8 of the stake money. An alternative argument could be how close was each person to winning when game got interrupted. A was just one game away from the win whereas B was 3 games away from the win. So the money be divided in the ratio 3 : 1, i.e., A's share be ¾ and B's ¼. Someone else could say let it be fifty-fifty as none has won, and it could happen that B wins all three next games and A not even a single. It all depends upon how one looks at the problem.

4. Simpson's Paradox

This paradox concerns calculation of values of certain statistical parameters of data.

Suppose two shipments of an item are received from two different suppliers, say A and B. Further suppose that the number of defective items received from each supplier in the two shipments are as under:

	Supplier	
	A	B
Shipment I	Good items = 600 Defective items = 500	Good items = 400 Defective items = 300
Shipment II	Good items = 300 Defective items = 600	Good items =500 Defective items = 900

When the quality control engineer inspected shipment I he noticed that whereas proportion of defective items received from supplier A was $\frac{500}{1100}$, i.e., $\frac{5}{11}$ it was $\frac{300}{700}$, i.e., $\frac{3}{7}$ from supplier B.

In case of second shipment proportion of defective items from suppliers A and B were respectively $\frac{600}{900}$, i.e., $\frac{2}{3}$ and $\frac{900}{1400}$, i.e., $\frac{9}{14}$.

Now since 3/7 < 5/11 and also 9/14 < 2/3 so it appears supplier B is sending better quality goods in both shipments.

Next suppose we combine the items of both shipments and then calculate the proportion of defective items received from the two suppliers. In this case we find whereas proportion of defective items from supplier A is

$$\frac{500+600}{1100+900} = \frac{11}{20}$$

it is $$\frac{300+900}{700+1400} = \frac{12}{21} \text{ for supplier B.}$$

Now $$\frac{11}{20} < \frac{12}{21}.$$

So goods received from supplier A are of better quality compared to goods received from supplier B. In a way we are forced to reverse our earlier decision. Why is it so?

(This is due to the fact that if $\frac{a_1}{b_1} < \frac{c_1}{d_1}$ and $\frac{a_2}{b_2} < \frac{c_2}{d_2}$, then $\frac{a_1+a_2}{b_1+b_2}$ need not be less than $\frac{c_1+c_2}{d_1+d_2}$).

Mind Teasers

1. Three dots of ink are flung randomly on a circular disc. What is the probability that all three will fall in the same half of the disc?
2. Eight slips of paper with numbers 1, 2, 3, 4, 5, 6, 7 and 8 written on them are placed in a bin. Slips are drawn one by one. What is the probability that the four slips drawn will be 1, 2, 3 and 4 in that order?
3. A coin is tossed twice. Given that the first outcome was a head, what is the probability that the second is also a head?
4. A person is walking along a road. He starts from a point (say O) on the road and takes one step at a time of one unit length. However, this step can be in the forward or backward direction (both with equal probability). He continues moving along the road like this till he reaches one of the posts A or B. Post A is located a units in the forward direction and post B is *b* units in the backward direction. What is the probability of his exit at B?

B — *b* — O — *a* — A

4.(a) Starting from (0, 0) an object moves in a coordinate plane via a sequence of steps each of length 1. The step is either to left or right or up or down with equal probability. What is the probability that object reaches (2, 2) in six or fewer steps?

5. **Ruin Problem:** Two players A and B play a game. Player A has *a* rupees and player B has *b* rupees. A coin is flipped. If it is head A wins and B pays A rupee one. If it is tail B wins and A pays him rupee one. The game continues, the two players flipping coin alternatively till one of them has exhausted his money. Find the probability of winning of A and B.

6. Two players A and B play a game. They throw a coin alternatively. The first to get a head wins. If A starts the game what are the probabilities of each winning the game?

7. If two dice are tossed together which number has the greatest probability of appearing as their sum?

8. A and B are two events in probability space with P(A) = 0.20 and P(B) = 0.40. What is the probability that at least one of these occurs given that (i) sample spaces of A and B are disjoint in S, (ii) A is a proper subset of B, (iii) $P(A \cap B) = 0.15$.

9. **Tea Tasting Lady:** A lady claims that by tasting a cup of tea, she can tell whether the milk or tea was first poured in the cup. Suppose she is presented 8 cups of which 4 have been prepared with milk being first poured in the cup and the remaining 4 in which tea was poured first. The lady is then asked to separate these into two groups, one in which tea was poured first and other consisting of cups in which milk was poured first (she is not told as to how many are of each type). Suppose the lady in reality has no such expertise and just guesses. What is the probability that she will correctly identify all the eight cups?

10. Eight persons attend a party. At the end of the party, three persons are chosen by lottery to receive first, second and third prize. What is the probability that person A receives first prize, person B second prize and person C third prize. Again suppose instead three persons are chosen randomly and given identical prizes. What is the probability that these chosen persons are A, B and C?

11. A teacher assigned 20 problems to the students for the test to be held next day. Five of these problems were chosen by the teacher for the test. A student had prepared only five problems out of these 20. What is the probability that (i) exactly those very five problems were there in the test, (ii) none of these five problems was in the test.

12. Ten persons gather for new year party. What is the probability that (i) at least two of them have same birthday, (ii) at least one has birthday on Jan 1 itself.

13. What is the probability that a randomly selected 5 digit number reads same both ways (backward and forward) (such numbers are called Palindromic numbers example 50205).

14. For every 1000 births, the number of persons who are alive at their 65th birthday is 746 (i.e., $P_{76} = 0.746$) and probability that a person who has just turned 65 will die in next five years is 0.160. What is the probability that a child born in this population will reach his/her 70th birthday. Also find the probability that a person who is 25 years now will be alive at 50 given $P_{25} = 0.95$ and $P_{50} = 0.85$.

15. A population was grouped according to blood group and it was observed that in this population 42% are of group A, 33% of group O, 18% of group B and 7% of group AB. Assume that a marriage has taken place in this population without checking blood group of the partners. What is the probability that the couple will have the same blood group?

16. **Question of Life and Death:** A criminal has been sentenced to death. To give him a chance of pardon he is given two identical jars and 100 white and 100 black balls. He is asked to put these balls in these two jars the way he likes. He is informed that after that he would be blindfolded and asked to randomly choose one of these two jars and pick a ball from it. If the picked out ball is white, he will be set free else executed. How should this unfortunate person distribute the balls between the two jars so that probability of drawing a white ball is maximized?

17. On Friday weathermen predict 50% chance of raining on Saturday, on Saturday they again predict 50% chance of raining on Sunday. What is the probability that it will rain this weekend?

18. There are 25 balls in all in two boxes. Out of these some are white and some black. One ball is taken out randomly from each box. The probability that both are black is 27/50. What is the probability that both are white?

19. Ram and Shyam have test on Wednesday. On Tuesday night, they were attending a marriage party till late night and so could not prepare for the test. They came late to the college and pleaded before the teacher that they became late for the test as a tyre of the car in which they were coming to the college got punctured and had to be replaced. The teacher was in no mood to accept this excuse lightly. He called Ram to his office and asked him which tyre had got punctured. After he replied, he was asked to sit in a side room. Then Shyam was called and asked the same question. What is the probability that their answers were identical?

20. **How to Change Sex Ratio:** Once a monarch wanted to alter the sex ratio of his subjects in favour of women as their number was going down. He ordered that every married woman could continue bearing children as long as these were females. The moment a woman gave birth to a male child, she was to stop bearing babies. Assuming that when the king issued this decree the number of males and females was equal, will this policy yield expected result in the long run?

21. Ram and Shyam speak truth only 1/3 times (i.e., out of three statements which either of them makes 2 are false and only one true). Ram makes a statement and Shyam says that Ram is speaking truth. What is the probability that statement made by Ram is true?

22. Mrs. X is mother of two children. You meet her in the street with a child who is her son. What is the probability that Mrs. X has both sons. One possible answer is ½ as the other can be a boy or a girl. The other possible answer is 1/3 as you know that Mrs. X has at least one son so possible combinations are (B, G), (G, B), (B, B). Which answer is correct and why?

23. A coin is flipped continuously till a head appears. If head does not appear upto 15th flip you win Rs. 100. However if a head appears before 15th flip you lose Rs. 100. Is it wise to play this game? What is the chance of your loosing? What is the expected pay off?

24. Mean height of 15 students in a class is 5′ – 5″. A new student whose height is 5′-9" joins the class. What is now the mean height of students in the class?

25. A class of 30 students was given two tests. Suppose the mean score and S.D. of scores of students in both the tests was the same. What can be said about the performance of students in the two tests. Does it imply that each student scored identical marks in both the tests? What if the same test was given to two classes of 30 students each and their mean score and S.D. were same?

26. Demographers wanted to study data of children of a small village hemlet comprising of 30 families. They visited each house and recorded the number of children in each family. It was found that all the children of the hemlet were attending the same school in the village. Demographers then went to the school and asked each child as to how many children his/her family had (including the child himself/herself). The following distribution of families and children was obtained

Number of children, x_i	0	1	2	3	4	5	6	7
Number of families, f_i	3	4	6	6	4	3	2	2

Calculate the mean (average) number of children per family based on this information.

27. A bug is at the origin of coordinates in a plane. It can jump from one lattice point to another (a lattice point is a point whose both coordinates are integers such as (1, 1), (2, 3), (3, 4), etc.), one jump per second. However from lattice node (m, n) it can jump to either $(m, n + 1)$ or $(m + 1, n)$ and both with equal probability. Where is the bug expected to be after 5 seconds. Are all these positions equally likely?

28. There was an advertisement for car tyres. Tyres which were guaranteed to last one lakh kilometers were to cost Rs. 4,500 per piece. Tyres that were guaranteed to last 2 lakh kilometers were to cost Rs. 6,000 per piece, and tyres which were guaranteed to last 3 lakh kilometers were to cost Rs. 7,500 per piece. Is it possible that company is in reality manufacturing only one type of tyres which can last n lakh kilometers (n = 1, 2 or 3) and marketing these under different brands? Assume tyres of each brand sell in different numbers. For what value of n is the profit maximized?

29. A bag contains four notes of denominations 10, 50, 100 and 500. A person is blindfolded and asked to draw these notes from the bag one by one and put them in boxes numbered 1, 2, 3 and 4 which are intended to contain notes of denominations 10, 50, 100 and 500 respectively. Person will be paid Rs. 100 for every note placed in the correct box. What is the expected amount of his winning?

30. A class of 30 students were given two tests in the same subject. Let x denote the score of a student in the first test and y in the second test. On finding the average score and S.D. of the scores in these two tests it turned out that $\bar{x} = \bar{y}$ and $\sigma_x = \sigma_y$ Does it mean that there is perfect correlation between the scores in the two tests (i.e., does it imply $P_{xy} = 1$ for each student)?

31. Some years ago a well known personality left Punjab and settled in Bhopal. A local newspaper remarked that by shifting to Bhopal he has raised the average income of both the states of Punjab and Madhya Pradesh. Is it theoretically possible? If yes, how? Considering this and keeping in view that average income of India is average of income of persons living in different states of India it appears that by mere reshuffling of the population one could increase average income of Indians. Is it really so?

Chapter 9
Calendars, Horoscopes and Astrology

1. Introduction

In this chapter we briefly try to highlight the role of mathematics in formation of calendars and horoscopes and the statistical nature of astrology.

2. Necessity of Keeping Record of the Passage of Time

Humans have always felt the necessity of keeping a systematic record of the passage of time so that posterity can know precisely as to the sequence in which a set of events occurred and precise time of their occurrence. It is also necessary so that public in general can know as to when an event or a festival happened in the past. It also wants to know in advance the date on which a particular festival is to be celebrated next.

Human civilization has taken help of the periods of revolution of moon around the earth and earth around the sun for this purpose. Whereas the period of rotation of earth about its axis is taken as the length of a day, the period of revolution of the moon about earth is regarded as the length of a month. Similarly, the period of revolution of the earth around the sun is taken to be the length of an year. For further convenience, a day is divided into 24 hours and each hour into 60 equal parts called minutes and each minute is further divided into 60 equal parts called seconds. Division of a hour into 60 minutes and a minute into 60 seconds of time roughly corresponds to division of the measure of 1 degree of an angle into 60 minutes and again 1 minute of angular measure into 60 seconds. For most of the civil purposes, second is the smallest unit of time. However for more precise record of time in games, a second is further subdivided into 10 and even 100 equal parts. For record of certain observations in experimental work requiring very high precision, we even hear time being recorded in nanoseconds.

3. Calendars

A calendar is an official record of the passage of time starting from a certain epoch back in time. We have several calendars in vogue in different religions and communities. However the calendar which is now most commonly used all over the world is Gregorian Calendar.

4. Gregorian Calendar

Gregorian calendar which is now being followed at international level had its origin in the west. Pope Gregory XIII is credited with having initiated it. It has as its basis period of rotation of earth about its axis as the length of the day and the period of revolution of the earth around the sun as the length of an year. A day is not considered as the time period between two successive sun rises at a place but the time period from one midnight to the next midnight which is 12 hours after the time the sun is at its highest position in the sky in the day. May be this was due to the fact that in most of Europe weather being foggy and cold most of the year it was not convenient to record exact times of rising or setting of the sun compared to recording the time of the day when sun is highest in the sky. Moreover in this calendar no specific importance is given to the period of rotation of the moon around earth except for dividing a year into twelve months.

In this calendar a new day begins at 12 midnight and an year has ordinarily 365 days except in the case of a leap year when it has 366 days. The start of the calendar is associated with Christ. Events which happened before Christ are referred to as B.C. (before Christ) and events which happen after Christ are referred to as AD (anno domini, after Christ). Thus April 12, 201 BC means 12th day of April of the year preceding 200 years before Christ. However April 12, 201 AD means the happening of event on 12th of April of year which started after 200 years had elapsed after Christ.

In Gregorian calendar an year is divided into 12 months which are usually of length 30 or 31 days except February which has 28 days making a total of 365 days. However in a leap year February has 29 days making a total of 366 days. Each year divisible by 4 is a leap year except when it is a century. A century is a leap year only when it is divisible by 400. Thus years 1864, 1896 being divisible by 4 were leap years but 1900 was not a leap year as it was century which was not divisible by 400. However year 2000 was a leap year as it is divisible by 400. Next century 2100 will again be not a leap year.

It is not that since inception calendar has been as it is now. It is in fact adaptation and improvement upon an earlier calender in use in Europe and called Julian Calender after Julius Caesar (45 B.C.) It has seen periodic changes for improvement. For instance it is said that when this calendar was introduced even though different months had different lengths, February had also 30 days. Also every fourth year (even if it was a century) was a leap year. Lengths of some months were later increased to honour ruling monarchs of the day. For example Augustus got a month named after him (August) and assigned it 31 days. Similarly Julius had a month named after him (July) which also has 31 days. To adjust the total number of days in the year to 365, two days were deducted from the month of February (which perhaps did not have any godfather) and was left with 28 days. In order to compensate it partially the additional day of leap year was allotted to it so that in a leap year it has 29 days.

Initially every year divisible by four was a leap year. This is in fact a necessary correction as the period of revolution of earth around sun is not exact number of days. It is roughly 365 and a quarter days. Keeping year of 365 days correction was made every fourth year for each quarter day lost in the previous three years. However as the observational techniques became more and more precise, it was realized that the period of revolution was not exactly 365.25 days, but in reality around 365.22 days. Making every fourth year a leap year was a bit over correction and it was noticed that such an over correction was resulting in adding 3 extra days in 400 years. Thus it was decided that even though normally every year divisible by 4 will be a leap year of 366 days, this rule will not apply to a century year. A century year will be a leap year only if it is divisible by 400 thus reducing total number of leap years in a period of 400 by 3. This correction was not adopted uniformly even in Europe. Some countries adopted it earlier some later. As a result in calendars of most of the European countries certain dates are missing depending upon when these corrections were adopted so as to make calendar become correct retrospectively. For instance England adopted it very late in 1752 on September 2. As a result the date on next day was not September 3 but September 14. Technically even with this correction a minor error is still left. However its effect will not be visible for centuries to come. The prime reason as to why it is necessary that the length of a year be same as the period of revolution for earth around the sun is that if it is not ensured then seasons will not keep pace with months. For example if we have winter in December and January months now it would have gradually shifted to other months if the correction had not been adopted.

5. Finding the Day of the Week on a Particular Date

We are sometimes interested in knowing the day of the week on a specific date. For example we may be interested in knowing the day of the week on August 15, 1947 when India became independent.

Keeping in view the pecularities of Gregorian calendar, the following simple rule enables us to determine the day of the week on a particular date.

1. There is a four hundred year cycle. In other words after every four hundred years the day of the week on a particular date of a month will be same as it was four hundred years earlier.
2. In an year there are 365 days. On dividing it by 7, we have 52 complete weeks leaving an extra day. However, since a leap year has 366 days it leaves 2 extra days after 52 complete weeks. Thus in one hundred years there are $1 \times 100 +$

24 = 124 = 5 (modulo 7) extra days if the hundred year is not divisible by 400. Thus in a century not divisible by 400 there are five extra days beside complete weeks. (However in a century divisible by 400 rule 1 applies).

3. We can calculate the number of extra days and complete weeks from the start of calendar till the preceding year as well as the number of extra days till the date under question in the year under consideration.
4. Adding all extra days till date remainder modulo 7 gives the day of the week. Remainder one is Monday, remainder two Tuesday and so on.

For example in case of August 15, 1947

(i) Number of extra days till 1600 — Nil

(ii) Number of extra days till 1900 — $5 \times 3 = 15 = 1$ (modulo 7)

(iii) Number of extra days from 1901 to 1946

$= 46 + 11 = 57 = 1$ (modulo 7)

(iv) Number of extra days from Jan 1 to August 15

$= 3 + 0 + 2 + 3 + 2 + 3 + 3 + 15 = 31 = 3$ (modulo 7)

Total number of extra days till August 15, 1947 = 1 + 1 + 3 = 5

Thus it must have been Friday.

6. Calendar Pecularities

How is it possible?

A boy was asked in a party as to how old he was. His reply was, 'I was 15 years old day before yesterday and will be 18 years old next year.' Is it possible?

Yes it is possible if his birthday was 31 December and he was asked the question on January 1. He was 15 years on December 30 and became 16 years old on December 31 of the previous year. He will be 17 years on December 31 of the current year. So he will be 18 years old December 31 of next year.

What is the probability that April 4, June 6, August 8, October 10 and December 10 in a year will fall on the same day of the week?

At first sight it might appear that there is very little probability of this happening. How can 5 different dates chosen in 5 different months fall on the same day of the week. However it will surprise you to know that it is hundred percent sure each year! This is because of the pecularities of the Gregorian calendar.

A closer look reveals that all these dates are exactly 9 weeks apart and hence fall on the same day of the week.

Another calendar related problem

Suppose we have the following calendar of a month

Sunday	Monday	Tuesday	Wednesday	Thursday	Friday	Saturday
		1	2	3	4	5
6	**7**	**8**	**9**	10	11	12
13	**14**	**15**	**16**	17	18	19
20	**21**	**22**	**23**	24	25	26
27	28	29	30	31		

Select a 3 × 3 square of any nine dates (say 7, 8, 9, 14, 15, 16, 21, 22, 23). Add 8 to the smallest date in the selected square and then multiply it with 9. In the present case it is $(7 + 8) \times 9 = 135$. Now multiply sum of entries in the middle row with 3, which in present case is $(14 + 15 + 16) \times 3$. We again get the same number 135. Surprising. Is it just a chance? No, it will always be so. The reason is as follows. The numbers in the middle row (or column) are averages of the corresponding three numbers in other two rows (or columns). In fact the entry at the centre (15) is average of all 9 numbers. So naturally their sum will be 9 times the sum of number at the centre (in this case 15) which will always be eight more than the first entry of the rectangle. Their sum will also be three times sum of the averages of the column entries which appear in this rectangle. Result will also hold even if instead of second row, sum of entries of second column (or sum of entries of a diagonal) of this rectangle are multiplied with 3.

7. Persons having same Birthday

Suppose a class has 35 students. What do you think is the probability that in this class there are two students who have the same birthday (same day and month but not year). Intuitively it seems that the probability of such a coincidence is very small. However it may be surprising to know that amongst first 35 presidents of U.S.A., there were two who had same birthday (James K. Polk, the 11th president (Nov. 2, 1795) and Warron G. Harding, the 29th President (Nov. 2, 1985). In fact one may be surprised to learn that it was not just by chance. In a group of 35 persons the probability that two of them will have same birthday is around 80%! Let us see how?

Suppose a person is chosen from this group and his birthday noted. Now the probability that the next choosen person will not have this birthday is 364/365. The probability that next chosen persons will not have the birthday of the earlier two chosen is 363/365 and so on.

Hence the probability that no two persons amongst the group of 35 have identical birthday is

$$\frac{364}{365} \times \frac{363}{365} \times \frac{362}{365} \times \ldots\ldots \times \frac{(365-34)}{365} = \frac{364 \times 363 \times 362 \times \ldots\ldots \times 331}{(365)^{35}}$$

So the probability that at least two persons have same birthday is

$$1 - \frac{364 \times 363 \times 362 \times \ldots\ldots \times 331}{(365)^{35}} \approx 0.81438$$

In fact whereas the probability of finding two persons in a group of 35 with same birthday is more than 80% it is more than 50% even in the case of a group of 25 persons. Arguing similarly the probability of finding two students in a class of 60 students who have same birthday is almost 99%!

It is not only in the case of birthdays. It can also be marriage dates or death anniversaries. It may be of interest to note that amongst first 35 presidents of U.S.A., Millard Filmore and H. Taft died on same day of the year (March 8). Again three presidents of U.S.A. have died on July 4 (John Adams, Thomas Jefferson and James Monroe).

(This should serve as eye opener for relying too much on intuition).

8. Friday the Thirteenth

Western world generally regards 13 to be an unlucky number. Many hotels in the west do not have room numbered 13. Thirteenth is not considered auspicious for starting a new project. Forming a group of 13 persons is generally avoided. Why is it so? It is said that 13 persons were sitting together for supper when Jesus Christ was betrayed which resulted in his crucification on a Friday. Do you think that thirteenth of a month falls on a Friday with more or less same regularity as on other days of the week? The answer surprisingly is No. In fact 13th of a month falls more frequently on a Friday as compared to the other days of the week (This fact was first reported in Monthly American Mathematical Society in the year 1933 by B.H. Brown). Let us see why it is so.

We know that Gregorian Calendar has a cycle of 400 years after which the day of week and date coincide. Now since there are $400 \times 12 = 4800$ months in a cycle of 400 years, 13th comes 4800 times in this cycle. The accompanying table summarises the frequency with which 13th appears on different days of the week in a cycle of 400 years.

Days of week	Frequency	Percentage
Sunday	687	14.313
Monday	685	14.271
Tuesday	685	14.271
Wednesday	687	14.313
Thursday	684	14.250
Friday	688	14.333
Saturday	684	14.250
Total	**4800**	

Clearly 13th comes a bit more times on Fridays than other days of the week.

9. Other Calendars

Gregorian calendar that is now more or less used internationally is not the only calendar in vogue. There are other calendars which are still in use. Different communities use these particularly to decide the dates of their festivals. We discuss here briefly some of these.

Hijri Calendar

Muslims usually follow Hijri calendar in fixing their festivals such as the month of Ramzan and Id. This calendar is purely a lunar based calendar in which length of a month is from one new moon day to the next new moon day. An year consists of 12 lunar months. No importance is given to the period of revolution of the earth around the sun. As a result a month is of 29 to 30 days (roughly 29.5 days). The reason why no importance was attached to revolution of the earth around the sun seems to be due to the fact that Arab world in which this calendar originated is located near equator. So not much variation in the position and time of rising and setting of the sun occurs from day to day. Hence no appreciable change was noticed in the duration of daylight from day to day over the year. On the contrary waxing and waning of the moon was easily noticed by a common man as the nights in Arab world are generally clear throughout the year. The beginning of this calendar is the date of Hijra in the life of Prophet Mohammad and that is why it is called Hijri calendar. As no care has been taken to adjust the length of 12 months of this calendar with the actual period of revolution of the earth around sun, dates of muslim festivals computed as per this calendar do not keep pace with seasons. That is why in some years we have Id in summer, then in spring, then in winter, and then in autumn. The cycle is repeated every 19 years. This period is known as Metonic cycle. After this interval sun and moon revert back to their original relative positions vis-à-vis the earth. In a metonic cycle whereas there are 20 years of 12 lunar months, there are 19 years of 12 solar months.

Vikrami Samvat

Hindus in north India usually follow Vikrami Samvat calendar for determining the dates on which different Hindu festivals fall in an year. Like Hijri calendar its base is also lunar month, the time of one revolution of moon around the earth counted either from one new moon to the next new moon or from one full moon to the next full moon. This duration is around 29½ solar days. Thus a lunar year of 12 months has $29.5 \times 12 \approx 354$ days in place of 365 days of solar year. Thus there is a deficit of around 11 days each year. In Hijri calendar no effort has been made to adjust this difference. However in Vikrami Samvat an effort has been made to periodically adjust this difference so that over the years same festivals fall in the same seasons. In fact Indian civilization where this calendar originated was essentially agriculture based. Necessity was therefore felt of informing the farming community of appropriate times for beginning or completion of certain agricultural activities such as sowing and harvesting by associating them with certain festivals which should fall arount the same time of the year every year. For example festival of Baisakhi is associated with the beginning of harvesting of wheat crop, Dussehra and Diwali with harvesting of summer crops of rice etc. and preparing of land for sowing of the next winter crop. To achieve this linkage of festivals with seasons help is taken of an extra month called **Malmas** which is added every third year or so. This adjusts the deficit of around 11 days each year. That is why we notice that whereas dates of hindu festivals as per Gregory calendar shift back for two successive years, in the third year these revert back. As a result over a long period of time we have festivals falling in the same seasons unlike muslim festivals whose dates are decided as per Hijri calendar and so keep on shifting from year to year, making same festival fall in different seasons over the years. Vikrami calendar is quite an elaborate affair in which details of times of rising and setting of sun, moon, planets and stars are given from day to day. According to this calendar the time of start of new day is determined by relative positions of earth and moon, and so is not same from day to day. However this is accounted for while preparing precise horoscopes. For normal civilian purposes whatever is the tithi at the time of rising of the sun is taken to be the tithi for that entire day. The starting point of this calendar is the reign of king Vikramaditya in the hoary past. New year of this calendar usually falls in the month of March. In March of 2011 we had the beginning of Vikrami Samvat 2068.

Saka Calendar

After India became independent in 1947, government of India thought of introducing a national calendar which could bring uniformity by superceding different calendars being followed by different communities in different parts of the country. To achieve this a special purpose committee was constituted. It consisted of eminent academicians and astronomers well versed in Indian calendars. It had as its members eminent physicists like Dr. Meghnad Saha and eminent astronomers like Dr. Gorakh Prasad. The committee after due deliberations submitted its recommendations to the government of India. Based on these recommendations the government formally approved the implementation of a national calendar called Saka Calendar. This calendar is an intermix of Gregorian and Indian calendars. Like Gregorian calendar a year of this calendar is also of 365 days except in the case of a leap year when it is of 366 days. Like Gregorian calendar every fourth year is a leap year except when it is a century. New year of this calendar always falls on March 22, the day of spring equinox when days and nights are of equal length throughout the world. Months are either of 31 days or 30 days. Second to six summer months are of 31 days duration whereas the remaining 7 months are of 30 days each (including February) making a total of 365 days in an year. March 22 of 2011 was the beginning of Saka Samvat 1933.

Unfortunately like national language Hindi, national calendar has also only ritualistic significance and is neither used by public at large nor the pundits. Even the government is not very serious about it. The only time we hear of it is in the morning on All India radio and Doordarshan. Just as English is now being used for all official purposes, Gregorian calendar is being used in all government dealings and pundits and astrologers still prepare the horoscopes and panchangs and Jantris based on Vikrami and other Indian calendars.

10. Horoscopes

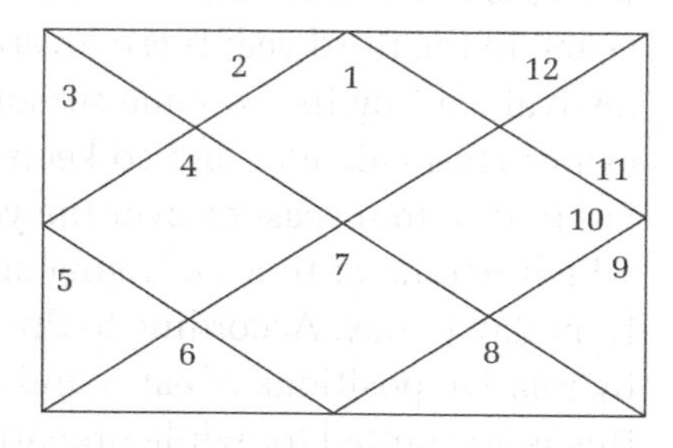

Horoscope of an individual is a chart giving the location of various important heavenly bodies such as the sun (सूर्य), the moon (चन्द्रमा), the planets mercury (बुध), Venus (शुक्र), Mars (मंगल), Saturn (शनि), Jupiter (बृहस्पति) and the ascending and descending nodes (राहु and केतु) (which are the points of intersection of the celestial equator with the ecliptic). Positions of these heavenly objects are plotted in a chart as shown in the accompanying figure. It has twelve cells. Each cell corresponds to a part of the ecliptic (path of sun relative to other stars in the sky traced by it in a year), the ecliptic having been divided into 12 parts each of which can be recognized in the sky by a cluster of easily recognizable stars in that part. These are called Rashis. These are Aries (मेष) March 21-April 20, Taurus (वृषभ) April 21-May 21, Gemini (मिथुन) May 22-June 21, Cancer (कर्क) June 22-July 23, Leo (सिंह) July 24-Aug. 23, Virgo (कन्या) Aug. 24-Sept. 23, Libra (तुला) Sept. 24-Oct. 23, Scorpio (वृश्चिक) Oct. 24-Nov. 22, Sagittarius (धनु) Nov. 23-Dec. 22, Capricorn (मकर) Dec. 23-Jan 20, Aquarius (कुंभ) Jan 21-Feb 19 and Pisces (मीन) Feb 20-March 20.

The planet which is in the rashi which is on the eastern horizon at the time of birth is placed in the first cell and remaining heavenly bodies placed in rashis as per their actual location at that time with respect to assigned rashis order. Each year Jantaris are published which indicate the positions of different planets, sun, moon and राहु केतु in appropriate rashis at each time of the day all the year around. Pundits who prepare the horoscopes take help of these charts. In fact motions of all the heavenly bodies being periodic, even though these positions change from day to day, month to month and year to year, these eventually start repeating. Astrologers and pundits use this fact to make horoscopes which are used for future predictions. In fact besides their use for future predictions, horoscopes are precise and authentic record of the time of birth of a person. This was basically how in the past births of individuals were recorded. Thus far it is all scientific even though astronomers now say that because of imprecision in measurement of observations in the past, the actual position of planets in the horoscopes as depicted by pundits using Jantaries is not in reality what it actually should be. In fact now astronomers say that there is gradual shifting of the points of intersection of ecliptic and equator (राहु and केतु). This is because pole star (which is usually regarded as the pole of ecliptic) is not exactly fixed (as it is commonly thought to be). It is in fact making a small circle in the sky around the true position of pole in a period of 26000 years. As a result we now have besides the earlier 12 known rashis, another rashi (a well defined cluster of stars) which has now become prominent on the ecliptic. As a result there are now 13 and not 12 rashis. However the pundits and astrologers preparing horoscopes are not willing to accept any change in whatever is mentioned in the scriptures.

In this regard I recall an incident which happened in the past while I was teaching at Roorkee University, Roorkee. Since childhood I have been fascinated by astronomy. In fact even my initial research work is in a related field. I also taught this subject at the University for several years. Those days we had in Roorkee a research centre, namely Structural Engineering Research Institute. Its director, Dr. Sharma, had research interests similar to mine those days. So we used to meet from time to time to discuss research related problems. Like me, Dr. Sharma had also a leaning towards astronomy and we often used to discuss ancient Indian astronomy vis-à-vis modern astronomy. Once after discussion with me regarding discrepancies in actual planetary positions and their positions as indicated in Jantaries, he wrote letters to pundits in Jaipur and Varanasi asking them as to why were they not indicating in Jantries more accurate positions of stars and planets (which are now readily available in nautical almanacs' that are published each year) in place of positions which they predict using earlier methods which have accumulated errors over the years. At first there was no response. However when he sent a reminder, he received the following reply, 'It would be better if you confine yourself to your own field of work and do not meddle in our affairs. We know how to make Jantaries.'

In fact the tragedy with our traditional methods of acquiring and preserving knowledge is that we accept whatever is written in scriptures as the last word and are not willing to accept that there may be scope for further improvement. This is the basic difference between scientific and religious approaches. In science we always feel that best is yet to happen and there is always a scope for improvement. However it is the reverse in religions. In religions it is believed that best happened at some epoch in the past when a prophet appeared and that was the last word. No improvement or modification in it is permissible.

11. Observational Astronomy in India

It is not only the ancient Greeks who took keen interest in observational astronomy. Observation of heavenly objects has also been a part of Indian ethos since Vedic times. Weather in India being moderate and sky generally clear for most part of the year, it has always been a pleasure and curiosity to observe sun, moon, planets and stars in the sky. Whereas majority is attracted to the sky just out of curiosity, there have also been astronomers and pundits who have been making keen observations of heavenly bodies to understand their movements. Records of such observations, their critical analysis and conclusions drawn are available in literature. Books are available on ancient Indian astronomy. One of these is (भारतीय ज्योतिष का इतिहासए लेखक डा० गोरख प्रसाद, उत्तर प्रदेश प्रकाशन) 1956). Even kings and monarchs in the past had been taking keen interest in observational astronomy. It is said that Mughal emperor Humayun used to take keen interest in observation of stars. For this purpose he often used to go to the rooftop of the library located in his fort in Delhi (now called Purana Qila which is near Zoological garden on Mathura Road). Once while he was observing stars from the roof top, he heard call for night prayers from the mosque in the fort. Hearing it he hurried downstairs to be in time to say his namaz. While descending the stairs in hurry, he missed one step of the stairs and fell down injuring himself. These injuries ultimately proved fatal.

A still later example is that of king 'Sawai Jai Singh' who founded the city of Jaipur. He reigned in 16th century towards the fag end of the reign of Mughal emperor Aurangzeb. Jai Singh was so much fascinated and committed to observational astronomy that he got observatories built in Jaipur, Benaras and Delhi. These observatories still exist. The one in Delhi is popularly known as Jantar Mantar. It is located near the Parliament House. Most of the observational instruments are built with bricks and plaster and are still intact and provide reasonably accurate observations even by modern standards. In fact the title 'Sawai' (one and a quarter) was confered upon him by Aurangzeb for his such types of deeds.

12. Astrology

Sometimes people confuse astrology with astronomy. Astronomy deals with observation of the positions of heavenly bodies at different times from different places, maintaining records of this data and its scientific analysis to draw conclusions of practical significance. However astrology essentially deals with the preparation of the horoscope of an individual at the time of his/her birth and its use for making predictions about his/her future. Since times immemorial humans have always felt an urge to know the future in advance and if possible to take precautionary steps to ward off and mitigate the ill effects of predicted adverse events. Different disciplines such as astrology, palmistry and neumrology etc. have evolved over the years with which some people claim to predict future and offer remedial measures for warding off predicted adverse effects.

In astrology, the astrologers claim that they can predict the future happenings in the life of a new born with the help of horoscope prepared at the time of his/her birth. Even in history we have accounts of futures of new borns being predicted this way. For example, it is said that when Gautam, the Buddha, was born astrologers had predicted that the child when he grows would either become a great monarch or would renounce the world to become a great spiritual leader. It was because of this prediction that Gautam's father tried to bring him up in a luxurious style befitting princes so that he would get engrossed in worldly pleasures and be not attracted towards spirituality. However that was not to be and ultimately he became a spiritual leader of world renown. Even now in modern times of science and technology a large majority have an inbuilt faith in astrology and its predictions. In fact its popularity is growing day by day. We now have several T.V. programmes where pundits predict each day the future of persons born in different rashis.

A question arises as to what is the authenticity of astrology and how can the future predictions made on its basis be relied upon? Are these exact and precise or should these predictions be taken with a grain of salt or totally ignored? Personally I feel that such predictions are to some extent statistical in nature. Observing that astronomers through their calculations could predict the exact date, time and duration of eclipses that was to happen in future, people started believing that such computations could also be used to predict the future happenings in lives of individuals as well. Being pressed for making such predictions it is possible that pundits started analyzing records of the happenings in the lives of persons born at same time of the year and collecting similar types of happenings in the lives of individuals born with identical positions of planets in their horoscopes. Based on such a data they might have started making predictions in the life of a new born based on the horoscope prepared at the time of his birth. Gradually this art must have got perfected and we now have well documented books on astrology.

Astrology has two components. One is the preparation of horoscope at the time of birth of a child. The other is prediction of the future happenings in the life of the child based on this horoscope. The first part is scientifically based. However the same can not be said about the second part. Can the future of a person be predicted based on his/her horoscope? Some believe it can be. However there is no rational or scientific base for it except that such predictions are to an extent statistical in nature based on the analysis of happenings in the lives of individuals with identical horoscopes in the past. Therefore I personally feel that such predictions should be accepted with a grain of salt and sould not be expected to be hundred percent correct. Just as when a weatherman says there is more than 90% chance that it will rain today and it may still not rain, same is true of predictions based on a horoscope. In fact as pointed out earlier in this chapter, in view of the inaccuracies in calculations which have accumulated over the years, even the positions of planets indicated by the pundits in the horoscope of a child at his birth are not their exact positions.

A still more perplexing part of reading of a horoscope by pundits is that they not only predict future but also offer remedies to offset the ill effects of unfavourably placed planets in the horoscope. Effect of remedies suggested seems to be more psychological in nature. After having performed these remedial rituals, a person feels that predicted ill effects will either not happen or their ill effect will get mitigated. And of course pundits charge hefty fees for preparing and reading horoscopes and for performing remedial measures. In fact this is their main source of income. By the way is there anyone whose all planets are in favourable position at the time of birth? Moreover if remedial measures do not work explanation is offered that ill effects were so strong that even remedial measures did not prove effective (just as in some diseases remedies do not work). It is not only in astrology but also in other fields of human activity that shrewd persons exploit the weaknesses of innocents and befool them for their self interest.

Personally I am of the view that predictions based on horoscopes should not be taken as a gospel. These can be regarded at the most as statistical predictions. It is not only so in the case of horoscope, same is also true in case of other trades of future prediction such as palmistry and numerology. Just as when weatherman predicts rain, it is advisable to take umbrella along with even though it may not become necessary to use it, same can be said of astrological predictions. However the dark side of it is that whereas a bright predicted future may make one lethargic with the result that it may not happen (even though it could have happened), prediction of an adverse future may make one lose hope and courage in advance thus making future still more bleak. Actually what should happen is that the prediction of a bad future should caution one to be well prepared to brace it, prediction of a bright future should goad one to work still harder to make it still brighter. As they say future is after all future and no one can predict it with hundred percent certainty.

Mind Teasers

1. In a certain year there were exactly four Fridays and exactly four Mondays in January. On what day of the week was 20th January of that year?
2. In USA, elections of president are held on first Tuesday after first Monday in November every four years. What is the earliest date of November on which election can fall? What is the latest date on which it can fall?
3. On which days of the week can first day of century fall?
4. On which day of the week 30th of a month falls most frequently?
5. On which day of the week (Saturday or Sunday) does new year fall more frequently?

Chapter 10
Problem Solving

1. Introduction

Problem solving is an important and vital constituent of mathematics. In fact the entire progress and development of the subject has been prompted by and is geared towards this end. It is one of the most exciting and useful application of mathematics. The subject is in fact primarily appreciated on this very account. Perhaps one of the main reasons as to why mathematics is not generally liked these days is because of the fact that now there is more stress on cramming statements, results and proofs, rather than in practical applications of derived results in solving problems from real life. In earlier days there used to be greater stress on application of techniques of mathematics in problem solving. It is said that in the past in Mathematics Tripos exam in U.K. there used to be one key question in each paper. Solution to this question required a judicious use of the learnt techniques in solving an unsolved challenge problem or a real life related problem. The student had the choice. He could either attempt the specified number of questions from the rest of the paper or just attempt the key question. Solution of this key question carried weight equivalent to solving correctly all the specified number of questions from the rest of the paper. Obviously only very bright students who had a clear grasp of the subject showed the courage to attempt it. It worked as a sort of challenge to the enterprising students. It is said that once an Indian student (Prof. Hemraj of Punjab University, Lahore) had gone to U.K. for Mathematics Tripos. Once during the exam, on looking at the paper he felt that perhaps he could crack the key question. He kept on brooding over it for a long time without writing anything in the answer sheet. Meanwhile other students were busy solving their papers. The invigilator thought that perhaps this Indian boy was unable to answer any question. When about half the time was over the boy sought the permission of the invigilator to go for toilet. He went to the toilet, smoked there for a while (smoking was in fashion those days) and returned. He just wrote something on the paper for a few minutes and came out of the examination hall after submitting his paper. The invigilator formed the impression that this student would fail. However when the result was declared that very student had topped in that paper scoring full marks as he was the only one who could solve the key question.

In earlier days there used to be a lot of craze for solving problems on one's own initiative. This is lacking these days. This is partly due to shift in stress in teaching of the subject and partly due to easy availability of help books which provide ready made solutions to questions. This kills the initiative and drive in the student. He just resorts to cramming facts and results without any critical appreciation. There is another instance of the past which shows the zest and craze for cracking unsolved problems. The incidence is again of Punjab University Lahore before partition. A student in M.A. mathematics who used to top the class was confronted with a problem which he could not solve. He worked over it for a few days but failed to solve it. He somehow felt that it should have been possible for him to solve it. He took it as a challenge and resolved that he would either find its solution by next morning 4 a.m. or would commit suicide! He worked on the problem all day and whole night but it would not yield solution. It was now almost 4 a.m. He took a knife and started climbing stairs of the hostel so as to kill himself at the roof top. While he was ascending the stairs, the solution to the problem flashed across his mind. He hurriedly came down and wrote the solution. Later on he became one of the renowned mathematicians of his times. (I won't recommend any one to go to that extent. Such types of geniuses are of course rare).

Another incident which highlights the importance of the role of mathematics in training individuals in solving problems which one faces in real life is the following one. Indian Administrative Civil Services (IAS), was previously known as ICS before partition. It holds a lot of fascination for young educated Indians. Many young persons try their luck after graduation and post graduation. The competition is quite tough. It is held in two parts : written papers followed by interviews. Now we have even preliminary exam before main written exam and marking is normalized. However it was not so earlier when actual marks obtained were considered. As a result a student who was good at mathematics stood greater chance of scoring higher marks. Once it is said that a mathematics background student topped the written part of the exam. When he appeared for the interview, the chairman of the interview board remarked, "Well gentleman you are a mathematician. However mathematics is not going to be of any help in administration. So you do not seem to be suited for the job." The candidate did not loose his cool and replied, "Well, Sir, how can it be said that mathematics will be of no help to me in discharging administrative responsibilities. In mathematics we have a set of rules. These have to judiciously applied in a given problem to obtain the desired result. The same is the case in administration. There also we have a set of rules and regulations which have to be applied to a practical real life problem to obtain desired results". The chairman was very much impressed by this answer. It is said that he was selected there and then without asking any further questions.

2. Problem Solving

Mathematics is closely associated with problem solving. There is no general rule which can be recommended as a guide to help an individual solve a problem at hand. It depends upon his background and his knowledge of various commonly used techniques for solving problems. Based on this the person has to make a judicious choice of the most appropriate technique for solving the problem at hand. In some cases more than one alternative technique may be applicable. In such a case he has to decide which is more appropriate. Sometimes after a problem has been solved using a particular method, it may be noticed that it could have been solved more easily and much faster using an alternative procedure. It is all part of the game.

The object of teaching mathematics to young students should be to inculcate in them the habit of problem solving. Given a problem they should be able to decide which would be the most appropriate method to use to solve it. Not all the problems can be solved. At a given moment there are always a host of unsolved problems. Some of the problems which are not solvable now may become solvable later when more appropriate techniques have been developed. That is how the subject has developed over the ages. The day when it becomes possible to solve each and every problem that is faced will be the end of progress of not only mathematics but perhaps even of human civilization. As they say even when confronted with a difficult problem that appears unsolvable, try and keep on trying and thinking over it. Some day it may get solved. The fun is not that much in getting the solution to a problem, but in trying to solve it.

An important job of a student of mathematics is to formulate mathematical model of a real life problem and then try to obtain its solution in a simple and efficient manner using mathematical tools at ones disposal. Solving a problem (be it of mathematics or of any other field) requires a suitable strategy to be adopted. If attempt is made to obtain its solution in an inept manner, it may look tedious and daunting. However, the same problem if attempted using an appropriate technique may yield solution easily and faster.

Here is a simple problem with a very simple solution which can be obtained fast. However the same problem looks complex and difficult to solve when an inappropriate strategy is used to solve it.

Example: In a single elimination tournament (once a match is lost, the team is out of the tournament), there are 25 teams competing for a trophy. How many games in all will be played till the winner team of the trophy is decided?

Solution: One way to solve this problem is to simulate the way the tournament will be actually played. As there are 25 teams, in the first round 12 teams will play other 12 teams, one team getting a by. The winning 12 teams from these twelve matches and the team which got by will play next round. Now we have 13 teams.

Of these 6 will play the other 6, one again getting a by. So thus far there have been 12 + 6 = 18 matches. At the end of round two we have 7 teams (six winning teams and one team that got by). These will play third round. In this round 3 teams will play other 3 teams, one team again getting by. Thus far there have been 12 + 6 + 3 = 21 matches. At the end of this round we have three winning teams and one that got by making a total of 4. So in the semifinal two teams will play the other two (no team getting by now) to give two winner teams which play the final match to decide winner of trophy. So in all 21 + 2 + 1 = 24 matches will be played before winner team emerges.

This is the normal way in which this problem will be generally solved. The answer is correct. However there is a more simpler and faster way in which the same answer could have been arrived at much faster. As there are 25 teams and there is to be only one winner team other 24 have to loose a match and for loosing a match a team has to play a game. So in all 24 matches will have to be played. So simple. Isn't it?

3. Problem Solving Techniques

In this section we present some simple techniques which can help in solving a variety of problems. Confronted with a problem one has to decide which will be the most appropriate technique to be used. Sometimes more than one technique may appear appropriate. The list of techniques being presented here is not exhaustive. There can be problems where none of these techniques may appear suitable. In such cases one has to think of some alternative technique. As mentioned earlier there are still some problems for which solution at the moment is not possible.

3.1 Properly Organising the Data

Given a problem it is essential that its data is properly organized before an attempt is made to solve it. In some cases it makes a lot of difference. Here is an example,

Example: find the value of

$$\left(1-\frac{1}{4}\right)\left(1-\frac{1}{9}\right)\left(1-\frac{1}{16}\right)\left(1-\frac{1}{25}\right)\dots\dots\dots\left(1-\frac{1}{225}\right)$$

Solution: One obvious way is to first simplify expression in each parenthesis to obtain

$$\frac{3}{4}\times\frac{8}{9}\times\frac{15}{16}\times\frac{24}{25}\times\dots\dots\times\frac{224}{225}$$

and then simplify $1\times\frac{2}{3}\times\frac{3}{2}\times\frac{3}{5}\times\dots\dots$

and proceed step by step to obtain the final answer.

However there is an alternative way to obtain the result much faster. This is based on properly reorganizing the data first. Given data of the problem can be reorganised as

$$\left(1-\frac{1}{4}\right)\left(1-\frac{1}{9}\right)\left(1-\frac{1}{16}\right)\left(1-\frac{1}{25}\right)\cdots\cdots\left(1-\frac{1}{225}\right)$$

$$=\left(1-\frac{1}{2^2}\right)\left(1-\frac{1}{3^2}\right)\left(1-\frac{1}{4^2}\right)\left(1-\frac{1}{5^2}\right)\cdots\cdots\left(1-\frac{1}{15^2}\right)$$

$$=\left(1-\frac{1}{2}\right)\left(1+\frac{1}{2}\right)\left(1-\frac{1}{3}\right)\left(1+\frac{1}{3}\right)\left(1-\frac{1}{4}\right)\left(1+\frac{1}{4}\right)\times$$

$$\left(1-\frac{1}{5}\right)\left(1+\frac{1}{5}\right)\cdots\cdots\left(1-\frac{1}{15}\right)\left(1+\frac{1}{15}\right)$$

$$=\frac{1}{2}\left(\frac{3}{2}\times\frac{2}{3}\right)\left(\frac{4}{3}\times\frac{3}{4}\right)\left(\frac{5}{4}\times\frac{4}{5}\right)\cdots\cdots\left(\frac{15}{14}\times\frac{14}{15}\right)\times\frac{16}{15}$$

$$=\frac{1}{2}\times\frac{16}{15}=\frac{8}{15}$$

3.2 Focussing on Right Information

In a problem a lot of information is given. For solution of the problem it is necessary that right type of information which is essential for the solution of the problem be searched and focused upon. We illustrate this with the following example.

Example: Suppose there is 1.5 litres of milk in a bottle. On the first day 100 c.c. of milk is taken out from the bottle and in its place 100 c.c. of water added to the milk in the bottle. Next day 200 c.c. of this mixture of milk and water is taken out of the bottle and 200 c.c. of water added. This process is continued for 15 days. What will be the ratio of the milk and water at the end of 15 days?

Solution: In this problem the vital information to be taken note of is as to how much water has been added to original 1.5 litres of milk in the bottle in 15 days.

This is 100 + 200 + 300 + ... + 1500

$$=\frac{15}{2}[2\times 100+(15-1)\times 100]$$

$$=\frac{15}{2}[200+1400]$$

$$= 12000 \text{ c.c. or } 1.2 \text{ litres}$$

Hence ratio of milk to water now is 1.5 : 1.2 or 5 : 4.

One may wonder as to why no attention has been given to the mixture being taken out each day. The reason is that concentration of milk and water in mixture taken out each day is same as their concentration in the mixture left behind in the bottle.

3.3 Logical Thinking

Sometimes a problem as posed looks daunting. However when its solution is presented one wonders as to why he/she could not think of it. It is exactly such problems which leave a dramatic effect upon us and help us in analyzing systematically similar situations in future. Here is one such example:

Example: There are three identical boxes. One contains two red balls, other two white balls and the third one red and one white ball. Boxes are labeled RR, WW and RW. However label of none of the boxes matches its contents. It is desired to determine actual types of balls in the labeled boxes by just picking one ball from any one of these three boxes and looking at its colour and then putting correct balls as per labels on the boxes.

Solution: At first sight the problem looks daunting. How is it possible to rectify the contents of all three incorrect boxes by just looking at the colour of one ball picked from one of these three boxes. However there is no need to panic. A way out is possible if we think a bit logically.

Since contents of none of the boxes match the label on them, the box labeled RW will be containing either only red balls or white balls. If we pick out a ball from the box labeled RW and notice its colour (say it is white), then the box marked RW will be having only white balls. Now the box which is marked WW must be containing red balls. It can not have mix of red and white otherwise the third box marked RR will contain correct type of red balls. The box marked RR will be containing mix of red and white balls.

3.4 Worst Case Scenario

In some problems analysis of the worst case scenario helps in finding solution to the problem much faster.

Example: Suppose there are 8 white, 6 red and 12 black balls in a drawer. What is the minimum number of balls which must be taken out so that we may have at least two balls of the same colour.

Solution: If we draw upto three balls these can be of different colours (one white, one red and one black). However if we draw 4 balls, we can have two balls of the same colour. So simple!

3.5 Do Not Overlook the Obvious

Sometimes a problem may look complex. However it is possible that its solution is obvious but is being overlooked. Here is an example.

Example: P is any point on the circumference of a circle whose centre is O. Perpendiculars PE and PF have been drawn from P on perpendicular diameters AOB and COD of the circle. What is the length of EF?

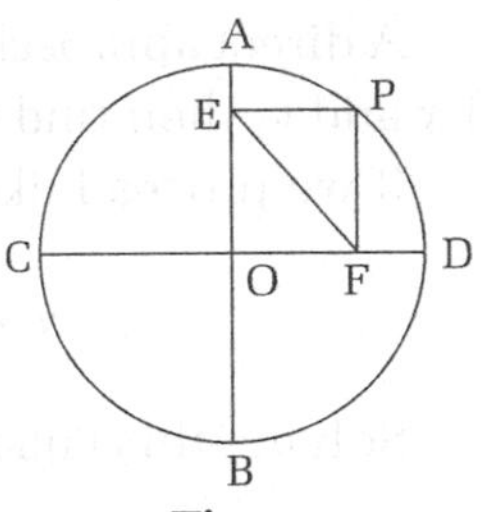

Fig. 1

Solution: One can obtain its solution using techniques of Euclidean geometry or coordinate geometry. However, there is a simple way of obtaining its solution also and this is as follows:

Since P is any point on the circumference of the circle, it can be A also. In that case foot E of perpendicular on AB will be A itself and foot F of the perpendicular on CD will be O. Thus EF = AO the radius of the circle! (So simple, isn't it?).

3.6 Looks Difficult but is Easy

Some problems at first sight look formidable and daunting. However if observed and analysed carefully, these may not be so difficult. Here is an example again from geometry.

Example: ABCD is a parallelogram and E lies on the side AB. EDGF is also a parallelogram such that C lies on FG. It is desired to determine the area of parallelogram EDGF given the area of parallelogram ABCD.

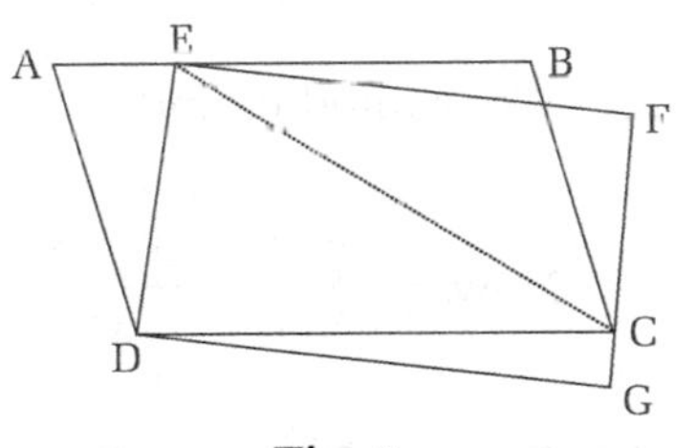

Fig. 2

Solution: At first sight the problem looks difficult. However it is not so. Join EC. Then ECD is a triangle on the same base as parallelogram ABCD with its vertex on side AB. Hence area of DEDC is one half of area of parallelogram ABCD.

Again parallelogram EFDG and EDC are on the same base ED with vertex of triangle lying on opposite side FG of the parallelogram. Thus area of DEC is one half of area of parallelogram EFGD.

Combining the two, area of parallelogram EDGF is same as area of given parallelogram ABCD.

3.7 Working Backwards

In some problems working backwards may yield solution much faster than working in forward direction. Consider for example the following simple problem.

Example: Sum of two numbers is 2 and their product 3. What is the sum of their reciprocals.

Solution: Formally stated in language of algebra let the two numbers be x and y, then we are given $x + y = 2$ and $xy = 3$ and we want to find $\frac{1}{x} + \frac{1}{y}$

A direct approach will be to solve simultaneously $x + y = 2$ and $xy = 3$ to get values of x and y, then find their reciprocals and add.

If we proceed like this, then substituting $y = 3/x$ in $x + y = 2$, we get

$$x + \frac{3}{x} = 2 \text{ or } x^2 - 2x + 3 = 0$$

Solving this equation, we obtain

$$x = \frac{2 \pm \sqrt{4 - 12}}{2} = 1 \pm \sqrt{-2} = 1 \pm i\sqrt{2}$$

So, $\quad y = 2 - x = 1 \mp i\sqrt{2}$

Hence, $$\frac{1}{x} + \frac{1}{y} = \frac{1}{1 \pm i\sqrt{2}} + \frac{1}{1 \mp i\sqrt{2}} = \frac{1 \mp i\sqrt{2} + 1 \pm i\sqrt{2}}{1 - i^2 2} = \frac{2}{3}$$

Answer has been obtained in a very complicated manner.

On the contrary had we tried to work backwards, we would have obtained the same answer in a much simpler way.

We wanted to find $\frac{1}{x} + \frac{1}{y}$

Now, $\frac{1}{x} + \frac{1}{y} = \frac{y + x}{xy} = \frac{2}{3}$ Ans. (using values of $x + y$ and xy as given in the data)

3.8 Do not Jump to Generalizations in Haste

On observing similar behaviour on a number of similar instances, one is often tempted to generalize the result. However, one must be cautious in this respect. Sometimes, it may be misleading. Here is an example.

Example: Observe the following pattern of natural numbers

$3 = 2^0 + 2$, $5 = 2^1 + 3$, $7 = 2^2 + 3$, $9 = 2^2 + 5$, $11 = 2^3 + 3$, $13 = 2^3 + 5$, $15 = 2^3 + 7$, ...,

$51 = 2^5 + 19$, ..., $125 = 2^6 + 61$, $129 = 2^5 + 97$, $131 = 2^7 + 3$, ...

This shows that an odd number can be expressed as the sum of some power of 2 and a prime number. The rule works for odd numbers up to 125. However, it does not work for 127 even though it again works for 129, 131, etc. Again it may not work at some later stage. So it will not be correct to generalize after looking at behaviour of odd numbers upto 100 that every odd number can be expressed as sum of some power of 2 and a prime number.

3.9 Use of Relativity

Concept of relativity is normally associated with Einstein who propounded theory of relativity. However relativity also occurs in other situations. For example the speed of a boat relative to the flowing water of a river. Its proper use often simplifies solutions. We will illustrate this with following example.

Example: A person while rowing his boat upstream drops a cork in the river and continues rowing upstream for the next ten minutes. He then turns around and is able to retrieve the cork after it has traveled 1 km downstream. What is the velocity of the flowing stream?

Solution: From the relativistic point of view, we are not bothered whether the stream is flowing or not. We are primarily interested in the separation and retrieving of the cork. Had the stream been stationary, he would have required as much time in rowing away from the cork as rowing to it. Now the cork has traveled 1 km in 20 minutes. So the speed of the stream is 3 km per hour.

3.10 Pigeonhole Principle

Pigeonhole principle is not so well known. However, it is a simple principle which helps in obtaining solutions to some problems (which otherwise look formidable) in a simple and elegant manner. The principle is as under:

If $n + 1$ objects have to be put in n boxes (pigeonholes), then at least one box (pigeonhole) will have at least two objects.

Principle is simple and its justification self-evident. We illustrate its use in solving problems.

Example 1: How many persons need be there in a town before we can be sure that there are at least two persons who have identical number of hair on their heads, assuming that the maximum number of hair which a person can have is 500,000.

Solution: Using pigeonhole principle, clearly 500002 (including totally bald persons).

Example 2: One million pine trees grow in a forest. It is known that no pine tree has more than 60,000 pine needles on it. Show that there are at least two pine trees which have same number of pine needles.

Solution: It is straight forward use of pigeonhole principle. Since it is mentioned that no pine tree has more than 60,000 pine needles, so even amongst 60001 pine trees there will be at least two which have identical number of pine needles. Hence out of one million pine trees there will definitely be at least two which have same number of pine needles.

Example 3: There are 50 faculty mail boxes in a school. One day 151 letters of faculty members were received in the school office and put in the mail boxes. One mail box has more letters than the others. What is the least number of letters that it can have?

Solution: It is possible that some faculty members may have not received any letter on a particular day. However that situation is excluded in this problem as it is mentioned that one mail box has more letters than others. Suppose all of them receive equal number of letters. Then each box will have 3 letters each and one letter will be extra. When it goes to a box this box will have one letter more than the other boxes. Clearly four is the least number of letters which a box must have so that it has more letters than the others.

Example 4: Show that amongst any six persons there are either three persons who know each other or there are three persons amongst whom one does not know the other two.

Solution: Solution of this problem is also based on pigeonhole principle though it is not as straight forward as in the earlier cases.

Choose any one of them and call him A and put the rest five in two groups I and II such that persons of one group (say I) know A but persons of the other group II do not know A. There will be at least three of the remaining five in one of these two categories. Suppose A has three acquaintances. If two of these know each other they form the required triplet. If none of them know each other then they themselves form the triplet. Similar argument holds if these three persons are not known to A.

Example 5: Show that there exists an integer whose decimal representation consists entirely of ones and zeros and which is divisible by 1987.

Solution: Consider 1988 pigeons numbered 1, 11, 111, ..., 111..111 (1998 times). Pigeonhole principle assumes that out of these there will be at least two which when divided by 1987 yield identical remainder. Their difference will be of type 111...111000...000 and this will be divisible by 1987.

3.11 Use of Logical Reasoning

Several real life problems get easily solved through logical reasoning. We present in this section some examples for illustration purpose.

1. How many cattle and how many attendants

Example: A herd of cattle alongwith attendant cowboys has 35 heads and 130 feet. How many cattle and how many attendants are there in the herd?

Solution: Whereas cattle and attendant have one head each, each cattle has four legs and each attendant two. So if number of attendants is x and number of cattle y, then

$$x + y = 35$$

and
$$2x + 4y = 130$$

Solving $x = 5$ and $y = 30$. Thus 5 attendants and 30 cattle.

2. To form heaps of equal weight

Example: Given a set of stones weighing 1 gm, 2 gm, 3 gm, ... 555 gms, arrange these 555 stones in two heaps of equal weight.

Solution: It may be noted that 1 + 555 = 556, 2 + 554 = 556, 3 + 553 = 556 and so on. So arrange the stones into heaps A and B as under

A 1, 554, 3, 552, 5, 550, 6, 548, ...

B 555, 2, 553, 4, 551, 6, 549, ...

Total weight of both heaps will be same. So to form such heaps, put first stone in A, last in B, last but one in A, second in B, third in A, last but second in B and so on.

3. Identifying counterfeit coin

Example: We are given 9 identical coins. Amongst these 9 coins one coin is counterfeit and weighs less than others. Find out this counterfeit coin using upto 3 weighings only.

Solution: Keep aside one coin and of the remaining 8, put 4 in one pan and the rest 4 in the other pan of the balance. If these balance, kept aside coin is counterfeit. Else, counterfeit coin is amongst the four which are weighing less. Now put 2 of these four less weighing coins in one pan and remaining two in the other. The pan which weighs less contains counterfeit coin. Now put one of these two less weighing coins in one pan and the other in second pan. Pan which is lighter contains counterfeit coin.

4. How many hand shakes

Example: Mrs and Mr Gupta hosted a party. They invited four couples to the party. On arrival, many, but not all, shook hands. No one shook hands twice with the same person. Also, no one shook hand with his/her spouse. Both host and the hostess also shook hands (but not with all). At the end of the party, Mr. Gupta asked each one as to how many hands he/she had shaken. Each person gave different answer. How many hands did Mrs. Gupta shake?

Solution: At first sight it looks a bit strange. How can one decide on the basis of available information as to how many hands Mrs Gupta shook. However a rational and logical analysis of the provided data yields the answer. Let us see how?

Let G denote Guptas and A, B, C and D other four invited couples. Since none shook hand with his or her spouse so none shook more than 8 hands. As the number of hand shakes of each person is different, so the number of hand shakes of each person will be a number between 0 and 8. Suppose A shook 8 hands. Then how many hands did Mrs. A shook? As everyone of the couples B, C, D and G has shook hands with Mr A, so each one of these has shaken hand at least once. As at least one of the persons must have shook 0 (no) hand it must be Mrs A.

Now eliminate Mr. A who shook 8 hands and Mrs. A who did not shake hand with any one (zero handshakes). Now we are left with couples B, C, D and G. They should have shaken hands between 1 and 7. Some one amongst these must have shaken 7 hands. Let it be Mrs. B. She has already shaken hand with A so she must have shaken hands with each member of the couples C, D and G (six handshakes) as she has not shaken hand with Mr. B. By now each member in the couples C, D and G has already made two hand shakes one with Mr. A and another with Mrs. B. Since one person must have shook hand only once, it must be Mr. B. Continuing like this we find that amongst the remaining couples C, D and G, one must have shaken six hands and the other only 2. Say these are Mr. C and Mrs. C. Then from amongst the remaining two couples D and G, one must have shaken 5 hands and one 3. Say these are Mrs, and Mr. D. Now only Mrs and Mr G are left and only 4 hand shakes are possible.

So Mrs G would have shaken 4 hands (What about Mr. G?).

(*Note in the solution we started with the assumption that it was Mrs. A who shook 8 hands. It could be Mr. A or Mr. or Mrs. B, C and D, but not Mr or Mrs G because it is mentioned that they shook some hands not all*).

3.12 River Crossing Problems

There are some interesting river crossing problems that need ingenious solutions based on logic. Some of these are:

1. To cross a water channel using two planks of smaller lengths

Example: A water channel 4 metres wide makes a turn of 90°. It is desired to cross it using two planks of lengths 3.5 metres each (without joining them at the ends). Will it be possible?

Solution: Yes. Arrange the two planks in the shape of T as shown in the accompanying figure.

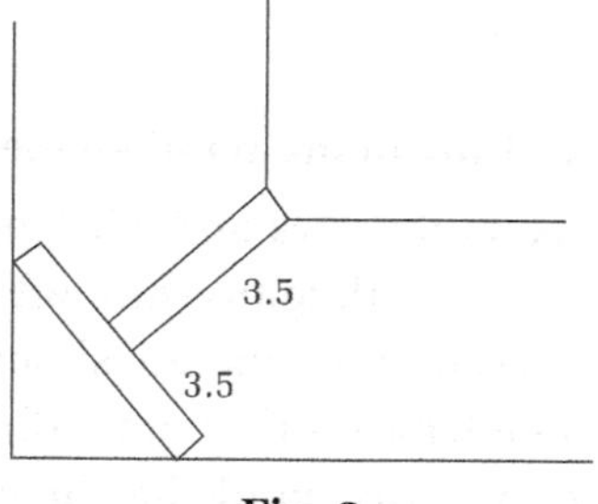

Fig. 3

2. Married couples on honeymoon

Example: Two married couples were out on honeymoon. They wanted to cross a river. A small boat was available which could ferry two persons at a time. No male wants his wife to travel with the other male and there is no oarsman to ferry the boat. How can these couples cross the river?

Solution: Denote couples by [H_1, W_1] and [H_2, W_2]. They can cross the river in three trips in the following manner:

Trip I	[H_1, W_1]	return trip [H_1]
Trip II	[H_1, H_2]	return trip [H_2]
Trip III	[H_2, W_2]	

3. How to take troops across the river

Example: A general has to take his troops across the river. There is one boat with two boys. The boat unfortunately is small. It can either carry two boys or one soldier. Yet the general is able to take his troop across the river. How? (Number of soldiers is unspecified).

Solution:

Trip I	[B_1, B_2]	return trip [B_1]
Trip II	[S_1]	return trip [B_2]
Trip III	[B_1, B_2]	return trip [B_1]
Trip IV	[S_2]	return trip [B_2]

and so on...

3.13 Age Related Problems

In this sub section we present some age related problems.

1. Age of the captain of the boat

Example: Captain of a boat is of age (say A). He has some children (say C). He has a boat whose length is L. Given that ACL = 32118 and that L is in exact feet and age of A is exact years (i.e., A, C, L are all integers), find the age of the captain.

Solution: At first sight it looks a strange type of question. How can the age of the captain be determined from a single numerical data? But a little thinking will convince that it is possible. Let us see how

It is given that ACL = 32118 and A, C, L are all integers.

Now possible factors of 32118 are

$$32118 = 2 \times 16059 = 2 \times 3 \times 5353$$
$$= 2 \times 3 \times 53 \times 101$$
$$= 6 \times 53 \times 101$$

So, clearly C must be 6, L 101 and age of A 53 years.

2. Age of self, father and son

Example: When I am as old as my father is now, I shall be five times as old as my son is now. However at that time my son will be eight years older than what am I now. At present the sum of my age and my father's age is 100. How old am I now and what are the present ages of my son and father?

Solution: It will be convenient to take help of algebra. Let my present age be x, then my father's age is $100 - x$. I shall be as old as my father (i.e., $100 - x$ years old) after $(100 - x) - x = 100 - 2x$ years.

Let s be the age of my son now.

Then according to provided information $100 - x = 5s$ or $x = 100 - 5s$. (i)

Also it is given that at that time my son will be 8 years older than what I am now. Therefore

$$s + 100 - 2x = x + 8 \text{ or } 3x = s + 92 \qquad \text{(ii)}$$

Solving (i) and (ii), $x = 35, s = 13$

Also, $100 - x = 65$

So, present age of my son is 13 years. My present age is 35 years and that of my father is 65 years.

3. Age of Ram

Example: Combined age of Ram and Shyam is 24 years. Ram is twice as old as Shyam was half as old as Shyam will be when Shyam will be thrice as old as Ram had been when Ram was thrice as old as Shyam had been. How old is Ram?

Solution: The statement of facts is quite intricate and not easy to comprehend. However, if we keep in view that the problem deals with four different epochs : present is (I), future will be (II), past was (III) and distant past had been (IV), then it is not that formidable as it appears. Let us see how

Let Shyam's age be x in epoch III. Then Ram was $3x$ in epoch IV. Now Shyam will be $9x$ in epoch II. Hence Ram was $4.5x$ in epoch III. Since age difference between them is $2x$, Shyam's age was $2.5x$ in epoch III.

Hence in the present epoch I, Ram's age is $5x$ and Shyam's age $3x$. As sum of their present age is 24,

so we have $5x + 3x = 24$ or $x = 3$

Hence the present age of Ram is 15 years and Shyam's 9 years.

4. Some Assorted Problems

In this section we present problems from different areas of human activity which can be solved using logic.

1. How far the worm crawls

Example: Two thick volumes of a book stand back to back in a shelf of an almirah in a library as shown in the accompanying figure. Hardboard covers of these two volumes are 2 cm thick each. Each volume has 500 pages having thickness of 8 cm. A bookworm started boring a hole from the first page of volume I towards the last page of volume II. How much distance the worm will crawl before it reaches last page of volume II?

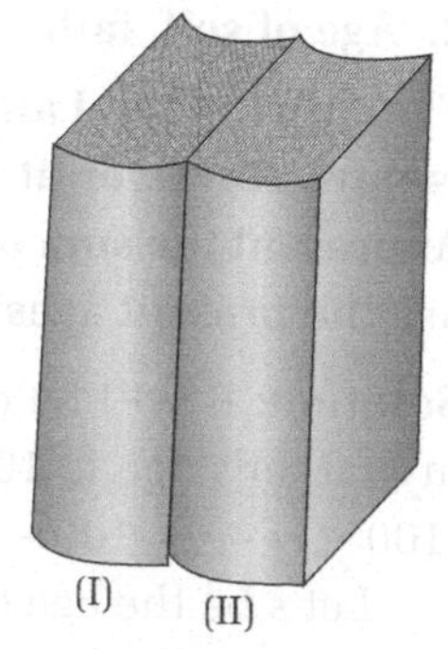

Fig. 4

Solution: One may be tempted to say $8 + 2 + 2 = 12$ cm. However that is not so. If we look at the figure carefully we notice that page

1 of volume one is separated from the last page of volume II by just the cover pages of two volumes. So the worm will have to crawl only 2 + 2 = 4 cm.

2. Volume of a taper top bottle

Example: A bottle has round bottom and straight sides. However it is tapered at the top and has a cap which can be screwed tight. The bottle is partly filled with water. Determine its volume using a ruler only.

Solution: At first sight it looks impossible. However there is a way out. First use ruler to measure the diameter of the base and then use it to calculate the area of the base. Suppose this area is A. Now use ruler to measure the height of water in the bottle. Let it be h. Then the volume of water in the bottle is V = Ah. Now turn the bottle upside down and measure the height of the air column in the bottle. Let it be h'. Then volume of air in the bottle is Ah'. Hence total volume V of the bottle is

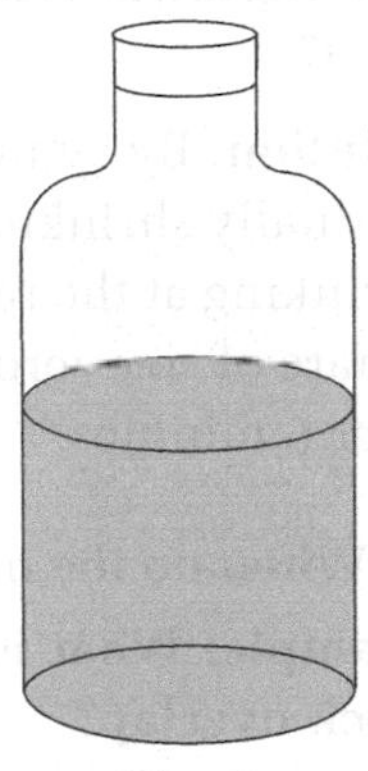

Fig. 5

$$V = Ah + Ah' = A(h + h')$$

3. Finding the centre of a circle

Example: A circle of radius between 2 and 4 inches is drawn on a piece of plane paper. Find the position of its centre given a plastic square sheet of side 10 inches.

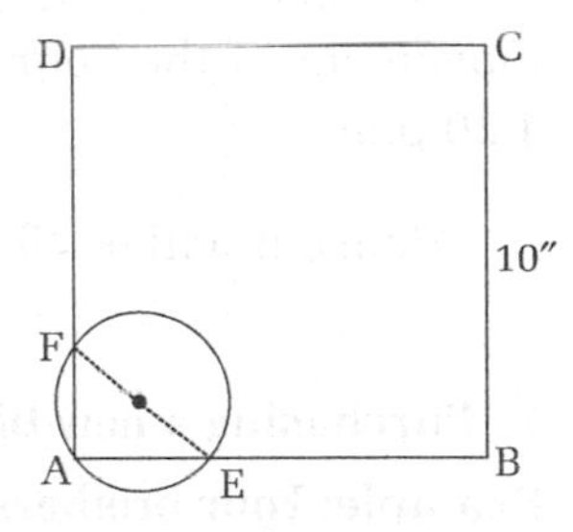

Fig. 6

Solution: This of course needs knowledge of results of plane two-dimensional Euclidean geometry. Place a corner say A of the square plastic sheet ABCD so that it touches the circle from inside. Let sides AB and AD of the plastic square sheet cross the circle at E and F respectively. Then EAF is a right angled triangle in the circle and hence EF is a diameter of the circle. Using an edge of the plastic sheet join EF. Similarly draw another diameter of the circle by placing corner A at some other corner of the circle. The point of intersection of these two diameters is the centre of the circle.

4. Using two containers to fill a third container

Example: We have two containers. One of 6 litres capacity and another of 4 litres capacity. These are full of water. We have another 3 litre container which is empty. Is it possible to fill it from the other two containers if only the following measures are permissible, (i) filling a container, (ii) emptying a container, (iii) pouring contents of one container into another container.

Solution: It is not possible with specified moves. As these moves mean just addition and subtraction of 6 and 4 both of which are even. By adding or subtracting two even numbers we can not create an odd number.

5. Where do the four runners meet

Example: Four runners are stationed at the four corners of a square ABCD of side 1 km each. They start running at the rate of 10 km per hour. At each moment, the runner A is running towards B, B towards C, C towards D and D towards A. When and where will they meet?

Solution: By symmetry at each stage they will be at the vertices of a square which is gradually shrinking in size. They will finally meet when its size becomes zero, its sides shrinking at the rate of 10 km per hour. So original square of size 1 km will shrink to a square of size zero in 1/10 hours or 6 minutes. So they will meet at the centre of square after 6 minutes.

6. When do the hands of a clock overlap

Example: What is the time after 4 p.m. when the hour and the minute hands of a clock overlap?

Solution: An immediate answer will be 4:20 p.m. However it is not correct because in 20 minutes after 4 p.m. the hour hand will have also moved a little ahead of 4.

In 12 hours whereas hour hand makes one revolution, minute hand makes 12 revolutions. Hence there are in all 11 overlappings in 12 hours. Hence in view of the movement of the hour hand, we have to apply correction of 12/11 to our answer of 4:20 p.m.

Hence it will be $20 \times \frac{12}{11} = 21\frac{9}{11}$ minutes past 4.

7. Purchasing a new bike

Example: Four brothers pooled their money to purchase a bike. First paid half of the sum paid by the other three. Second paid one third of the sum paid by the other three. Third paid one quarter of the amount paid by the other three. Fourth paid Rs. 13,000. What is the price of the bike and how much did each of them pay?

Solution: Let the money contributed by each brother be x_1, x_2, x_3 and x_4 respectively. Then as per information supplied

(i) $x_1 = \frac{1}{2}(x_2 + x_3 + x_4)$ or $2x_1 - x_2 - x_3 - x_4 = 0$

(ii) $x_2 = \frac{1}{3}(x_1 + x_3 + x_4)$ or $x_1 - 3x_2 + x_3 + x_4 = 0$

(iii) $x_3 = \frac{1}{4}(x_1 + x_2 + x_4)$ or $x_1 + x_2 - 4x_3 + x_4 = 0$

(iv) $x^4 = 13000$

So, $2x_1 - x_2 - x_3 = 13000$ using (iv) in (i)

$x_1 - 3x_2 + x_3 = -13000$ using (iv) in (ii)

$x_1 + x_2 - 4x_3 = -13000$ using (iv) in (iii)

Solving $x_1 = 20{,}000$, $x_2 = 15{,}000$,

$x_3 = 12{,}000$ and $x_4 = 13{,}000$

So first contributed Rs. 20,000, second Rs. 15,000, third Rs. 12,000 and fourth Rs. 13,000.

Price of bike is Rs. 60,000.

8. Which job to accept

Example: A person receives offers of two jobs: Job 1 gives him annual salary of Rs. 10 lacs with annual increment of Rupees 2 lacs per annum. Second job offers him salary of Rs. 5 lacs for six months with six monthly increment of Rupees 50,000. Which of the two jobs he should prefer?

Solution: Let us compute his annual earnings from the two jobs.

First year: Job 1 = Rs. 10,00,000,

Job 2 = First six months Rs. 50,0000, next six months Rs. 55,0000

Total in one year = Rs. 10,50,000

Second year: Job 1 = 12 lacs

Job 2 = 60,0000 + 650000 = 12 lacs 50 thousand

So the second job is better, as each year second is paying more. So, he should prefer Job 2.

9. Fuel Saving Devices

Example: A car is equipped with three fuel saving devices, say A, B and C. Device A by itself saves 25% fuel, device B 45% and device C 30% fuel. How much fuel will be saved if all devices operate simultaneously?

Solution: Equipped with devices A, B and C separately car uses only 75%, 55% and 70% of fuel to travel same distance as it would travel without using any of these fuel saving devices. So when all devices are used simultaneously, it will need only $0.75 \times 0.55 \times 0.70 = 0.28875$ parts of fuel it uses normally.

Hence there will be a saving of 1 – 0.28875 or 0.71125, i.e. 71.125%.

10. Spider catching a fly sitting on the opposite wall

Example: A room has a floor measuring 8′ × 10′. Ceiling is 9′ high. A spider is perched midway on one of the 8′ × 9′ wall just 6″ below the ceiling. A fly is sitting on the opposite wall midway 6″ above floor. The spider wants to quietly sneak to the fly to catch it walking along the walls, floor and ceiling. What can be its shortest path.

Solution: If we dissemble the room of Fig. 7, we get Fig. 7(a).

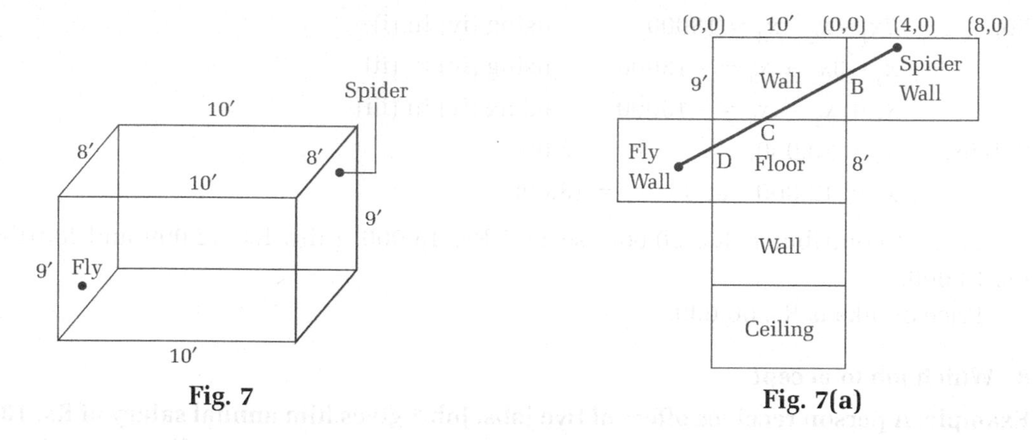

Fig. 7

Fig. 7(a)

The spider should travel along the straight path connecting it to a point B on the junction of this wall with the neighbouring wall, then from there it should travel straight to C, a point on the join of that wall with floor. Then from C to D on the floor to its junction with wall on which fly is sitting.

(This problem illustrates that it may not always be advisable to tackle the problem as it is presented. Sometimes it may be convenient to redraft it).

11. Determining the ages of the children and the house number

Example: A salesman knocked at the door of a house and requested the housewife to buy his product. The housewife said, "I shall buy your product if you can correctly answer my following question: I have three kids. Their ages add upto 13. The product of their ages is the number of the house which you see on the door. Can you tell me their ages? The salesman looked at the house number, thought for a while and said, "Madam, I am sorry. The information supplied by you is insufficient." The housewife thought for a while and smiled and said, "O.K. I give you some more information. My eldest daughter takes lessons in Piano." Whereupon the salesman gave the correct ages of the children. Can you tell the ages of the children and house number?

Solution: At first sight it may appear strange as to how provided information suffices to determine the ages of the children and the house number. However, the salesman was shrewd. He had to break 13 into three integers so that their product was the house number. He tried to find these by trial and error. If the number of the house was 48 then the ages of the children could be 2, 3, 8. Why could he not arrive at an answer in the first instance. This must have been due to the fact that he must have been getting two sets of answers. In case of 48 there is no confusion in forming its triplets whose sum is 13. As he had confusion, the house number must have been such that two possible triplets of 13 were possible, whose product was house number. A little thinking tells us that this number is 36. In this case we have two sets of triplets of 13 whose product is 36. These are: (1, 6, 6) and (2, 2, 9). So initially the salesman had the confusion as to which of these was correct answer. However when the housewife informed him that her eldest daughter took lessions in Piano, he became sure that the correct answer must be (2, 2, 9) and not (1, 6, 6).

12. Three friends visiting fair with their pet monkey

Example: Three friends, Ram, Shyam and Mohan went to a fair taking along with them their pet monkey. In the fair they purchased a basket of bananas which they carried alongwith on their way back home. As they started back for their village late, they decided to spend the night under a tree. While everyone was asleep, Ram got up. He divided bananas into three equal portions and kept one part for himself. However while dividing bananas into three equal portions he found one banana extra. This he threw towards the monkey who was also awake at that time. Monkey ate it instantly. After keeping aside his part of bananas, he again went to sleep. After sometime Shyam woke up. He repeated the same exercise. He divided left over bananas into three equal portions, found one banana extra which he threw to the monkey. After keeping apart his portion he also went to sleep. After some time Mohan woke up. He also repeated the same exercise. In the morning when they woke up, they divided the remaining bananas into three equal parts. Again one was left which was given to the monkey. What could be the number of bananas in the baskct to begin with given that these were less than one hundred.

Solution: In the morning the minimum number of bananas which could be there in the basket when they woke up in the morning is four. Suppose when they woke up there were 4 bananas in the basket so that each got one and 4th was given to the monkey. This is 2/3 of what Mohan equally divided.

So when Mohan woke up there were $4 \times \frac{3}{2} + 1 = 7$ bananas.

This is again 2/3 of what Shyam divided equally.

Hence when Shyam woke up there were $7 \times \frac{3}{2} + 1 = 11.5$ bananas.

This being fraction is not acceptable. Same thing happens if each had got 2 bananas in the morning and thus there were 7 bananas in the morning. In this case

Mohan when he woke up found $7 \times \frac{3}{2} + 1$ (which is again a fraction and not acceptable).

If we continue like this we find 7 is the least number of bananas which each must have got in morning. In that case total bananas in the morning were $3 \times 7 + 1 = 22$.

Mohan when he woke up found $22 \times \frac{3}{2} + 1 = 34$ bananas.

Shyam when he woke up found $34 \times \frac{3}{2} + 1 = 52$ bananas.

Ram when he woke up found $52 \times \frac{3}{2} + 1 = 79$ bananas.

So, total number of bananas in the basket must have been 79.

Mind Teasers

1. A sheep can eat grass of a field in one day. A cow will be able to eat the same grass in half a day. How long will the grass lost if both eat together?
2. There is a herd of cattle in a barn. There are also some attendant boys. Head count reveals 35 heads and 130 feet. How many cattle and how many attendant boys?
3. Three men can cut the crop of a field in 2 days. Same field can be cut by 2 women in three days. How many days will it take to cut the crop of this field if one man and one women work together?
4. Ten persons are sitting around a table. Rs. 10,000 are to be distributed amongst them so that each person receives average of what person on his right and left neighbours receive. How much does each person receive. One obvious answer is Rs. 1000 each. Is there some other answer also?
5. Two teams played each other in a decathlon (an event having 10 games). In each game the winning team gets 4 points and the loosing team one point. In case of a draw both the teams get 2 points each. At the end of the event, both teams had scored points whose total was 40. How many draws were there? One obvious solution is all the ten games ended in draw. Is there some other solution?
6. Dolly, Paul and Amit, each drew 2 cards from 7 cards numbered 2, 4, 6, 8, 10, 12 and 14. Dolly says that the sum of numbers on her cards is 20. Paul says that sum of numbers on his card is 10 while the sum of number of Amit's cards is 14. Which card is left?
7. If every boy in a class buys a chocolate and every girl a sandwich, they spend Rs. 10 less than if each girl buys a chocolate and each boy a sandwich. There are more boys in the class than girls, and each sandwich costs rupee one more that a chocolate. How many boys and how many girls are there in the class?
8. Suppose one keeps on folding a very thin muslin handkerchief of thiskness 1/10 of a millimeter. Then each fold doubles the thickness but its area becomes one half of the original area. If folding continues 32 times, what will be the thickness of the muslin sheet?
9. Five workers in a factory receive a total of 1.5 lacs per month as wages. However their wages are not equal. Each wants to buy a motorbike costing Rs. 32000. Show that at least two will have to wait for the pay of next month.
10. Is it possible to choose three persons from a group of seven persons of unequal ages such that their combined age exceeds 140 given that the sum of the ages of seven persons is 322 years.

11. Out of 101 coins, one is counterfeit (either of more or less weight). What is the minimum number of weighings needed to find it and decide whether it is heavier or lighter?

12. A road connecting two towns in a hilly region goes totally uphill in one direction and down hill in the other. A bus travels at the speed of 40 km per hour uphill and 60 km per hour down hill. It takes the bus four and a half hours to make the round trip with a halt of half an hour in between. What is the distance between the towns?

13. Twenty five crates of apples are delivered in a store. Apples are of three varieties. However each crate contains apples of same variety. Show that amongst these 25 crates there will be at least 9 crates which have apples of same variety.

14. A bag contains beads of two colours black and white. What is the smallest number of beads which must be drawn from the bag without looking at the colour of the beads so that there are at least two beads of the same colour?

15. In a drawer there are 8 white, 6 red and 12 black balls. What is the minimum number of balls which must be taken out to be sure that we have at least two black balls.

16. Given 12 integers from 1 to 20, no two identical, show that this group of integers will have 2 integers whose difference is divisible by 11.

17. Show that in a group of 5 persons, there must be at least two who have identical number of friends in the group.

18. Show that in a group of 6 persons, there are either 3 who know each other or 3 none of whom knows the other.

19. Show that there exists a power of 3 which ends with digits 001 in decimal notation.

20. Three volumes of a book are 5 cm thick each. These are placed on a table one above the other. A bookworm starts penetrating from the first page of bottom volume. It penetrates 1 cm each week. How long will it take the worm to reach the last page of the topmost volume?

21. If in problem 4(1), in place of two, there are three volumes of identical size placed pack to back, how far will the worm have to crawl to reach from first page of volume III to first page of volume I?

22. Each morning a shopkeeper purchases an item for Rs. 100 and sells it by the evening for Rs. 110. Do an analysis to justify that he is making a profit of 10%. Again justify that he is making a profit of 3650% on money invested.

23. Suppose a person is walking in rain. Will it be advisable for him to walk steadily or run fast at steady speed to minimize the amount of rain water falling on his body?

24. A grasshopper jumps along a straight line. His first jump takes him 1 cm, second 2 cm, third 3 cm and so on. However each jump can be either to right or left. Show that after 1985 jumps he can not be at the point from where he started.

25. A rectangular town has 8 streets running north-south and 6 streets running east-west. In how many ways can a person starting from the south-west corner reach the north-east corner driving in east and north directions only.

26. In the problem of friends going to fair with monkey (problem 12 of section 4), is the total number of bananas received by each friend same? If not who gets maximum bananas?

27. In the same problem of friends going to fair with their pet monkey, is another solution possible in which the number of bananas purchased in the fair is less than 300 but more than 100?

28. Find the actual length of the shortest path to be traveled by the spider in problem 10 of section 4.

29. Suppose in place of containers of sizes 6 and 4 in problem 4 of section 4, we have containers of sizes 9 and 4 litres. Will it now be possible to use these to fill a container of 6 litres as desired there?

30. Clock hands coincide at 12 O'clock. When next do they coincide?

Chapter 11
Mathematics and Computers

1. Introduction

No present day book on mathematics will be complete without reference to computers. Mathematics and computers are intimately inter-related. Computers are changing daily lives of not only ordinary persons but are also showing their impact on mathematicians as well. Mathematics has helped in a fundamental way in the design and working of computers and now computers are forcing mathematicians to change the way of their thinking. The impact of computers on mathematics has been thus a two way process. If it has added to the power of computing its very origin and the way it works is based on mathematical concepts.

2. Contribution of Mathematics in Development of Computers

The first fundamental idea on which the design of computers is based was introduced by George Boole in 1854 in the form of 'laws of thought'. In his famous treatise, Boole presented the algebraic formulation of the semantic behaviour of simplest linguistic particles such as conjunction and negation which is today known as Boolean Algebra. The mathematical treatment and the laws governing the thinking process was carried further by Frege and Russell who successfully extended it to entire logic. Later on 'artificial intelligence' tried to extend (but with limited success) to the formalization of thought even beyond logical and rational domain.

The second significant concept which contributed to the development of computers was given by Alan Turing. Starting with the logical calculus of Frege and Russell, he demonstrated in 1936 that there can not exist a procedure for deciding whether any given formula of logical calculus is valid or not. In other words it is impossible to mechanize the semantics of logical reasoning in a way similar to what had been achieved for its syntax. To prove this impossibility, Turing introduced the notion of an abstract machine capable of executing all possible formal calculations and showed that such a machine can not solve decision problems. Turing machine was in a way theoretical blue print of a computer.

To physically implement such a machine required one more idea. This was born through the collaboration of a neurophysiologist and a mathematician namely Warren McCullock and Walter Pitts. They could develop in 1943 a simplified model of the working of human thought process in the form of an electrical network called 'neural network'. They showed that an electrical network could be built up to mimic the rudiments of the working of human thought process. What these networks could achieve is precisely what is known as Boolean Algebra.

An electric computer is nothing but practical implementation of Turing's machine combined with McCullock and Pitt's neural network. The later provides the former with a brain capable of making most elementary logical decisions. As a result a computer can execute all possible calculations excluding decision problems which need logic of a higher order.

These developments resulted in the construction of first generation computers: the Electronic Numerical Integrator and Calculators (ENIAC) directed in U.S.A. by Von-Neumann. Automatic Computing Engine (ACE) developed in Great Britain under the supervision of Turing. Both projects took place in 1950's. Thus computer is the outcome of mathematical research carried out in the first half of twentieth century.

3. Use of Computers in Mathematics

Computers are now playing an active role in helping mathematicians solve many of their problems which were earlier considered unsolvable. Major contribution of computers in mathematics is their vast computational power. In fact construction of computers had been stimulated by the desire to speed up and automate vast amount of calculations required in second world war efforts which both Turing and Von-Neumann had experienced first hand, Turing while doing counter intelligence work and Neumann during construction of atom bomb. Use of computers in performing large computations accurately and fast is even to this day one of the most widespread use of this machine and this is implied by its very name 'computer'.

3.1 Impact on Pure Mathematics

Ability to carry out large computations at high speed has its impact even on pure mathematics. For example, people have been using computers to search large prime numbers. Record holder at the end of 20th century was the prime number $2^{6,972,593} - 1$. This number is almost two million digits long! Recently researchers have identified a prime number $2^{57,885,161} - 1$. It has almost four million digits more than the previous record holder. This discovery was a part of Great Internet Merscene Prime Search, a distributed computing project designed to search for a particular type of prime numbers.

In 1940, Fermat had conjectured that all numbers of the type $2^{2^n} + 1$ where n is a positive integer, are prime. It is true for $n = 0, 1, 2, 3, 4$ for which corresponding numbers are 3, 5, 17, 257 and 65,537 and all these are prime. For $n = 5$, the corresponding number is $2^{2^5} + 1 = 2^{32} + 1 = 4{,}294{,}967{,}297$. Now it is not easy to check whether this number is prime or not. For it to be prime it has to be ensured that it can not be written as the product of two factors. A systematic check by a computer showed that it can be factored. In fact $4{,}294{,}967{,}297 = 641 \times 6{,}700{,}417$ and hence Fermat's conjecture is not true in general as it does not hold for $n = 5$. For record no other Fermat type prime numbers beyond $n = 4$ are known. In 1990 combined powers of one thousand computers made it possible to show that number for $2^{2^n} + 1$ for $n = 9$ has factors like $n = 5$.

We have another interesting problem labeled RSA129. In this problem computer scientists Ron Rivest, Adi Shamir and Leonard Adilman while working on encryption of messages became interested in finding two prime numbers whose product is a 129 digit number. In 1993 a group of more than 600 computer enthusiasts and academicians used Internet to coordinate their efforts to search for such two prime numbers. It took them more than a year to find such two prime numbers. One of these numbers has 64 digits and the other 65 digits! (This is from the book 'Road Ahead' by Bill Gates). Discovery of such numbers has the possibility of their use in development of unbreakable codes for encryption of messages.

3.2 Four Colour Problem

Four colour problem dates back to 1852 when Francis Guthrie while trying to colour map of counties in England noticed that just four colours were sufficient to colour a map in a manner which ensured that no two adjacent counties had the same colour. He asked his brother Fredrick if that would be true for every map. In other words is it possible to colour every map with just four colours so that none of its adjacent regions that share common boundary lines (but not just common points) receive the same colour?

Fredrick communicated this conjecture to a famous mathematician of the times, namely DeMorgan. However no authentic proof was coming although every known map satisfying the specified conditions could be coloured with four colours.

It was in 1977 that two mathematicians, K. Appel and W. Haken could formally establish it using very powerful computer of the day by considering all possible formulations for such maps and showing that it would be never necessary to use more than four colours to colour a map. In fact it took them thousands of hours of machine time to exhaustively check all possibilities. This would not have been possible in the absence of computers.

1997 also saw the first proof of a mathematical theorem performed by a computer without any human help. In 1933 Herbert Robins had conjectured that a certain system of three equations was an axiomatization of the theory of Boolean algebras. Robin's conjecture was proven true by a computer program written by William McCume and LarryWos.

3.3 Applications in Applied Mathematics

Computers are producing noticeable impact in applied mathematics as well. Earlier most of the mathematical models developed for real life problems could not be analysed analytically. This generally resulted in introduction of assumptions and approximations to bring the mathematical model of the problem in a format which could be analysed using available analytical techniques. Such solutions however provided only an insight into the nature of the possible solution but not the actual solution of the problem.

These days when it is realized that it is impossible to obtain analytic solution to the mathematical model of a real life problem, its numerical solution is obtained using computers for different sets of input data. This gives a feel of the nature of its possible solution. The numerical solutions provided by the computer are actual and not approximate solutions as no approximations are introduced in the mathematical model. A number of efficient and fast computational algorithms such as genetic algorithms, swarm optimization technique, ant colony method, etc. are now available for obtaining numerical solutions to a vast variety of mathematical models of real life problems. These techniques are efficient, accurate and fast and take their cue from how nature works.

In fact in many cases instead of obtaining numerical solutions (or in addition to the numerical solution) it is possible to view the graphical behaviour of the solution for different sets of input data directly on computer screen using simulation. This has enabled mathematicians to analyse a host of complex problems which were earlier considered intractable. This forms a part of the subject which is now known as 'Theory of Chaos'. Its name not withstanding, the subject consists in the study of systems which may not in reality be chaotic but are so complex and complicated that they appear to be chaotic at first sight.

Simulation of weather is one such example. Weather of a place gets effected by several factors, including what may be happening even in other continents and far off oceans and seas. Most famous metaphor regarding chaotic systems such as weather is the term **'butterfly effect'**. It is said that a butterfly flapping its wings in one continent may cause a tornado in another continent! Computers have made analysis and impact of such types of effects possible. This has resulted in generation of more accurate short term weather predictions.

Great strides have been made in computer graphics. Besides their commercial applications, they are now also playing important role in mathematics by providing visual aids. Computers have lead to the discovery and visualization of surfaces which would have been hard to picture even with mind's eye. The most well known of such computer image curves are **'fractals'**. These were originally discovered in the beginning of 20th century. However they came into prominence only in 1980's when use of computers had become widespread.

4. Computer Games

One of the most common and most popular use of computers these days is in form of computer games. A variety of computer games are now available. There are not only one person games but even two person games such as chess and cards. In these games whereas one person is the player himself/herself, the opponent is the computer. Softwares have been developed and built on the basis of which computer plays an opponent as good as an experienced player would play (and in many cases even better). Chess competitions have been organized in which some of the renowned chessmasters of world were pitted against computers and in many cases computers could even defeat them!

5. Limitations of Computers

With the vast computing powers which new generation computers possess, many feel as if a computer can compute everything. However that is not really so. Turing's definition of an algorithm divides numerical functions into two classes : computable functions and non computable functions. This division is only a first approximation. There are several functions which are computable in theory but not in practice. For instance an algorithm whose execution time is greater than the life of the universe or even the duration of human life can not be in reality considered executable even if it is so theoretically. From application point of view only those algorithms whose execution time is sufficiently fast have relevance. In 1965, Edmonds and Cabham proposed a distinction between algorithms which can be run in polynomial time and those which can not be run in polynomial time. Execution time of an algorithm is measured by the number of steps performed by a computer in its execution. A quadratic algorithm does not require more than 100 computational steps on 10 digit numbers and not more than 10000 steps on 100 digit number and so on.

Of course actual computation time depends to a large extent on the type and power of the computer on which the algorithm is run. However if an algorithm requires polynomial time on one computer it will also need polynomial time on every other computer. In other words, the difference between various computer models and their different implementations effects the execution time of a polynomial time algorithm by a polynomial factor. Change of machine does not alter the polynomial time nature of the algorithm. To be executable in polynomial time is thus an intrinsic and not a contingent characteristic of an algorithm. In case the execution time of an algorithm is not polynomial, it does not mean that it can never be executed in polynomial time but there will be instances in which it will not be so.

Class of problems for which a polynomial time solution exists is denoted by P. Class of problems which are not executable in polynomial time is denoted by NP (not polynomial). NP class of problems although not necessarily solvable in polynomial time have almost that property in the sense that for any candidate of such a class, one can verify in polynomial time whether suggested solution is indeed a solution or not.

The difference between P and NP problems is the following. For a problem belonging to the class P it is necessary that there exists a method which can find its solution in polynomial time. However in the case of a problem belonging to NP class it is enough to verify whether a solution of the problem in polynomial time exists or not. For example to verify that a given telephone number is of some person or not is relatively easy. We have to just look up the number of telephone against the name of the person in the telephone directory. Therefore this problem belongs to class P. However to find the person whose telephone number is the given number from a telephone directory (where telephone numbers of persons are listed in alphabetic order) is much more difficult. This is to an extent NP hard problem. Similarly, it is easy to verify that 641 and 6700417 are factors of 4294967297. However the converse problem of checking whether 4294967297 is a prime number or is a composite number having factors is much more difficult. This is in a way NP hard problem. The well known traveling salesman problem in which one is given a map of a number of cities connected to each other by roads along with their lengths, and one is required, to find the path of minimal length which will enable a salesman to visit each city once and only once and return back to the starting city is NP hard.

Stephan Cook, Richard Karpand and Leonid Levin have shown an interesting property which is possessed by all NP hard problems. They have shown that finding a polynomial time solution for any of these NP hard problems would automatically ensure that polynomial time solutions exist for every NP hard problem and distinction between P and NP problems will disappear. The question whether P and NP problems are two different classes of problems or a single class is a challenging one and one of the most famous open problem of theoretical computer science.

6. Dangers of Indiscriminate Use of Computers

Computers are changing day to day life of not only ordinary persons but even mathematicians in a fundamental way. However as happens in the case of new techniques and technologies, not all change is for good and the use of computers in mathematics is no exception. Whereas computers have helped mathematicians in solving some complex problems which were earlier considered unsolvable, it has brought lethargy and in some cases even misuse of computers. Inherent dangers of mindless use of computers are illustrated by the following episodes which show how an indiscriminate reliance on the power of computers may hinder rather than stimulate human thought process.

As mentioned earlier, in 1640, Fermat had conjectured that all numbers of the type $2^{2^n} + 1$, n a positive integer are prime and could verify it for $n = 0, 1, 2, 3$ and 4. For

$n = 5$, the number $2^{2^5} + 1 = 4{,}294{,}967{,}297$ is so large that it is not easily possible to check that it has no factors and is prime. A systematic manual search to check if factors are possible for this large number is not possible. It was only computer which could check that this number is not prime and has factors. In fact, 4,294,967,297 = 641×6700417.

It was again computer which could show that even for $n = 9$, $2^{2^n} + 1$ is not prime. A computer can of course be used to check it for even larger integer values of n. As mentioned earlier till date for no power of n other than 0, 1, 2, 3 and 4, $2^{2^n} + 1$ has been found to be prime. Checking a number to be prime by checking for its all possible factors is pure waste and misuse of computer. Mathematicians have been able to show that in order to check whether a number N is prime or not it is just sufficient to check if any number less than $\sqrt{N}$ is its factor or not. For example to check if 103 is prime or not we need not check if it is divisible by 2, 3, 4, ..., 10, 11, ... up to 102. We need only check if it is divisible by any number less than $\sqrt{103}$, i.e., any number from 2 to 10. Systematic use of logical reasoning considerably shortens the number of factors to be checked. In fact not only that in the case of 4,294,967,297, Euler had shown theoretically that even such a systematic search was also not necessary. It is enough to consider whether it can have factors of the type $64k + 1$, $k = 1, 2, 3, ...$ or not. In fact the factor 641 of 4294,967,297 is indeed of type $64k + 1$ for $k = 10$.

Non availability of computers had thus forced Euler to shift the problem from a mere accounting problem to a problem of higher mathematics. Even though computers could verify that $2^{2^n} + 1$ is not prime for $n = 5$ and $n = 9$, such verifications do not produce any interesting result of general type.

Another instance of this nature of which I personally had the first hand experience in 1970's is the following. Those days I was in England under British Council Exchange of Younger Scientists programme. One day I had gone to a shopping mall for shopping. I made some purchases and placed the purchased items on the counter for preparation of bill. Even in those days most of the big shops and stores in England had automatic calculating machines. The lady at the counter entered into the machine prices of the purchased items one by one and presented me the slip of paper showing the amount of money to be paid by me. The machine had calculated the amount 5 pounds 50 pence. However, mentally I had calculated that the amount should have been 15 pounds plus a few pence. I asked the lady to recheck. She remarked 'How can it be incorrect? It has been calculated by machine (she thought I was expecting the bill to be of lesser amount). I replied 'Alright, I will pay the amount demanded but may be you will regret at the end of the day.' Those days 10 pounds was a big amount (almost her one day salary). When I made this remark she re-entered the prices of the items in the machine. It now produced a bill of 15.50 pounds. She was awestruck. She thanked me and remarked as to how could I make such a long calculation mentally fast and that too accurate. The reason was that in our school days we were not exposed to use of calculators and computers and were trained to make large calculations mentally.

These episodes and Wittgenstein's philosophical investigations warn us that all progress looks bigger than what it in reality is. Consequences of the use of computers in daily life and mathematics are both beneficial as well as harmful. Benefits need not be unduly exaggerated. Their indiscriminate use kills the initiative. Computers must be used when absolutely necessary. As they say **computer is a good servant but a bad master**. Computers have great potential for performing large computations and that too fast. However they work as per instructions provided. They do not have any initiative and decision capability of their own. A computer has not the capability of self-consciousness of humans. It can not say, 'This is I'. The day it happens humans will become subservient to it.

PART II

Mathematics Based Diversions and Recreations

Those who regard mathematics as a dull and boring subject will find certain aspects of the subject highlighted in the subsequent chapters of this book quite interesting, fascinating and even entertaining. These are available generally in the form of games and puzzles and other mathematics based diversions. These have evolved over the centuries to entertain. No doubt some of these tax the mind to some extent but this teasing of mind is not without a purpose. It provides relaxation and joy to mind when a puzzle gets solved. It is said that famous mathematician Leibnitz used to spend a lot of his time on peg-jumping puzzles. Einstein's book shelf was stocked with books on mathematical games and puzzles. Creative thoughts generated by such mathematical games and puzzles have lead many good scientific minds to important and path breaking discoveries. To a common man these provide relaxation and recreation.

This part of the book consists of five chapters. Chapter 12 is devoted to conventional types of games and puzzles. Sudoku puzzle which is a creation of the last quarter of twentieth century is presented in Chapter 13. Arithmetic operations based puzzles are presented in Chapter 14. Chessboard related problems are discussed in Chapter 15. Finally Latin squares and Magic squares are presented in the Chapter 16.

Chapter 12
Games and Puzzles

1. Introduction

In this chapter we present some games and puzzles which a person with elementary knowledge of mathematics and agile mind should find no difficulty in solving. We start with games and then present puzzles. There is no hard and fast rule to distinguish a game from a puzzle. The division is just arbitrary. Deviating from the convention followed in the previous chapters we do not provide solutions to puzzles in the text so that the reader can think and muse over these in search of solution. Solutions are provided in hints and answers component of this chapter.

2. Single Person Games

In this section we present certain games which need only a single person for playing. You may ask your friend to act as instructed or do it yourself.

1. Make a total of 1000 using sixteen fours only.
2. Cards with numbers 7, 8, 9, 4, 5, 6, 1, 2 and 3 are laid in a row on a table in the order specified. Arrange these in ascending order: 1, 2, 3, 4, 5, 6, 7, 8 and 9 using only operations in which it is permitted to choose one or more consecutive cards and rearrange them in reverse order.
3. Given a rectangle of size 3×9, cut it into 8 squares (all squares need not be of same size. However lengths of sides should be integers).
4. Trace the following four shapes on a piece of paper or board and then cut these out. Now arrange these to form letter T.

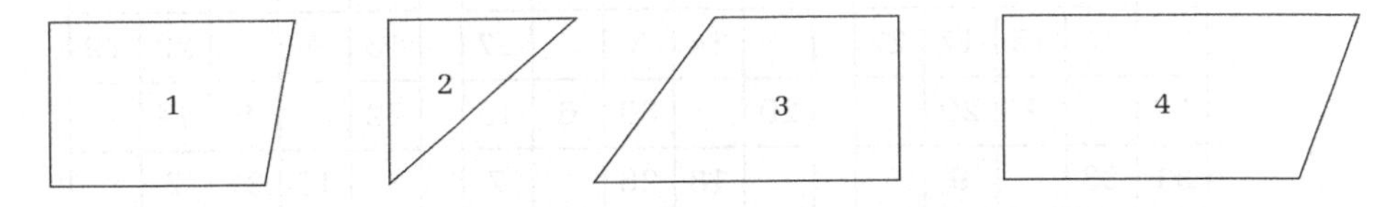

5. Given following 9 dots which are placed randomly in a plane. Join all these nine dots using only four straight lines. Each next line should start from where the previous line ends.

6. In the following two figures made with six match sticks, there are two dots which are outside the two tub shaped figures. Move four matches so that the dots come inside tub shaped figures.

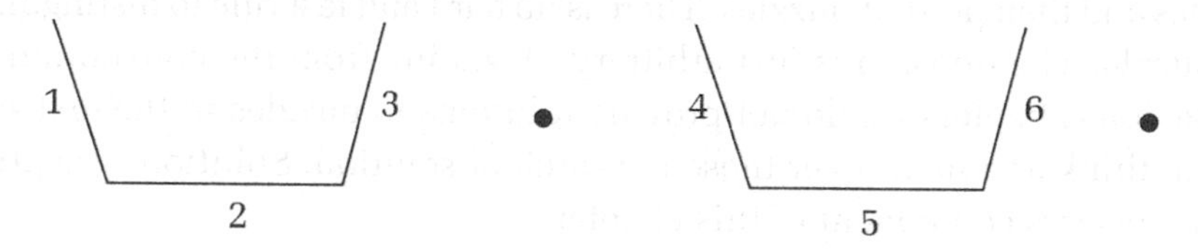

7. Sixteen matchsticks are arranged to form figure shown below. Move two matchsticks to change its shape into four squares.

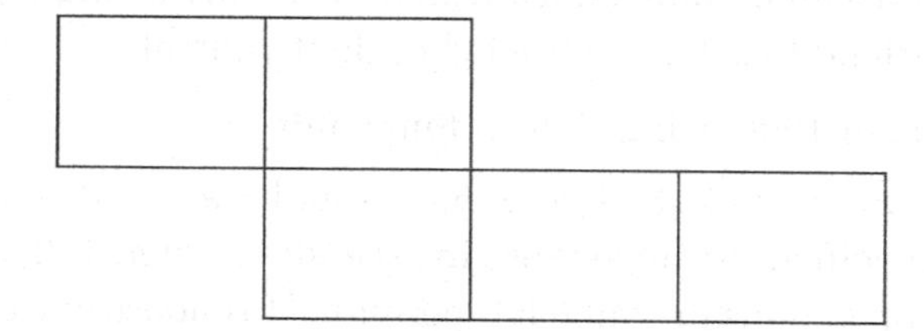

8. **Telling the number thought by another person:** Here is an interesting method to tell the number thought by some other person. Prepare the following five cards with number 1 to 31 written in squares as under:

1	13		25	
	3	15	17	27
19		7	29	
31	23		9	
	5		21	11

2	11		23	
	14	3		27
20		30	6	15
	18	26		7
10			19	31

4			12	21
13	5		23	29
22		6	14	
	15	31	7	
28		20		30

8		13		26
	9	27	14	
15			10	28
	24	11		31
29	30		12	25

16		21	27	
	17	29		22
	24		20	30
23		18		25
28	26		31	19

Ask your friend to think of a number between 1 and 31 and then inspect these five cards to tell which of these cards have that number. After a short pause you can tell the number he chose by just adding the top left entries of the cards indicated by your friend. (Think why it works).

9. Oznam's Problem: Take out all the four aces, four kings, four queens and four jacks from a deck of playing cards and ask your friend to arrange these 16 cards in a square of four rows and four columns so that each row and each column has all the four card values as well as all the four suits. (This arrangement is known as Latin square of order four. We shall discuss more about Latin squares in a subsequent chapter). Is an arrangement also possible where besides the above two conditions, the two diagonals also satisfy these conditions. If yes, make it.

10. A 3×3 table has all zero entries. Change it to

4	9	5
10	22	12
6	13	7

by increasing at a time each number in a 2×2 square by one.

11. Numbers 1 to 16 are written in a 4×4 table as shown below in (a). Obtain if possible table shown in (b) from it by making following moves: All numbers in a row can be increased by one or all numbers in a column can be decreased by one.

1	2	3	4
5	6	7	8
9	10	11	12
13	14	15	16

(a)

1	5	9	13
2	6	10	14
3	7	11	15
4	8	12	16

(b)

3. Two Person Games

In this section we present some games which require moves by two players.

1. **Game of Mora**

   Ancient game of Mora is played thus. There are two players. Both simultaneously pick a number which is either 1 or 2 or 3. This number is to be guessed by the other player. If both the players guess right or both guess wrong, the game is a draw and no payment is to be made. However, if one guesses right and the other wrong, then the player who guesses wrong pays the player who guessed right a sum equal to the number guessed and number picked by the other person. What is the best strategy for playing this game?

2. **Two Person Zero Sum Game**

   Each of two players (say A and B) have a choice of choosing a number between 1 and 4 without revealing it to the opponent. They have decided for the following pay off for the player A.

A \ B	1	2	3	4
1	4	–2	–4	–1
2	3	2	–1	1
3	2	3	2	–2
4	–1	–3	–3	2

   In other words if A chooses 2 and B chooses 1, then A gets Rs. 3. However, if A chooses 1 and B chooses 3, then A gets –4, in other words A will have to pay B Rs. 4. What is the best strategy for each player? (Such a game is called **two person zero sum game** as gain of one player is same as loss of other player making a total zero. The amount which person pays to the other is called the **value of the game**. Optimal strategy for each player is one in which the player maximizes his gain or minimizes his loss).

3. **Race to 50**

   Two players A and B play a game alternately. Each can add any number from 1 to 6 to the current total. Game starting with initial total as zero. One who is the first to add a number which brings total to 50 wins. Here is an example:

Round	A adds	B adds	Total
1	4	–	$0 + 4 = 4$
	–	5	$4 + 5 = 9$
2	3	–	$9 + 3 = 12$
	–	6	$12 + 6 = 18$
3	5	–	$18 + 5 = 23$
	–	2	$23 + 2 = 25$
4	6	–	$25 + 6 = 31$
	–	4	$31 + 4 = 35$
5	1	–	$35 + 1 = 36$
	–	3	$36 + 3 = 39$
6	6	–	$39 + 6 = 45$
	–	5	$45 + 5 = 50$

So B wins. What is the best strategy for a player to play this game?

4. **Game of Noughts and Crosses**
This is a popular game amongst the children. In this game a square is subdivided into 9 or more subsquares by drawing equal number of horizontal and vertical lines. Two players take turns in putting noughts (O) and crosses (X) (one always putting a nought and the other always a cross). The player who is the first to get his symbol in an entire row or a column or a diagonal wins. In the following two squares, the player putting naughts wins in first square. However no player wins in second square.

X	O	X
X	O	X
	O	

X	O	X
O	O	X
X	X	O

What is the best strategy that a player should adopt while playing this game?

5. **Removing marbles from a box**
A box has 300 marbles. Two persons take turns in removing marbles from the box. On his turn, each player can remove from the box not more than half of the marbles currently in the box. The person who can not remove the marbles from the box loses. What is the recommended strategy for a player so that he wins?

6. **Removing chips from a pile**
   A game is played between two players. They begin with a pile of 30 identical chips. For a move, a player may remove from 1 to 6 chips. They remove chips alternatively and this process continues till the last chip is removed. The player who removes the last chip wins. Devise winning strategy for such a player.

7. **Putting plus or minus sign between numbers written on blackboard**
   Numbers from 1 to 20 are written on a blackboard in a row. Two players take turns in putting plus or minus sign (whatever they like) between these numbers. When signs between all the numbers have been placed, the resulting expression is evaluated performing indicated plus and minus operations. First player wins if answer is an even number. Second wins if answer is odd number. Is there any winning strategy?

8. **Joining points on the circumference of a circle**
   Twenty points are marked on the circumference of a circle. Two players take turns in joining these points (two at a time) with straight lines which form chords of the circle. However newly drawn chord should not intersect any of the earlier drawn chords. The player who can not do so loses. Is there some winning strategy?

9. **Game of Nim**
   There are three heaps of stones. Two players say A and B make their moves in turns. They can take out as many stones as they like (but from the same heap). The player who picks the last stone wins. Is there some winning strategy?

4. Puzzles

In this section we present some simple well known puzzles. It is suggested that the reader should first himself/herself try to figure out their solutions before looking at the solutions provided.

1. Find the smallest number of digits which can generate all other integers from 1 to 40 by just adding two or more of these.

2. **More milk or more water**
   There are two tumblers. One is full of milk and the other full of water. Three table spoons of milk are poured from tumbler of milk into the tumbler of water and well stirred. After that three table spoons of water from this tumbler are put back into the tumbler of milk and well stirred. Which is greater: percentage of milk in tumbler of water or percentage of water in the tumbler of milk?

3. **When was the container half-full**
A number of bacteria are put in a glass of water. One second later each bacteria divides into two. Next second again, each resulting bacteria divides into two and so on. After one minute the glass is full of bacteria. When was the glass half full?

4. **When did his wife pick him up**
A person arrives back from office by train. His wife picks him up from the railway station at 6 p.m. One day he was free from office early and arrived at the station at 5 p.m. by an early train. As his wife had not reached the station to pick him up, he started walking home. He met his wife enroute and arrived home 10 minutes earlier. How long had he walked before his wife picked him up?

5. **How many sons and how many daughters.**
A person left behind Rs. 2,109,000 with the instructions that each of his sons should receive three times the amount which each of his daughters receive and each daughter must receive twice as much as their mother. If his wife's share came out to be Rs. 11,1000, how many sons and daughters he had?

6. **Centigrade and Fahrenheit readings identical.**
At what temperature, centigrade and Fahrenheit readings are identical?

7. **Missing digits.**
A scrap of paper was found in a dustbin. It was a bill which read : number of items 72, cost Rs. *679.*0 The thousand place and first digit of paise were illegible. What can these be?

8. **Burning ropes to determine time**
There are two ropes of uniform section and of equal length. When ignited at one end, each takes one hour to burn completely. Using these two ropes how can time period of 45 minutes be determined?

9. **How long the tyres of a car last**
A car has a stand-by wheel for emergency use besides the four wheels. A car owner uses all the five wheels equally by interchanging the fifth wheel with one of the four working wheels from time to time. Such a car which uses five tyres equally has traveled 20,000 km. For how many kilometers has each tyre been used?

10. **How many brothers and how many sisters**
A family has 7 children, some boys and some girls. Each boy in the family has as many sisters as brothers. However, each girl in the family has twice as many brothers as sisters. How many boys and how many girls are there in the family?

11. Number of steps in the escalator

While walking down an escalator, I noticed that if I walk down 26 steps, I require 30 seconds to reach the bottom. However, if I walk down 34 steps, I need only 18 seconds to reach the bottom. How many steps are there in the escalator?

12. How many cats and how many mice

Some cats killed 999919 mice, each cat killing equal number of mice. How many cats were there and how many mice did each cat kill given that each cat killed more mice than the total number of cats that were there.

13. How many stamps

One day a person went to a post office and placed Rs. 75 on the counter and asked for two rupee stamps six times as many as for 1 rupee stamps. For the rest he asked for 5 rupee stamps. How many stamps of each denomination did he buy?

14. What is the number

Unit place digit of a two-digit number is smaller than its ten place digit number by 4. If the number is divided by the sum of its digits, the quotient is 7. Find the number.

15. Problems on perfect square numbers

(i) Arrange all digits from 1 to 9 to form four perfect square numbers (without digits repeating).

(ii) Arrange digits from 1 to 9 to form the smallest and largest possible square numbers.

(iii) Find four numbers such that the sum of every two of these numbers as well as the sum of all four numbers are perfect squares.

16. What is my house number

I live in a long street. Houses on one side of my house are numbered 1, 2, 3, ... However the numbers of houses on one side of my house add upto same number as the sum of the numbers of houses on the other side of my house. What is the number of my house?

17. Number of the car plate

An old car number plate consists of 5 different digits. It got turned upside down with order of digits reversed. However even when upside down, the digits still formed a number which could be read. However, this number exceeded the original number by 78633. What is the actual car number on the plate?

18. How many marbles

Manu had some marbles. He said to his friend : "If you give me one of your marbles, I shall have twice as many marbles as you will then have." The friend replied, "If you give me one marble, I shall have as many marbles as you will then have." How many marbles Manu and his friend have?

19. How many days did domestic help work

Mr. Prakash engaged Shri Ram Lakhan as a domestic help at the salary of Rs. 3000 p.m. with the condition that an amount of Rs. 150 will be deducted from his salary for each day he would not turn up for the work. At the end of the month (of 30 days), Ram Lakhan got nothing. For how many days was he absent from his duty?

20. How many diagonals

How many diagonals are there in a (i) hexagon, (ii) in a convex polygon having n sides.

21. Which job is more lucrative

Which job is more lucrative, a job with Rs. 2 lacs to be paid annually or a job with one lac to be paid every six months with a raise of Rs. 30,000 per year in the first case and a raise of Rs. 10,000 every half year in the second case?

22. Dimensions of a square board

There are three square boards. One is five square feet more than the second and second is five square feet more than the third. What are the dimensions of the three boards?

23. How many triangles

How many triangles are there in this figure?

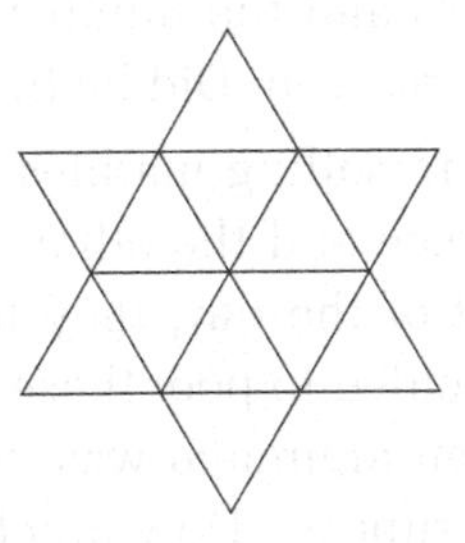

24. Dimensions of Canvas

What are the smallest dimensions of canvas which would allow a painter a margin of 4" at the top and bottom and 2" on each side with available painting space as 72 square inches.

25. To paint half window black

A room has $4' \times 4'$ square window. Half of its area is to be painted black in such a manner that unpainted clear part of the window is still a square measuring 4 feet from side to side. Is it possible, if yes, how?

26. Enclosing the largest possible area

Find the largest possible area that can be enclosed by four sides of lengths 20, 16, 12 and 10 units.

27. Heights of brothers

Next door to me live four brothers who have different heights. Their average height is 64 inches. Whereas the difference between the heights of two brothers is just two inches, the difference between the heights of other two brothers is 6 inches. What are the heights of the four brothers?

28. Gain or Loss

(i) A farmer had two cows. He sold them for Rs. 12,000 each making a profit of 20%, on the sale of one cow and a loss of 20% on the sale of other cow. Did he gain or lose in overall transaction and how much?

(ii) A person bought an item from the market for Rs. 600 and sold it to his friend for Rs. 700. After some time the person bought back the same item from his friend for Rs. 800. After some time, it was again bought back by the friend for Rs. 900. How much overall profit did the first person make?

(iii) A person said to me, "Heads I win, tails I lose." I bet half of money in my pocket on each throw of a coin. He threw a coin and lost and paid me half of the money in his pocket. He tossed the coin again and won and got back half of the money in his pocket. He continued tossing the coin several times and the number of times he lost was equal to the number of times he won. Did he finally gain or lose?

(iv) Two persons were selling marbles in market. One was selling three marbles for a rupee and the other two marbles for a rupee. One day towards the end of the day, they had equal number of marbles left unsold. They decided to pool these together and sell these for five for two rupees. (Their argument was that 2 for a rupee and 3 for a rupee meant 5 for two rupees). How much was the gain or loss to each one of them?

29. Average speed

While driving to office in the car in the morning as there is generally rush, my average speed is 40 km p. h. However on way back at night my average speed is 60 km p.h. What is my average speed in the round trip?

30. How much faster is the mail train

A mail train and a goods train are running on parallel tracks in the same direction. The mail train which is faster than the goods train overtakes it and takes three times as long to cross it as it would have taken if moving in the opposite direction. How much is the mail train faster than the goods train?

31. Age related problems

(i) When I am as old as my father is now, I shall be five times as old as my son is now. At that time my son will be 8 years older than what I am now. At present the sum of my age and my father's age is 100. How old are we now?

(ii) Mohan lives one fourth of his life as a boy, one fifth as a youth, one third as a middle aged person and has already spent 13 years of his life as an old person. How old is Mohan now?

(iii) A lady remarked, "If you reverse the digits of my age, you get the age of my husband. He is senior to me. Difference between our ages is 1/11 of the sum of our ages. How old are we?"

32. A squirrel climbing a circular pole

A squirrel climbs a circular pole spirally making one complete circuit in 4 feet (i.e., it rises 4 feet while completing one circle). If the top of the poles is 16 feet high and the pole is three feet in circumference, how many feet has the squirrel traveled when it reaches the top of the pole?

33. Josephus Problem

Forty one persons enter a suicide pact. They arrange themselves in a circle and make rule that starting with a specific person every third person will be killed. The process will continue till only two persons are left. Will the person from whom the counting started ultimately escape? Is it possible for two persons to enter into a pact and so position themselves that it is they who ultimately escape getting killed? How will the results get affected if instead of persons arranging themselves in a circle they arrange themselves in a row?

Next two puzzles are from Bhaskaracharya's Lilavatti

34. Lilavatti-I

Which is the number which when multiplied by 3 then increased by ¾ of the product, then divided by 7, next diminished by one-third of the quotient, then multiplied by itself diminished by 52, square root found, 8 added and finally divided by 10 gives answer just 2?

35. Lilavatti-II

From a swarm of bees, square root of half of the number of bees in it flew out of a Jasmine bush and 8/9 of the swarm of the bees remained behind. One female bee is flying about a male bee. What is the total number of bees in the swarm?

Mind Teasers

1. A 3 × 3 table is filled with numbers. It is permitted to increase each number in any 2 × 2 square by one. Is it possible to use this operation repeatedly to obtain the following table starting with all initial entries as zero?

4	9	5
10	18	12
6	13	7

2. Number 485 is written on black board. It is allowed to either double the written number or erase its last digit. How can 14 be obtained using these two operations.

3. Ram and Shyam have unequal number of marbles. Ram says to Shyam "If you give me one marble, I shall have twice as many marbles as you will then have". Shyam replies, "If you give me one marble, than both of us will have equal number of marbles". How many marbles each of them has?

4. A lady visited three temples in the morning one after another. She started the visit with a basket having some flowers(between 10 and 15) in it. She offered some of these in the first temple that she visited. After coming out of the temple, she purchased as many more flowers as she then had. Next she visited the second temple and again offered there as many flowers as in the first temple. After coming out of this temple, she again purchased as many flowers as she then had. Finally she visited the third temple and offered there also as many flowers as in the first two temples. When she came out of the third temple there were no flowers left in her basket. How many flowers she had in the beginning and how many flowers she offered in each temple.

5. A school bus is considered overcrowded if there are more than 50% of allowable children in the bus. Children ride in several buses to a summer camp. Which is greater: percentage of overcrowded buses or percentage of children riding in overcrowded buses?

6. There are 9 coins out of which one is counterfeit and weighs more or less than normal coins. Is it possible to find out this counterfeit coin using two weighings only?

7. A rail track runs parallel to a road until a bend in the road brings it to a level crossing. A cyclist rides along this road to his place of work every day at a constant speed of 12 km p. h. He normally meets the train which travels in the same direction at the level crossing. One day he was late in starting for his office by 25 minutes and met the train which was running right time 6 km ahead of the level crossing. What is the speed of the train?

8. One day a person started from city A towards city B at noon. Another person started from city B towards city A at 2 p.m. They crossed each other at five past four. At what time did they arrive at their destinations if the distance between A and B is 24 km and what were their speeds?

9. Product of three consecutive integers when divided by each of them yields sum of quotients as 74. What are these integers?

10. Numbers 25 and 36 are written on the black board. At his turn a student writes the positive difference between the numbers already written on the board in case such a difference does not already exist. Loser is one who can not write the number. Who is the loser if two students take part in it.

11. There are three containers which can hold 19 cc, 13 cc and 7 cc of water. 19 cc container is empty. However 13 cc and 7 cc containers are full. Can we measure 10 cc using these three containers?

12. A shopkeeper is offering two items at discount. One for Rs. 350 whose actual price is 8/7 of this and the other for Rs. 300 whose actual price is 7/6 of this. Which is a better bargain?

13. B is sitting on right of A and C to the left of A. D is sitting to right of C but left of B. How are they seated?

14. There are more adults than boys, more boys than girls and more girls than families. If no family has fewer than three children, what is the smallest number of such families?

15. A family of four persons lived in a remote village near a river bank. The family consisted of the farmer, his wife, his aged mother and a small child. One night there was flood in the river and flood water entered the helmet. The water was rising and they had no place nearby where they could go for safety. So they decided to ascend the roof of the helmet. They had a wooden stair which was not wide and strong so only two persons could ascend or descend it at a time with a light. They had only one torch light. So they decided to ascend two at a

time. One would stay at the top and other would return with the torch light. In view of their age and physical condition, whereas the farmer needed just one minute, his child 2 minutes, his wife 5 minutes and his mother 10 minutes to ascend or descend. How should they plan their ascent to the roof so that the entire operation is completed in shortest possible time to avoid any one of them being drowned in flood waters.

Chapter 13
Sudoku

1. Introduction

Sudoku is a form of puzzle which has become very popular in the last few years. Like all great puzzles that catch popular fancy, Sudoku is based on a simple idea. In its most popular form, it has a 9 × 9 square grid with two horizontal and two vertical heavy lines which divide it into 9 boxes, each box of size 3 × 3 grids.

2	4		8	1				3
5	8	6	3				1	2
				5	2	8	6	
8			5	4	3		9	
4	1	5	6	2	9	7	3	8
				8	1	2	4	5
	5		2	3			8	1
3	6	8			5			
7					8			

Fig. 1

Some digits from 1 to 9 are initially given in the grid and one is asked to fill the remaining cells of this 9 × 9 grid with numbers from 1 to 9 so that each digit from 1 to 9 appears once and only once in each column and each row as well as in each of the nine 3 × 3 boxes. It is that simple.

It is a single person game. No arithmetic operations of addition, subtraction or multiplication are involved. It is a game of pure logic. The game is in fact addictive. Once a person gets a hang of it, solving Sudoku becomes a fun and good pastime. Experts try to find techniques for cracking a Sudoku puzzle quick and fast.

2. Origin of Sudoku

As the name implies, Sudoku came to the world from Japan. However some westners credit Howard Garns of United States with inventing this game. Originally it was called 'Number Place'. It first appeared in Dell Pencil Puzzles and World Games magazine in May 1979. Dozens more followed in Dell Publications. Surprisingly Howard Garns who is credited with its invention was not a mathematician. He was an architect. He was seventy four when he created his first 'Number Place' puzzle. He died in 1989. By that time, the game had not become that popular.

After Gans stopped contributing his 'Number Place' puzzle, other puzzle makers started creating their own 'Number Place' puzzles for Dell. It is said that in 1980's an editor of 'Nikole Puzzles' magazine of Japan saw one of these on his visit to U.S.A. He took it back to Japan and published it in his magazine uner the caption 'Sudoku'. It became a hit in Japan. In 1997 a retired judge from Newzealand saw a Nikole book of Sudoku puzzles and started making his own Sudoku puzzles. Beginning November 2004, the Times Magazine of London started publishing these and they became instant hit. These days most of the leading daily newspapers in U.K., U.S.A. and even India publish Sudoku puzzles. In fact people all over the world have fallen prey to it. Books on Sudoku puzzles have also now appeared. In the words of Christian Science Monitor, 'Logic has not been this much fashionable since the days of 'Rubik Cube'. According to Daily Telegraph Sudoku puzzles are to the first decade of twenty first century what Rubik Cube was in 1970's.

3. How to Solve a Sudoku Puzzle

Solving a Sudoku puzzle involves pure logic. No guess work or calculations are needed. In fact guess work is not even desirable. A few simple basic facts and logic and experience usually help.

Take for example the Sudoku puzzle of Fig. 1. For easy reference, we first label the nine 3 × 3 boxes as I to IX as shown in Fig. 2.

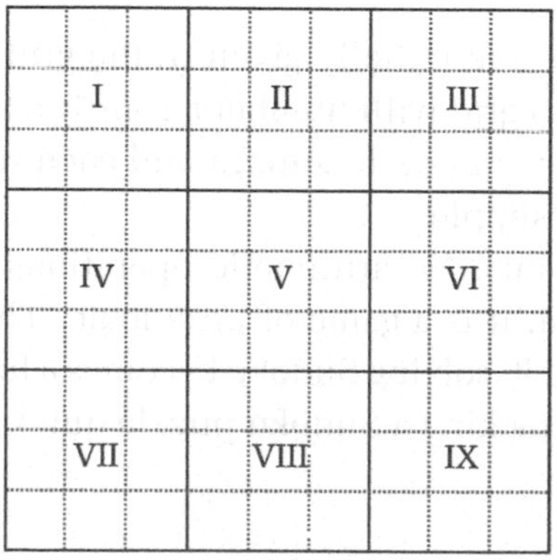

I	II	III
IV	V	VI
VII	VIII	IX

Fig. 2

There are several ways to get started. For instance we observe that boxes VIII and IX have already 8 in third and first row. So box VII is to have 8 in second row. Luckily there is only one vacant slot in second row of box VII. So this slot has to be filled with 8.

Now we look at the left three vertical boxes I, IV and VII, 8 appears in column 3 in box VII, and column two in box I. So it has to be in column one in box IV. But there are two vacant slots in row one of this box. So 8 can be either in row one or row three of the first column of box IV. However row three has already 8 in box V. So it can not be in row three. Hence 8 is in row one of first column box IV.

Now look at boxes I, II and III. Each of these has 8. Box I in row two, Box II in row one and box III in row three. So 8 is not to appear anywhere else in these three boxes.

Next consider number 5. It appears in row one in box V and row two in box IV. Hence it must appear in row three in box VI. Now there is only one slot vacant in row three of box VI. So 5 must appear there. Thus 5 appears in row 6 and column 9 of box VI.

Now box IX does not have 5. It can not be in row one or row two as 5 is already in row one of VII and row two of box VIII. So it has to be in row three of box IX. But all the three slots in this row of box IX are vacant. However 5 can not be in column three of this box as 5 is already in column three of box VI. So it has to be either in column one or column two of row three of box V. But which of these can not be decided at the moment. So we leave it there for the moment and look for some other clue.

For a different kind of logic, consider fifth row. It has entries 3, 4, 5, 6, 7 and 8. Missing numbers are 1, 2 and 9. 1 can not be in 5th or 6th column for it is already there in these columns. So it has to be in column 2. Thus 2 and 9 are to appear in columns 5 and 6 of row 5. Now 2 can not be in column 6 as it is already there. So 2 is to be in column 5 and 9 in column 6 and one in column 2. Thus row 5 is now completely filled.

Box V has now only two empty squares where numbers 4 and 7 are to appear. 4 can not be in empty square in 6th row as it is already in 6th row. So 4 is in empty square of 4th row of box V and 7 in empty square in 6th row.

Now consider status of number 3 in boxes I, II and III. 3 is already in first row of box III. So it can not be in empty squares of row 1 of boxes I, II. As row 2 of box I has no empty square so 3 is in row 3 of box I and row 2 of box II. Now 3 is already in columns 5 and 6. So 3 can only be in column three in box II. So 3 is at crossing of row 2 and column 4 in box II. Position of 3 in square I can only be in column 2 or column 3 of row 3 as 3 is already in column 1. However its exact location can not be as yet decided.

Next consider row 6. It now already has 1, 2, 4, 5 and 8. So 3, 6, 7, 9 are missing. 6 can not be in column 2 or 3 as it is already there. So it must be in column 1. However exact location of 3, 7 and 9 can not be decided as yet.

Now consider column 1. In this column 1 and 9 are missing. 1 can not be in row 7 as it is already there. So 9 has to be in row 7 of column 1 and hence 1 in row 3 of column 1.

Now 1 is in the column one of box I, column two of box IV; hence it has to be in column three of square VII. It can not be in its row 7 as 3 is already there. So it has to be in row 9.

Now box VII has two empty squares in which numbers 2 and 4 have to appear. 4 can not be in column 1 so it must be in column 3 and hence 2 in column 2.

Now entries 6 and 7 are missing in row 7. 7 can not be in column 7 so it has to be column 6 and thus 6 in column 7.

Again entries 4 and 6 are missing in column 6. 6 can not be in row 2. So it must be in row 1 and hence 4 in row 2.

Now 7 and 9 are missing in box II. 7 can not be in column 4. So it has to be in column 5 and 9 in column 4.

Now only 9 is missing in row 2. It must be in column 7 of row 2. Also 1 and 4 are missing in column 4. 1 can not be in row 9 so it has to be in row 8 and 4 in row 9.

Now looking at column five, 6 and 9 are missing. 6 can not be in row 8. So it must be in row 9 and 9 in row 8.

In box VI, 1 and 6 are missing. 1 can not be in column 9. So it has to be in column 7 and consequently 6 in column 9.

Again 2 and 7 are missing in row 4. Now 2 can not be in column 2 so it must be in column 3 and 7 in column 2.

Now 3 and 9 are missing from row 6. However decision regarding their exact location still can not be taken. But 3 and 9 are also missing in column 2. 9 can not be in row 3 as it is already there. So it has to be in row 6 and thus 3 in row 2. This leaves only one vacant slot in box IV and it has to be 3.

Box I still has two missing entries 7 and 9. As 9 can not be in row 3 it must be in row 1 and 7 in row 3.

By now boxes I, II, IV, V, VI, VII and VIII are completely filled. Box III has only one missing entry in row 3 which has to be 4. Now box III has two missing entries 5 and 7. As 7 can not be in column 7 so 5 is in column 7 and 7 in column 8 of box III. Six entries are still missing in box IX. As 5 can not be in column 7 or column 9 it has to be in column 8. As it can not be in row 8 also, it must be in row 9 of column 8.

Now there are two missing entries (3, 4) in column 7 and two (3, 9) in row 9 of this box. Clearly common missing entry 3 must be in grid at crossing of 7th column and 9th row. Of the remaining two, i.e. (4 and 9), 4 has to be in a vacant slot of column 7 and 9 in a vacant slot of row 9. Thus 4 is to be in grid (8, 7) and 9 in grid (9, 9).

Now only two entries are missing in row 8. One in column 8 and the other in column 9. Also entries (2, 7) are missing in columns 8 and 9. Thus 2 is in column 8 and 7 in column 9 of row 8.

Now, the puzzle is completely solved. The result is depicted in Fig. 3. Circled entries are some of the entries which have been determined using logic.

We have presented here a method for solving Sudoku puzzle. However the line of argument presented here is not unique. Alternative and perhaps even more effective approaches may be possible. It depends upon the experience and expertise of the concerned individual as to how fast a puzzle gets solved. It is a matter of pure logic. In fact now even computer softwares have been developed to solve Sudoku puzzles.

2	4	9	8	1	6	5	7	3
5	8	6	3	7	4	9	1	2
1	3	7	9	5	2	8	6	4
8	7	2	5	4	3	1	9	6
4	1	5	6	2	9	7	3	8
6	9	3	7	8	1	2	4	5
9	5	4	2	3	7	6	8	1
3	6	8	1	9	5	4	2	7
7	2	1	4	6	8	3	5	9

Fig. 3: Solution to Sudoku puzzle of Fig. 1

4. Level of Difficulty

At first sight it appears as if a Sudoku puzzle in which lesser number of initial entries are provided will be more difficult to crack compared to a Sudoku puzzle in which more entries are provided initially. However, that is not always so. Often it is noticed that even though at initial stages it is easy to decide about unknown entries, it becomes more and more difficult to fill up the missing entries as the number of missing entries decreases. Often one gets stuck up towards the end. Sudoku puzzles are often categorized by the designers as level I, II, III, IV, etc. depending upon how difficult it will be to crack the puzzle.

In the exercise part of this chapter we present 20 Sudoku puzzles for the interested reader to try and solve these. In our view first five of these are easy (level I), next five of moderate difficulty (level II), next five demanding (level III) and last five challenging (level IV). However this categorization is not very rigid and one should not get scared if puzzle is of level III or IV. Sometimes an alternative way of reasoning may make a puzzle categorized easy as difficult and vice versa. The solutions to these puzzles are also provided. However once a Sudoku puzzle is solved, checking the solution is simple. The reader is advised not to look at the solution before the puzzle gets solved. Otherwise it looses its charm.

5. Variation in the Size of Sudoku Puzzle

With Sudoku puzzles gaining wider acceptability and popularity, Sudoku puzzles of sizes other than the standard 9 × 9 size have also been framed. The most common amongst these being Sudoku puzzles of size 5 × 5, 6 × 6 which are often meant for children. Puzzles of sizes more than 9 × 9 have also been designed. However 9 × 9 size remains the most popular.

6. Designing a New Sudoku Puzzle

At first it appears as if a Sudoku puzzle can be designed arbitrarily by drawing a 9 x 9 square mesh with two equispaced horizontal and two equispaced vertical partitions which divide it into 9 boxes each having nine squares and then randomly writing some numbers between 1 and 9 in some of these squares. However that is not so simple. Care has to be taken to suitably place selected numbers between 1 and 9 in different grids of these boxes so that when missing entries are filled a solution is possible which ensures that each row and each column and each of the nine boxes has numbers 1 to 9 with no number repeated and no number missing. Moreover such a solution should not only exist but be also unique.

6.1 Creating New Puzzle from a given Sudoku Puzzle

Given a Sudoku puzzle it is easy to create new puzzles from it. We can interchange positions of provided entries of two rows amongst first three rows or next three rows or last three rows. In other words interchange entries of two rows of boxes I, II, III or two rows of boxes IV, V and VI or two rows of boxes VII, VIII or IX. More than one sets of rows can be interchanged simultaneously.

Alternatively we may interchange positions of provided entries amongst first three or middle three or last three columns (in other words interchange columns within boxes I, IV, VII or II, V, VIII or III, VI, IX). A judicious mix of the two is also possible.

7. Samurai-Sudoku Puzzle

This is a more extended version of standard 9 × 9 size Sudoku puzzle. It consists of five overlapping standard 9 × 9 size Sudoku puzzles. One such Samurai-Sudoku puzzle is presented in Fig. 4.

9						4	1						6	4	7					
5			9	1		2						1			4			6	5	2
8		1	7		3			6						9	6		1			
	5					1	4									7			2	
	7		4			6						6	5		2				3	
		9	3		5			2					9		3		5	1		
					4	3		1	9	5			2			9	3		8	
	9			5		7	6			1					8			2	1	
	8	4		7					4					8	5					
								7			4			9						
						6	4		1			2								
								8		2		6								
				7	4	8		3					6				1		3	9
3		5	6		1		7		5	6		3			2	5		6		
		4				5						8		7			9	2		
6						3								8		7			2	
1	3	8	7		6	2		9				7				6				
				3										4			8		9	7
	6	2				1		5						5		8		4		2
7		1		8	3	9	6							6				5		
				6								2		3		1	5			8

Fig. 4: Samurai-Sudoku puzzle

The standard rules of the Sudoku puzzle apply to each of these overlapping puzzles. In this enlarged version, the number of unknown entries is now more. However the number of provided hints is also more. It is in no way more difficult to crack a Sudoku-Samurai puzzle as compared to a standard Sudoku puzzle. However time required for obtaining the complete solution is comparatively more. Solution to the Samurai-Sudoku puzzle of this section is not provided and is left as an exercise. Some more Sudoku-Samurai puzzles are also provided in the exercises for the interested reader to try his hand. One can also try to solve each of the five superimposed Sudoku puzzles as Sudoku puzzles separately and see if each of these is separately solvable or not and whether their independently obtained solutions are identical with their super-imposed solutions in Sudoku-Samurai.

Mind Teasers

1. Solve the following Sudoku Puzzles

(1)

		6		4		5		2
7	5	8	3			9	1	4
1				5			3	6
4	3	2			6			
					4	1	6	
				9	7			3
8		5					9	7
2			9	8		6		
9		3	4	7	5	2	8	

Level of difficulty I

(2)

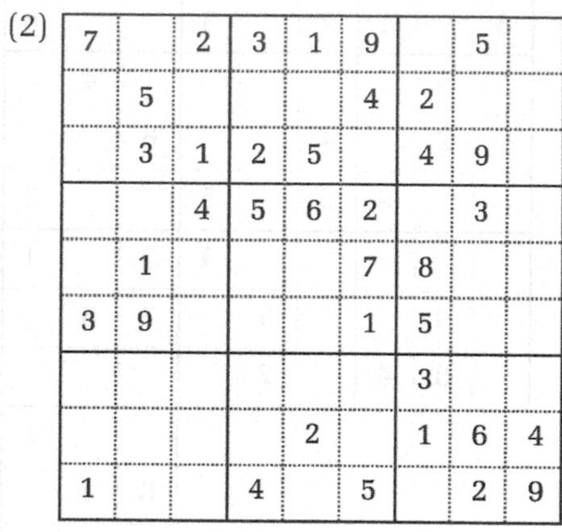

7		2	3	1	9		5	
	5				4	2		
	3	1	2	5		4	9	
		4	5	6	2		3	
	1				7	8		
3	9				1	5		
						3		
				2		1	6	4
1			4		5		2	9

Level of difficulty I

(3)

8				3		2		6
2	3	6		4			7	
7			2		9			
1		3	9		6	4		
			3	2			5	
4	2		1		7			3
5		9	6				8	
			4		8	3		5
		7				6	1	4

Level of difficulty I

(4)

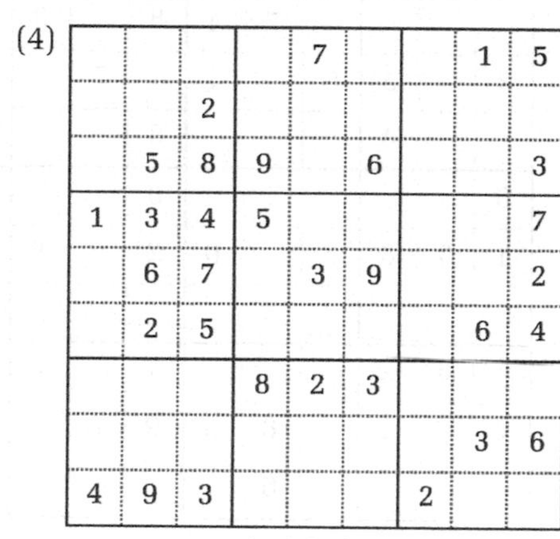

				7			1	5
		2						
	5	8	9		6			3
1	3	4	5					7
	6	7		3	9			2
	2	5					6	4
			8	2	3			
							3	6
4	9	3				2		

Level of difficulty I

(5)

			7		6			
	1	7		9				
	6		3	5				7
1	9		2	3				6
7			6			2		
	8	2		1	7	9		
	3		1			5		
		1	9				2	3
		5			3		8	

Level of difficulty I

(6)

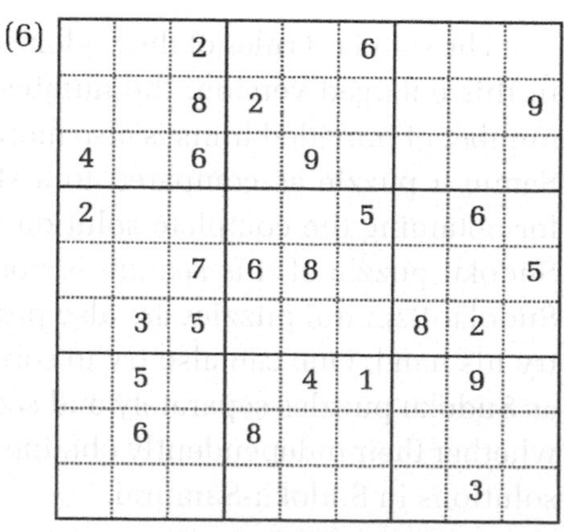

		2			6			
		8	2					9
4		6		9				
2					5		6	
		7	6	8				5
	3	5				8	2	
	5			4	1		9	
	6		8					
							3	

Level of difficulty II

(7)

							5	2
				2			1	
	8				4			
								6
9					8	5		3
		6	5	3				1
	4	2						
	7	5		8	6			
3	6		1			7		5

Level of difficulty II

(8)

					9	8		1
3				4			9	
		5			8			
6					3	1		
				2		9		6
		7	1				4	
		4				7	5	
					6			2
8	2							3

Level of difficulty II

(9)

					8		7	
3					9	4		
	5	4			1			
	1	7					9	
	6			4				1
9		2		3	5			
		9						7
				7	6	8		
								6

Level of difficulty III

(10)

		1	4					
	9					3		
	4			7			1	
	1	2			8	5		
	8							2
			5	6				4
				1	3		9	
							5	
		3		9		2		

Level of difficulty III

(11)

			8		5	2		
8						4		1
7		5		4				9
			1			7		2
	4							
		6	4	3				
	3		9		6			
	1							
							8	

Level of difficulty III

(12)

		9	1					
	2				5	8		
8						7		6
					1			
	4	1		9	8	2		
	7			5		4		
	1		3					
					6			
3		8		1			2	

Level of difficulty III

(13)

9						4		3
2				7		6		
							8	5
	1		2					
6			7	5				
		7		4	3			
4							7	
	9			2	1		5	
	5	1				8		

Level of difficulty III

(14)

						1		4
			8				7	
	2			3	6			
	4				3			
		3			4			7
				9		8		
	5	7			1			
9						5		
8							6	

Level of difficulty III

(15)

9				2				
8					4			
		6	5				4	7
		1				3		5
						1		
	3	5			6			8
	7		1	3		2		
					7	9		4
								1

Level of difficulty III

(16)

	2							5
				8				9
	1			7	4			2
2							9	
8				5				7
		5					4	
9	4	3			1			
			7			2		8

Level of difficulty IV

(17)

2		7					3	
				4				
		9	5	1				8
			8					5
6		1	7		4			
				6				
			3					
5						9	6	4
				7			1	3

Level of difficulty IV

(18)

		7	9					1
		3				4	6	7
					8		2	
4				5				3
					3		7	
	3		4		9	2		
		6			4			
		4	1				8	
2	1							

Level of difficulty IV

(19)

2					7	9	5	1
6			5		2			
8	2	4					6	
	1			6				9
	3						2	
			1					2
	4							7
				8	9		3	

Level of difficulty IV

(20)

		4					9	8
	5			7	4			
	3		9	8			5	
7		2			5		1	
5			7	9	3			
						8		4
			1					
					2		3	

Level of difficulty IV

2. From Sudoku puzzle number 20 create three new Sudoku puzzles by
   (i) interchanging entries of column 4 and 5,
   (ii) shifting entry 3 provided in row 9 of column 8 to row 8 of column 9,
   (iii) shifting entry 1 given in row 8 of column 4 to row 8 of column 5.

   and obtain their solutions in case these exist.

3. Solve Samurai-Sudoku puzzle given in section 6 of this chapter. Also solve the superimposed five Sudoku puzzles seperately. Are the solutions of these separate Sudoku Puzzles same as the solutions obtained in super-imposed Samurai Sudoku.

4. Solve the following two Samurai-Sudoku puzzles:

(1)

3		5				4								2			4	5	7	8
2			3		8	5		7					8	3	1		5		9	
		7	6	5									5		8					
1	3		8	7								6	7	8			2			5
		4	5			2		6									8	6	1	2
6				4		3	7							1		4				
		6	9			7	2	4			8	1	6					9		7
4			1	6			5		6		7	2	4		6	7	1		5	
	8	3			5			9	5			8			2			1		
						3	4						8							
						1			3	9	5	4	7	2						
								2			4									
	7	5	1	9	4	2		6	6	5	1							1		
	9						8							3		8	7		4	
		1		3				7			3	6		8		9	2	5	7	
4	5	3	8		2	7									6		4		5	
					5		2	3					3					2		
		8			1		4					2	6	7	9	5			1	8
			4		6	1		2				3		1		6	8	9		5
	8			1		3	6								3				6	
6	1			5								8				4			3	

(2)

1	7		9			2	4					2		8	4	7		5	3	
4		9	1	2		3	7							5					8	
														6	8	5			4	
		1				7	6					6	9		3					
2			8		1	5											8	2		4
	9	4	2		5	8						5	8		1	4		3		9
9					2	6			2								3			7
3					9	4		8	3	9		7	2	1			4	6	9	3
5		6		8	3	9		7	5		8						2			
							4	6	8		7	9								
								1	6				7							
											1		4	5						
	4			8			6	9			5	2				1		6		9
6			7		9		3		7		9							7		
7	1							5				1		7	6	8			5	
			5	7				6				8			3			2		
8		6	9					7				7	5		9	2		1	6	
1	9			4		3	5						1				7		9	
2			8		7	5								5			6			
4			3									6					8		3	4
9		5	1			8	7	2				3		8	1		2		7	6

Chapter 14
Arithmetic Based Puzzles

1. Introduction

In the previous chapter we considered Sudoku puzzle in which we had to find the missing numbers in cells of a 9 by 9 array. Finding these missing numbers keeping in mind the restrictions imposed in the puzzle was a matter of pure logic. It did not involve any knowledge or use of any basic arithmetic operations of addition, subtraction, multiplication or division. Puzzles being presented in this chapter also require finding missing digits. However, in situations that involve use of one or more basic arithmetic operations. For solving puzzles of this chapter one needs besides logic, the knowledge of basic arithmetic operations also.

We start with simple situations where one is required to find missing digits in an arithmetic operation in which some digits are provided and rest are missing.

2. Finding Missing Digits in Arithmetic Operations

Example 1: Find missing digits in the following addition operation

```
 * * *
 + * *
------
10 * 2
```

Solution: Clearly we have to find one three digit number and one two digit number whose sum is 10*2, where * digit in the result also has to be determined. This can have more than one answer. One of these is

```
  945
+  87
-----
 1032
```

Example 2: Find the missing digits in the following product operation.

```
  * * 2
  × * *
-------
  * * 8
9 * *
-------
* * 6 *
```

Solution: Again this problem can have more than one possible solution. One of these is

$$\begin{array}{r} 132 \\ \times 74 \\ \hline 528 \\ 924 \\ \hline 9768 \end{array}$$

Example 3: Find missing digits in the following division operation.

$$\begin{array}{r|l} & * * 3 \\ \cline{2-2} * * 9 & 6 * 8 * * * \\ & * * * 2 \\ \cline{2-2} & * 9 * * \\ & * * 4 * \\ \cline{2-2} & * * 4 * \\ & * * * * \\ \cline{2-2} \end{array}$$

Solution: According to division rules, 2 in unit place as a result of multiplying * * 9 with a number from 1 to 9 shows that first digit on left of quotient has to be 8. Similarly 9 in unit place of divisor and 3 in unit place of quotient shows that last digit of dividend must be 7 as there is no remainder. Arguing like this step by step we get the following answer.

$$\begin{array}{r|l} & 853 \\ \cline{2-2} 749 & 638897 \\ & 5992 \\ \cline{2-2} & 3969 \\ & 3745 \\ \cline{2-2} & 2247 \\ & 2247 \\ \cline{2-2} \end{array}$$

It is clearly a matter of judicious choice of missing entries, which when filled with appropriate digits, arithmetic operation looks justified. We present in the exercises to this chapter some more such types of problems.

3. Mathdoku Puzzles

A Mathdoku puzzle is essentially adaptation of Sudoku puzzle which involves arithmetic operations. It generally consists of a 5 × 5 square. Each row and each column has five cells. Numbers from 1 to 5 are to be placed in cells in such a way that each row and each column contains digits from 1 to 5. Like Sudoku no number is to be repeated in any row or column. Moreover each bold outlined group of cells contains a hint consisting of a number and one of the mathematical symbol of four basic arithmetic operations of +, –, × and ÷ . Numbers in this group of cells are to be the result of applying the mathematical operation represented by this symbol to the digits contained within the domain. The solution to puzzle is to be arrived at logically and is unique.

Example: Consider the following Mathdoku puzzle.

9+	4÷		10+	
		5÷		
1–		20×		2×
4–	20×	6×		
			1–	

Solution: In this case we have to find three numbers from 1 to 5 in the first three group of cells (one in first row and two in the second row) so that their sum is 9. Similarly, ratio of numbers in the second and third cell of first row is to be 4; sum of 4th and 5th cell in first row and 5th cell in second row is to be 9 and so on. So besides ensuring (as in Sudoku puzzle) that each number from 1 to 5 appears once and once only in each row and each column, one has also to ensure that selected numbers satisfy arithmetic operations specified in each subgroup of cells. The solution to the puzzles is given in the following table.

9+ 3	4÷ 1	4	10+ 2	5
4	2	5÷ 1	5	3
1– 2	3	20× 5	4	2× 1
4– 5	20× 4	6× 3	1	2
1	5	2	1– 3	4

We present in the exercises to this chapter some Mathodoku puzzles. Try to solve them. It is a good pass time and mental exercise particularly for the school going children. Try and have fun. Solutions to these are provided at the end.

3.1 Creating New Mathdoku Puzzle

Like Sudoku puzzles new Mathodoku puzzles can be generated from a given Mathdoku puzzle. One may change the arithmetic operation to be performed in one or more domains keeping in mind the entries. Or redefine domains and arithmetic operations to be performed in these domains or more generally interchange one or more rows and then re-specify arithmetic operations to be performed in each domain keeping in view the new entries.

4. Alpha-numeric Puzzles

In such puzzles, in an arithmetic operation unknown digits are represented by letters of alphabet. A repeated letter of alphabet means the same digit repeating. One is required to find digits which these letters must represent so that arithmetic operation is justified. We illustrate with some examples.

Example 1: Find appropriate digit values of the letters of alphabet used in the following addition problem.

$$\begin{array}{r} \text{CAAA} \\ +\text{AAB} \\ \hline \text{BAAC} \end{array}$$

Solution: Puzzle can have more than are possible answers. One of these is

$$\begin{array}{r} 1999 \\ 992 \\ \hline 2991 \end{array}$$

Example 2: Find digit values of alphabet letters used in the following product operation

$$\begin{array}{r} \text{AB} \\ \times\text{AB} \\ \hline \text{ACC} \end{array}$$

Solution: A possible solution is

$$\begin{array}{r} 12 \\ \times 12 \\ \hline 144 \end{array}$$

Is there an other solution also possible?

Example 3: Find digit values of alphabet letters used in the following product operation

```
  AB4
 × C
-----
 1404
```

Solution: A possible solution is

```
  234
  × 6
-----
 1404
```

Example 4: Find digit values of alphabet letters used in the following division operation.

```
      ABC
    ------
CD | BABD
     A8
    ------
      DB
      DA
    ------
       C4
       CD
    ------
```

Solution: Clearly, D has to be 4 and difference of B and A has to be C.

The answer to this Puzzle is

```
      231
    ------
14 | 3234
     28
    ------
     43
     42
    ------
      14
      14
    ------
```

5. Crypto-arithmetic Puzzles

Originally any mathematical problem involving basic arithmetic operations of addition, subtraction, multiplication or division in which digits are replaced by letters from alphabet was called **cryptarihm problem**. The terminology was coined by J.A.H. Hunter is mid 1950's. Later on they also came to be known as **crypto-arithmetic or alphamatic problems.** Solving an alphametic problem is generally not that easy. At times it becomes so frustrating that the solver throws up his/her hands in desperation. However perseverance and correct use of basic rules of arithmetic generally helps in seeing one through. Once an alphametic problem is cracked (solved) it gives one immense pleasure and joy. Here is an example.

Example:

$$\begin{array}{r} \text{SEND} \\ +\text{MORE} \\ \hline \text{MONEY} \end{array}$$

This is a famous cryptography problem. In July 1924 issue of Standard Magazine, H.E. Dueny, (one of the greatest puzzle devotees) introduced several verbal arithmetic puzzles. This was amongst those puzzles.

The puzzle is not easy to crack. In fact if trial and error approach is adopted it will take in general a very long-time to crack it. However a systematic approach keeping in view basic arithmetic rules helps to a large extent. Let us see how

(i) Since the sum of two four digits numbers is a five digit number, this sum must be less than 20,000. Hence only possible value of M is 1.

(ii) Now since M is 1 so MORE is less than 2000 and as SEND also has to be less than 10,000, so their sum must be less than 12000. Therefore O is either 0 or 1. Since M is already 1, so O is 0.

(iii) Now MORE is less than 1100. If SEND is less than 9000 then MONEY will be less that 10,100 implying N=0 which is not possible as O is already 0. Hence send is at least 9000 and therefore S is 9. We have thus far

$$\begin{array}{r} \text{SEND} \\ +\text{MORE} \\ \hline \text{MONEY} \end{array} \longrightarrow \begin{array}{r} \text{9END} \\ +\text{10RE} \\ \hline \text{10NEY} \end{array}$$

(iv) Now unit place contains two distinct digits D and E which are to have values from remaining digits (2, 3, 4, 5, 6, 7 and 8). So if D + E is less than 10, then D + E = Y. However, if D + E is more than 10 then D + E = 10 + Y.

(v) Consider first the possibility that D + E = Y. Applying same logic N + R = E or N + R = 10 + E. Again E + O = N and as E and N are distinct N + R ≠ E but N + R = 10 + E with one carry over. Thus N = E + 1. Substituting N = E + 1, N = R + 10 + E yields R = 9. However, 9 has already appeared. Thus D + E ≠ Y but D + E = 10 + Y.

Hence 10's column now is 1 + N + R = E which is at least 6 and at most 16. If it is 6, then there is no carry over and a contradictory conclusion is reached in hundredth place. So E + 1 = N and 1 + N + R = E + 10. Therefore, 1 + E + 1 + R = E + 10 yielding R = 8 and E + 1 = N and D + E = Y + 10. Moreover these digits are to be between 2 and 7. As D + E is at most 13, so Y is 3 or 2. If Y = 3, then D + E is 13 so D is either 6 and E is 7 or vice versa. However, neither of these alternatives is acceptable as former implies N = 8 (and R is already 8) while later implies N = 7 and D is already 7. So Y ≠ 3. Hence Y = 2.

(vi) Thus D + E = 12 with available digits 5 and 7. If E = 7 then N = 8 which is not possible as R is assigned 8. Also E = 5 implies N = 7. But 7 is assigned to D. So Y $\neq$ 3. Hence Y = 2 and D + E = 2 or 12. Now D + E cannot be 2 so D + E = 12 which available digits 5 and 7 can produce. But for E = 7, N = 8 which is not possible So E = 5 and hence D = 7.

Thus finally we have

$$\begin{array}{r} \text{SEND} \\ +\text{MORE} \\ \hline \text{MONEY} \end{array} \longrightarrow \begin{array}{r} 9567 \\ +1085 \\ \hline 10652 \end{array}$$

5.1 Some Useful Hints

A few useful hints which can help in cracking such crypto-arithmetic problems are:

(i) If the summand or sum is known to be divisible by an even number, then letter in units column of that number cannot be an odd digit 1, 3, 5, 7 or 9.

(ii) If the summand or sum is a perfect square, then the letter in its unit column cannot be 2, 3, 7 or 8.

(iii) If the summand or sum is known to be a prime number, then the letter in its unit column can only be 1, 3, 7 or 9.

(iv) If the summand or sum is known to be divisible by 5, then the letter in unit place will be either 0 or 5.

(v) If the summand or sum is divisible by 10, then the letter in unit place has to be 0.

(vi) Always keep in mind the possibility of a carry over.

(vii) Whenever possible try to determine constraints on the magnitude of digits which can represent the letter.

(viii) If same alphabet letter appears in the same column of the summand and the sum, then remaining entries in that column plus carry over from previous column (if any) must add to a number whose first digit is zero.

Although majority of crypto-arithmetic puzzles are based on addition operation some involve other arithmetic operations also. The strategy to be adopted in cracking such puzzles is also similar. We present some crypto-arithmetic puzzles in the exercises of this chapter. Try to solve them, it will be a fun.

6. Alpha-numeric Equations

Just like algebraic equations, we also have what are known as **alphabetic** or **alpha-numeric equations**. These are essentially crypto-arithmetic puzzles where instead of writing unknowns in the format of normal addition, these are expressed in form of an equation whose right hand side is the result of addition of terms on the left.

For example one could write

$$\begin{array}{r} \text{SEND} \\ +\text{MORE} \\ \hline \text{MONEY} \end{array}$$

in the form of an equation as

$$\text{SEND} + \text{MORE} = \text{MONEY}$$

Example: Consider the alphabetic equation.

$$\text{NUDE} + \text{NOT} + \text{NOR} = \text{C} \times 10^4$$

Keeping in mind that each distinct alphabet letter has to be some distinct number from 0 to 9 and the sum on left has to be a multiple of 10^4. It is noticed that for N = 8, U = 3, D = 5, E = 0, O = 2, T = 4, R = 6

Equation yields, $8350 + 824 + 826 = 10000 = \text{C} \times 10^4$

So $\quad \text{C} = 1$

So the solution is N = 8, U = 3, D = 5, E = 0, O = 2, T = 4, R = 6 and C = 1.

Mind Teasers

1. Find missing digits in the following arithmetic operations.

(i)
$$\begin{array}{r} * * * \\ + * * * \\ \hline 1 * * 1 \end{array}$$

(ii)
$$\begin{array}{r} * 2 * \\ - * * * \\ \hline * 1 \end{array}$$

(iii)
$$\begin{array}{r} * * * \\ \times * 2 \\ \hline * 8 * \\ * * * \\ \hline 4 * * 0 \end{array}$$

(iv)
$$\begin{array}{r} * 3 * \\ \times * 4 \\ \hline * * * \\ * * * \\ \hline 9 * 6 8 \end{array}$$

(v)
$$\begin{array}{r} 7 * * \overline{) * * 8 8 9 7} \\ * * * * \\ \hline 3 * * 9 \\ * 7 * * \\ \hline * * * * \\ * * * * \\ \hline \end{array}$$

2. Solve the Following Mathdoku Puzzles

(i)

2×		7+		75×
40×	3÷	8×		
			2÷	
	60×	15×		4÷
		2×		

(ii)

2÷	24×			5÷
	60×	14+		
1–			3+	
		2×	5+	1
6+				

(iii)

12×	8×		3–	40×
	8×			
		2–	4+	
1–				1–
7+		4÷		

(iv)

12×		3–		7+
3+	8+	60×		
				2÷
3–	4–		36×	
	2÷			

(v)

8+		80×		5×
	3+			
10+	10×		4+	
	1–	5+		6+
		3÷		

(vi)

1–		3+		25×
11+		24×		
	3×			5+
8+		4–	48×	

(vii)

7+	20×		15×	8×
	5+			
5÷		6+		
3–	3×		8×	2–

(viii)

6+	8+	3÷	1–	
			3+	9+
3×	1–			
		3–		
11+			3÷	

(ix)

12+	2–		11+	2×
		5×		2–
2–	3+		48×	
	2×			

(x)

4+	2÷		45×	20×
12×		5×		2÷
2–	9+		24×	
	2×			

3. Find appropriate positive integer values between 0 and 9 for the alphabet letters appearing in the following alpha-numeric arithmetic operations (a repeated letter means same numerical digit).

(i)
```
 ABB
+CBB
----
 DDD
```

(ii)
```
  A6B
 +5C7
 ----
 D332
```

(iii)
```
  AB5
 +8CB
 ----
 CC59
```

(iv)
```
 AB
-BA
---
 27
```

(v)
```
 ABCD
-BADC
-----
  855
```

(vi)
```
 AB
 ×A
---
8B8
```

(vii)
```
 AB
×BA
---
ACA
```

(viii)
```
 ABCD
  × 9
-----
 DCEA
```

(ix)
```
     A
    AB
   ABC
 +ABCD
 -----
  2600
```

(x)
```
   ABC
  ×BAC
  ----
    **C
   **A
  **B
  -----
  **C*C
```

(xi)
```
        *D
ABC ) C*AAD
      ****
      ----
       CB**
       ****
       ----
```

4. Solve the following (a bit more complex) crypto-arithmetic puzzles involving multiplication and division operations.

(i)
```
   ABC
 × BAC
 -----
   **C
  **A
 **B
 -----
 C**DC
```

(ii)
```
   CAN
 × CAN
 -----
  CCDC
  ***
 **N
 -----
 CDD*G
```

(iii)
```
    CAN
  × CAN
  -----
    ***
   N**N
  **C*
  ------
  N**CCC
```

(iv)
```
       ATOM
     × ATOM
   --------
      *****
     *****
    *****
   *****
   --------
   MOT***OM
```

(v)
```
    ERROR
  ×    OR
  -------
   ****O*
  **OO*
  -------
  MISTAKE
```

(vi)
```
        THE
SHE ) ***ES
      SHE
      ----
      ****
      *E**
      ----
       ***
       ***
       ---
```

(vii)
```
        RAT
RAT ) ROBBOT
      *AAR
      ----
      B*A*
      **AA
      ----
       ***
       ***
       ---
```

5. Solve the following alphabetic puzzles involving addition operation.

(i)
```
  ONE
 +ONE
 ----
  TWO
```

(ii)
```
  ONE
+FOUR
-----
 FIVE
```

(iii)
```
 SEVEN
+EIGHT
------
TWELVE
```

(iv)
```
  WRONG
 +WRONG
 ------
  RIGHT
```

(v)
```
  LETS
 +WAVE
 -----
 LATER
```

(vi)
```
 TERRIBLE
+NUMBER
---------
THIRTEEN
```

(vii)
$$\begin{array}{r} \text{TWELVE}\\ \text{ELEVEN}\\ \text{SIXTY}\\ +\text{SEVEN}\\ \hline \text{NINETY} \end{array}$$

(viii)
$$\begin{array}{r} \text{FIFTEEN}\\ \text{FIFTEEN}\\ +\text{FORTY}\\ \hline \text{SEVENTY} \end{array}$$

(ix)
$$\begin{array}{r} \text{FOURTEEN}\\ \text{SEVEN}\\ \text{TEN}\\ +\text{TEN}\\ \hline \text{FORTYONE} \end{array}$$

(x)
$$\begin{array}{r} \text{TOPSY}\\ +\text{TURVY}\\ \hline \text{YEARS} \end{array}$$

(xi)
$$\begin{array}{r} \text{ASK}\\ \text{THE}\\ +\text{RIGHT}\\ \hline \text{THING} \end{array}$$

6. Solve the following alpha-numeric equations.

(i) AYE + AYE + AYE + AYE = YES + YES + YES

(ii) FIFTEEN + FIFTEEN + SEVEN + SEVEN + SEVEN + TEN = SIXTYONE

(iii) Solve the following alphabetic simultaneous equations

AB + CD = EG

GF + HO = CF

CD + HO = AOD

Chapter 15
Chessboard Related Problems

1. Introduction

Chessboard provides a field of play for a number of games both ancient and modern like chess and its variants such as checkers and droughts. These have been source of recreation to humanity since times immemorial. In fact the game of snakes and ladders and even the word game scrabble are its variants. These games are played on boards. These boards come in different sizes, 8×8 for chess, 8×8 and 10×10 for checkers, 10×10 for snakes and ladders and 15×15 for scrabble. In fact games such as Bau and Owari, which are also widely played in various forms under different names use non-square boards such as 4×8 and 2×6. In some board games no special colours are given to the individual squares. However in some games colour of an individual square is important. The familiar black and white (or black and yellow or black and red) colours alternate in squares of games of chess and checkers.

Board games are in a sense a metaphor for life itself. Chess is often associated with war between opposing forces, having king, queen, knights, camels, rooks, etc. The objective is to conquer the enemy by trapping its king but at the same time protecting one's own king. Ladders and snakes game is sometimes associated with an individual's sole's effort to achieve salvation (nirvana) while facing in life from time to time ups (ladders) and downfalls (snakes). Omarkhayam, the well known Persian poet, saw in such games reflection of our lives as mere pawns in a game run by 'The Master of the Show'. Shakespeare in his well known play, 'As You Like It' also remarks: 'All the world is a stage and all the men and women mere players'.

However this chapter is not about games such as chess and checkers which are played on boards, but about the game board itself and the mathematics which arises from such an apparently simple structure. We start the chapter with a well-known puzzle commonly known as 'Guarini's problem'.

2. Guarini's Problem

This chess board based puzzle dates back to 1512 A.D. It involves four knights, two white and two black which are placed at the four corners of a 3 × 3 chessboard as shown in Fig. 2.1 (W denoting white and B black). It is desired to interchange the positions of white and black knights using only knight's permissible moves.

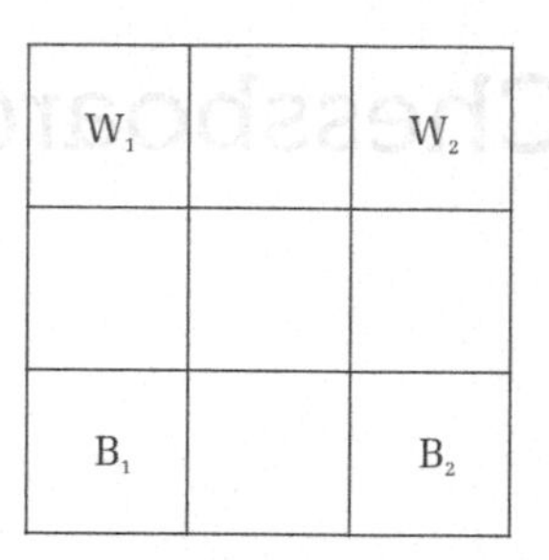

Fig. 2.1

A knight can move on a chessboard by first moving two squares in horizontal (right or left) or vertical direction (up or down) and then moving one square to right or left (if original move was in vertical direction) and one square up or down (if original move was in a horizontal direction).

This seems to be a simple puzzle which may be solved by trial and error (try). However it is a bit harder than what it looks at the first sight. Solution to this puzzle can be obtained in a more systematic manner by observing its underlying basic structure. In Fig. 2.2, the possible execution of knight's moves is exhibited explicitly. In Fig. 2.3, the structure of possible moves is exhibited by drawing lines that connect any two squares of the board between which knight can move.

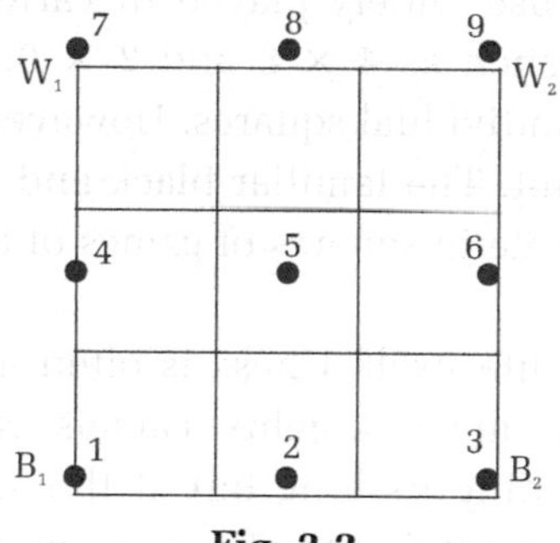

Fig. 2.2

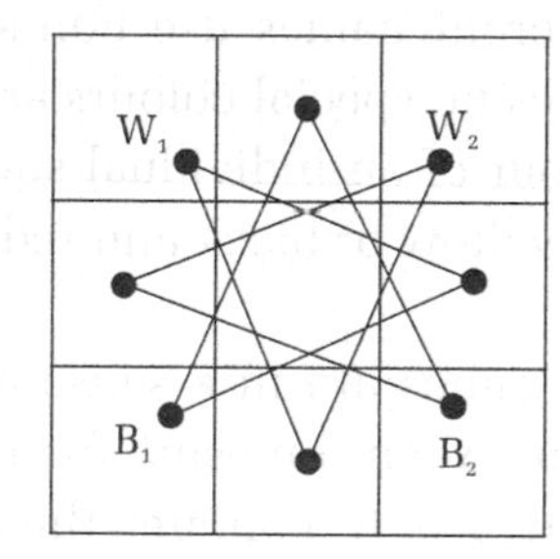

Fig. 2.3

The solution to the puzzle is now clear. Knights can march around all in one direction. One possible solution is:

$$B_1 \to 6 \to 7, B_2 \to 8 \to 1 \to 6 \to 7 \to 2 \to 9$$
$$W_1 \to 6 \to 1, W_2 \to 4 \to 3$$

This graphical solution can now be translated back to original board.

This is a nice illustration of the way in which mathematical abstraction sometimes helps in solving intricate problems. In Fig. 2.3 we are not bothered about original chessboard or how the knights move. The general process of transforming a realistic problem into an abstract diagram and then analyzing its behaviour is what is now known as '**Graph Theory**' in mathematics. This subject is now quite developed and extensively used in solution of different types of realistic problems arising in subjects such as computer science.

3. Knight's Tour Problem

In a knight's tour problem, one has to design a tour for a knight on a chessboard so that it visits each square of the chessboard once and only once (i.e., exactly once). The tour is called a '**closed tour**' if at the end of the tour, the knight returns to its original position from where it started. In case the tour finishes at a square which is different from the one from where it started, then it is called an open tour. Unless specified otherwise, a tour generally means a closed tour.

The knight's tour problem is very old. In fact it dates back to the beginning of chess in 6th century in India. It has a long and rich history. The smallest chess boards on which a knight's tour is possible are of sizes 5 × 6 and 3 × 10. The smallest chess board on which an open tour is possible is of size 3 × 4. These tours are shown in Figs. 3.1, 3.2 and 3.3 respectively. Numbers in squares indicate the sequence of moves.

5	26	1	16	11	20
2	15	4	19	30	17
25	6	27	12	21	10
14	3	8	23	18	29
7	24	13	28	9	22

Fig. 3.1: Knight's tour on 5 × 6 board

26	29	2	21	8	23	6	17	14	11
1	20	27	24	3	18	9	12	5	16
28	25	30	19	22	7	4	15	10	13

Fig. 3.2: Knight's tour on 3 × 10 board

1	4	7	10
12	9	2	5
3	6	11	8

Fig. 3.3: Knight's open tour on 3 × 4 board

Eminent mathematician Euler has done extensive work on knight's tour problem. One particularly attractive knight's tour on an 8 × 8 chessboard is presented in Fig. 3.4. What is specially interesting about this tour is that Euler first makes an open tour of lower half of the board starting from location 1 and ending at location 32. He then repeats exactly the same tour in a systematic fashion in the upper half of the board in such a way that these two solutions can be joined together to yield a tour of the entire chessboard.

We have seen that 5 × 6, 3 × 10 and 8 × 8 chessboards have knight's tours. A question that arises is that does every size chessboard has a knight's tour if not closed at least open. The answer is no. We have seen in section 2 in Gurani's problem that in the case of 3 × 3 chessboard, centre can not be reached from any other square by knight's moves. A more interesting and perhaps much more surprising instance is 4 × 4 board. This is shown in Fig. 3.5. The reason for knight's tour not being possible in this case is that there are closed loops amongst squares 1, 7, 16 and 10 and squares 4, 6, 13 and 11. So these can not form a part of a larger tour of the entire board.

.	62			35			
.						34	
63							
.		64					33
1	26	15	20	7	32	13	22
16	19	8	25	14	21	6	31
27	2	17	10	29	4	23	12
18	9	28	3	24	11	30	5

Fig. 3.4: Knight's tour on a 8 × 8 board

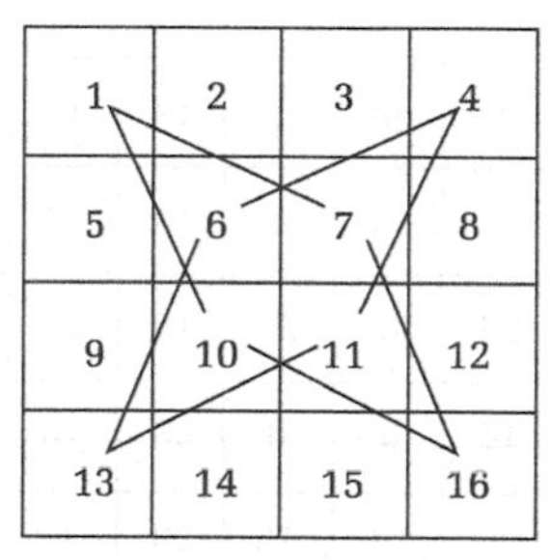

Fig. 3.5: Why no Knight's tour is possible on 4 × 4 board

The question that are there chessboards of other sizes which also do not have knight's tour will be answered later after we have discussed colouring of chessboard in the next section.

4. Colouring of Chessboard

Thus far we have not bothered about the colour of the squares of a chessboard assuming all the squares to be identical. However actual chessboards have squares of alternating colours (usually black and white. Omarkhayam regarded these as representing night and day).

Consider a 5 × 5 chess board as shown in Fig. 4 in which colours of squares alternate.

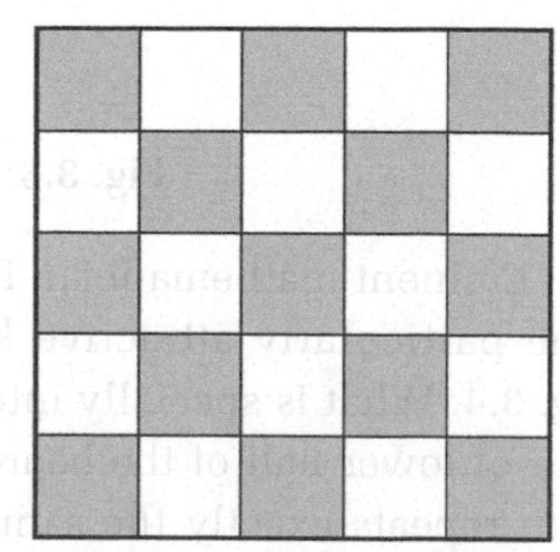

Fig. 4: Open knight's tour on 5 × 5 chessboard

In such a chessboard knight's tour alternates between squares of one colour and the other. Thus for a complete knight's tour, the number of white and black squares must be equal. However a 5×5 chessboard shown in Fig. 4 does not have same number of squares of each colour. There are 13 black and 12 white squares. So a knight's tour is not possible on 5×5 chessboard.

The same argument holds for any chessboard of odd size such as 13×7 whose both dimensions are odd.

5. Open Knight Tour

It is possible that on a chessboard a closed knight's tour may not be possible but an open knight's tour may be possible. For example we have just seen that in the case of a 5×5 chessboard a closed knight's tour is not possible. However an open knight's tour may be possible and indeed that is so. An open knight's tour in case of a 5×5 chess board is shown in Fig. 5. As usual numbers in squares indicate the move number, in which that square gets visited.

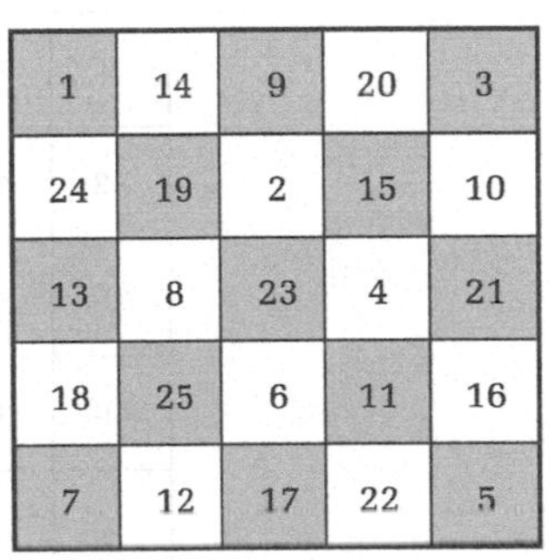

Fig. 5: Open knight's tour on 5×5 chessboard

5.1 Impossible Knight Tours

We have seen that a knight's tour is impossible on a 4×4 chessboard (even though its both dimensions are even). It is also not possible on a chessboard whose both dimensions are odd. In fact knight's tour is also not possible on a number of other boards. For instance if board has only one or two rows or columns. With one row (or column) knight's move is not at all possible. With two rows it can move only in one direction and thus gets stuck at the far end of the board. In case of boards with three rows the smallest board on which knight's tour is possible is of dimensions 3×10. Thus it is not possible on boards of sizes 3×4, 3×6, 3×8 even though one of their dimensions is even.

6. Generating a Knight's Tour

Generating a knight's tour on a chessboard of specified size is not that easy. As the tour nears the end, the available options fast dwindle and often it becomes clear that knight can not continue to move further. There is a genuine feeling of pleasure and exhilaration when a knight's tour is finally discovered.

Efforts have been made to discover some systematic way of finding a knight's tour on a given chessboard even when knowing that such a tour exists. Amongst the earliest known solutions to the knight's tour problem is an open tour of 8×8 chessboard by

De Movire. This open tour is shown in Fig. 6, numbers in the square indicating the move number. This also indicates a technique for generating a knight's tour. This technique is simple but usually effective.

34	49	22	11	36	39	24	1
21	10	35	50	23	12	37	40
48	33	62	57	38	25	2	13
9	20	51	54	63	60	41	26
32	47	58	61	56	53	14	3
19	8	55	52	59	64	27	42
46	31	6	17	44	29	4	15
7	18	45	30	5	16	43	28

Fig. 6: De Movire's open tour of 8×8 chessboard

The technique is to start from the outer ring of the squares and cycle around the outside of the board coming inside only when it becomes absolutely necessary. In the above tour first 24 moves are in outer two rows. At 25th move we venture inside and then immediately return to outer ring at 26 and continue.

It may be noted that in De Movire's tour shown in Fig. 6, a closed tour became impossible as soon as we landed in square numbered 12 because once squares 2 and 12 have been visited there is no way to return to square 1. So for finding a closed tour care should be taken of corners and their connecting squares.

Exactly one hundred years after Euler published his contribution to knight's tour, an Irish mathematician Hamilton marketed in 1859 a board game called 'Icosian Game'. In this game the object was to complete a tour of the vertices of the graph of dodecahedron having been given first five vertices to be visited. Because of this game, any closed tour in a graph that visits each vertex of the graph only once is called a 'Hamiltonian Cycle'. So in mathematical terminology what we are looking for in a knight's tour of a chessboard is to find a Hamiltonian cycle in the associated graph.

7. Tours with Chess Pieces other than Knight

Thus far we have considered only knight's tours on a chessboard. However there are other pieces of chess such as rook, bishop, queen and king. One can also consider the possibility of tours by these pieces. We present in this section examples of tours by rook, bishop, queen and king.

7.1 Rook's Tour

A rook can move in a horizontal or a vertical direction and it can move along these directions as many squares as it likes. Hence finding a rook's tour on an 8 × 8 chessboard is not that difficult. One possible rook's tour of 8 × 8 chessboard is shown in Figure 7.1. This is a closed tour.

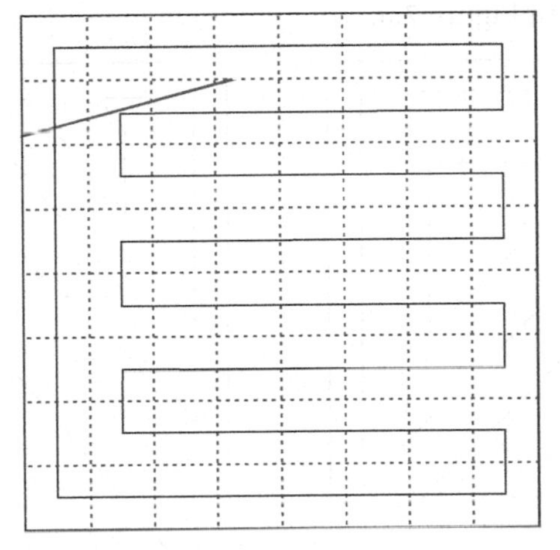

Fig. 7.1: Rook's tour

Alternative tours are also possible.

7.2 Bishop's Tour

A bishop can move along a diagonal as far a it likes. If a bishop passes through a square once it can pass through the same square again later while moving along another diagonal. Alternately it enters a square along one diagonal, turns and leaves along another diagonal. In such a case it need not visit the same square again Fig. 7.2.

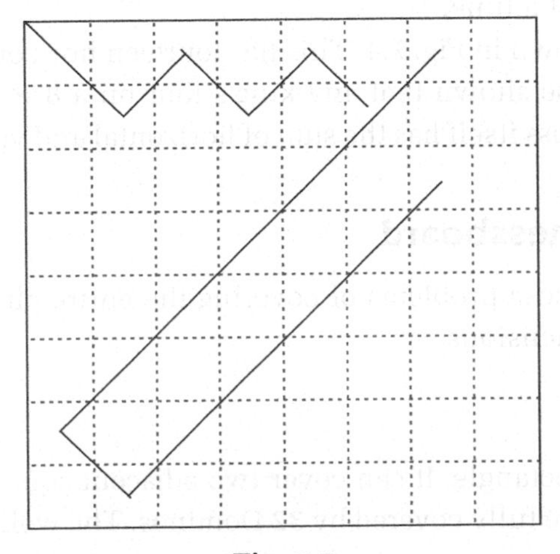

Fig. 7.2

7.3 Queen's Tour

The queen is by far the most powerful piece on a chessboard. It combines in its moves the strength of a rook (which can move any distance it likes horizontally or vertically) along with that of a bishop (which can move any distance that it likes along a diagonal). So it is possible to determine several alternative tours of a queen on 8×8 chessboard. One of these is shown in Figure 7.3.

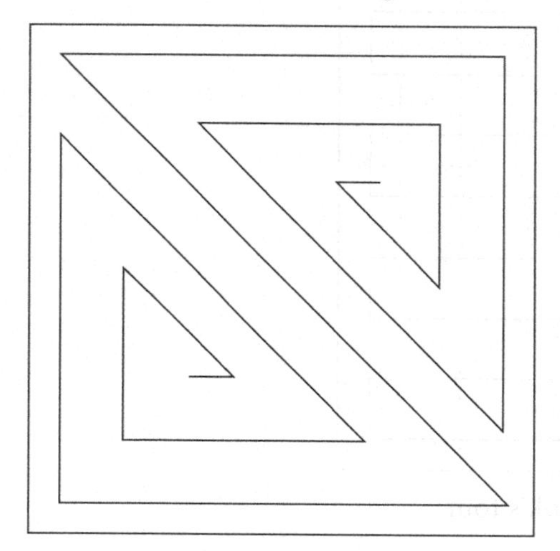

Fig. 7.3: A queen's tour

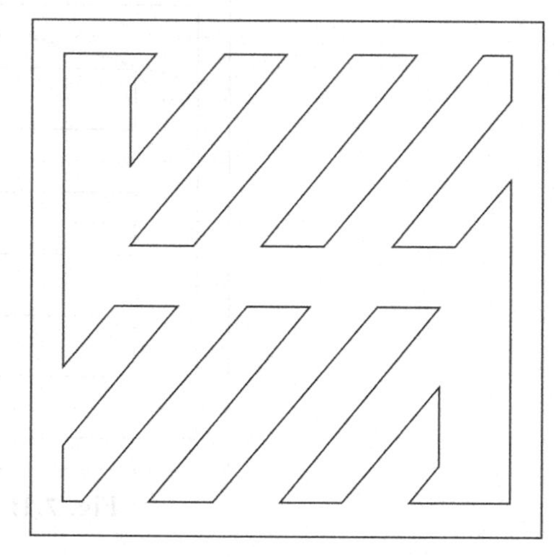

Fig. 7.4: A king's tour with fourteen horizontal moves and fourteen vertical moves

7.4 King's Tour

A king can move in a horizontal or vertical direction or along a diagonal as it likes but only one step at a time. Thus movements of king are similar to those of a queen but restricted to one step at a time.

A king's tour is shown in Fig. 7.4. This has fourteen horizontal and sixteen vertical moves. In fact it can be shown that any king's tour on a 8×8 chessboard in which king's path does not cross itself has the sum of horizontal and vertical moves at least 28.

8. Covering a Chessboard

In this section we discuss problems of covering the entire chessboard with identical pieces of specified dimensions.

8.1 Domino's Puzzle

A domino is a 1×2 rectangle. It can cover two adjacent squares of a chessboard. An 8×8 chessboard can be fully covered by 32 Dominos. The well known Domino puzzle is as follows.

Suppose we remove two diametrically opposite corner squares (both white or both black) of an 8×8 chessboard. Can the remaining 62 squares be covered with 31 domains? At first sight it appears that it should be possible. But it is not (you can try). This conclusion is based on the analogy of this problem with a knight's tour on 8×8 coloured chessboard whose opposite corners are removed. Fig. 8.1

Fig. 8.1

As the removed squares are of the same colour (say white) so we are left with 30 white and 32 black squares. Clearly since a domino covers one white and one black square it is impossible to cover 62 squares of which 30 are white and 32 black with 31 or any other number of dominos.

What makes this domino puzzle exciting is that people usually try to analyse the problem geometrically and feel it should be possible to cover the truncated chessboard as number of squares is still even. On the contrary if same side corners (one white and one black) are removed covering will be possible. It will also be possible even if all the four corners are removed.

Gomory has established this fact in a more general form in the form of a theorem known after his name.

Gomory's Theorem: If we remove one black and one white square from anywhere on an 8×8 chessboard, then it is possible to cover the remaining board with dominos.

8.2 Domino Covering Games

There are several games which can be played based on domino covering of a chessboard. For example two players take turns in placing a domino on a chessboard. The player who can not place a domino on the chessboard loses. However for the game to be of interest, the chessboard on which this game be played should be of odd by odd size. In case of even by even chessboard, the second player can always win by playing symmetrical to the moves of the first player. In case of an odd by even chessboard, the first player can always win by playing the first move at the very centre of the board.

8.3 Covering with Trominos

As an extension to dominos suppose we want to cover a chessboard with Trominos (1×3) in place of dominos (1×2). Trominos can also be L shaped two squares in one direction and the third in perpendicular direction. However we shall consider here only straight trominos in which all the three squares are in a straight line. Since the number of squares in an 8×8 chessboard is 64, it would not be possible to cover it with trominos. However if we remove one square from anywhere, it should be possible as then we are left with 63 squares which can be covered with 21 trominos.

Using standard labeling as in coordinate geometry (1, 1) for the left hand bottom square and (8×8) for the top right hand square, no matter wherever a (1×3) tromino is placed, one coordinate remains constant for the three covered squares and the other coordinates are three consecutive integers. Now if one square is removed from the board, then the remaining board can be covered by trominos only if removed square has coordinates each of which is divisible by 3. Thus only possible missing squares are (3, 3), (3, 6), (6, 3) and (6, 6). If any one of these is removed, it will be possible to cover the remaining board with trominos otherwise not (Try).

8.4 Tetraminos

Tetramino is a figure consisting of four squares. It can have several shapes. Five of these are:

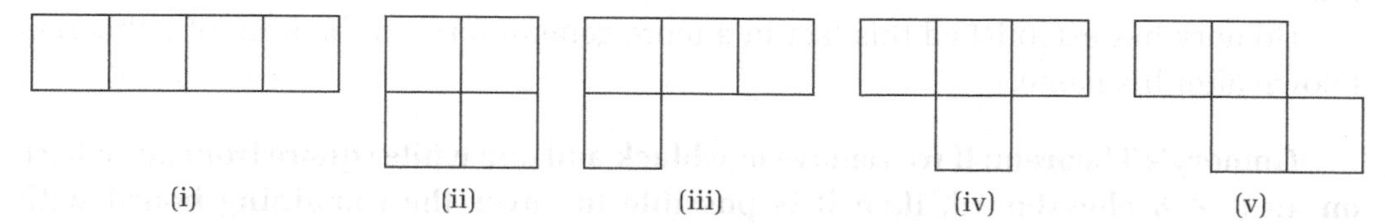

Fig. 8.2

(iii) and (iv) look different when turned over.

As regards covering of 8 × 8 chessboard with tetraminos, if we consider tetraminos of type (i) (i.e., 1 × 4) we can easily cover the chessboard with 16 of these. Similarly sixteen 2 × 2 square tetraminos of type (ii) can also cover the chessboard. What about other three types of tetraminos? Two possible ways of covering an 8 × 8 chessboard with L type tetraminos of type (iii) and T type tetraminos of type (iv) are shown in Figures 8.3 and 8.4 respectively.

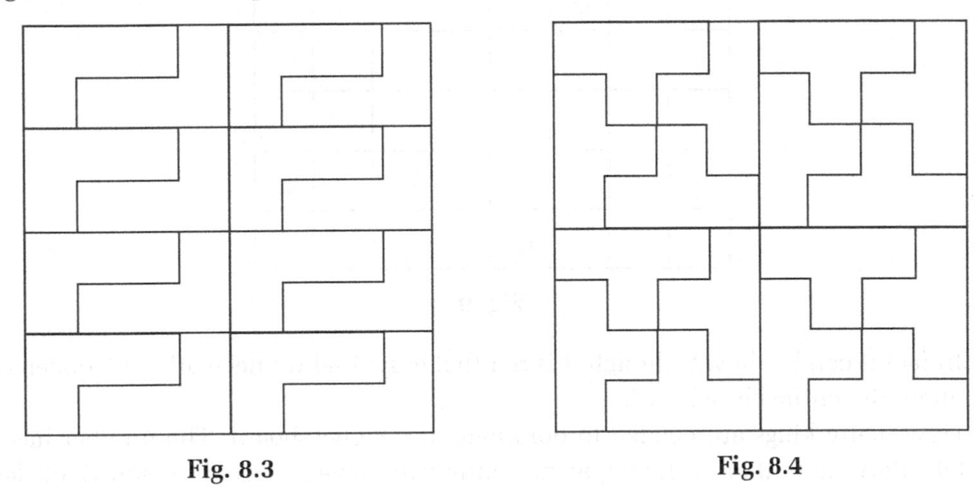

Fig. 8.3 **Fig. 8.4**

However Z type tetraminos (type (v)) can not cover an 8 × 8 chessboard (Try).

9. Dominance Problem

In dominance problem we are interested in determining the least number of pieces of the same type (all kings, all queens, etc.) so that through their permissible moves, they can cover the entire chessboard. Such domination problems arose in chess in analogy with earlier time war game plans. However these are now finding practical use in a variety of fields through graph theory.

As mentioned earlier, queen is by far the most powerful piece in chess. It combines the strength of a rook (which can move any distance it likes in a horizontal or vertical direction) with that of a bishop (which can move any distance it likes diagonally). Depending upon which square it is placed, a queen can dominate (or cover) from 22 to 28 squares of the chessboard.

A question that naturally arises is that what is the least number of queens and their locations so that they dominate the entire chessboard (i.e., one of such queens can move to reach any square of the chessboard).

For example four queens placed at squares indicated in Fig. 9 dominate the entire chessboard except two shaded squares. Their locations are indicated by writing Q in squares of Fig. 9. As only two squares are left uncovered, these can be covered by suitably positioning fifth queen.

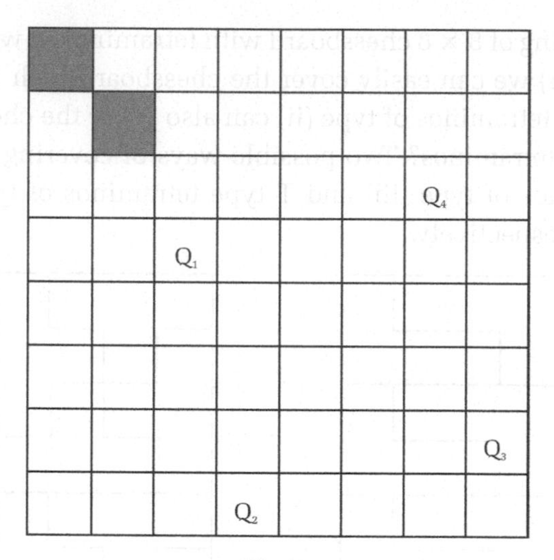

Fig. 9

In fact it can be shown (though it is not that easy) that we need atleast 5 queens to dominate the entire chessboard.

How many kings are needed to dominate 8 × 8 chessboard. The number has to be definitely more than 5. Like queen, a king can move in any direction right, left, up, down or diagonalwise but only one square at a time. Thus a king can cover at the most 3 × 3 = 9 squares (including its own position) in its immediate neighbourhood. In fact at least 9 kings are needed to cover the entire board (Try to find their locations).

A rook is regarded as the weakest of all chesspieces. It can move any distance right or left, up or down in the row or column in which it is located. So a rook in fact controls or dominates as many as 16 squares including itself (much more than the squares a king controls). But still we shall need 8 rooks to control the entire chessboard (find their positions).

10. Independence

Concept of independence is somewhat opposite to that of dominance. We may want to know as to what is the maximum number of pieces of the same type which can be placed on a chessboard so that no two of these can attack each other. We call a group of pieces independent if none of these can attack the other in just one move.

Let us see what is the maximum number of knights that can be independent of each other. We know that in one move, a knight on a black square can move only to white square. So if we place even 32 knights on all black (or all white squares) they will not be able to attack each other (Try for other types of pieces).

11. Dissection Problem

There is a long history of geometric dissection problems in recreational mathematics. We discuss here one of these related to chessboard. An 8×8 chessboard has got broken into eight pieces as shown below. The problem is to put these pieces together to reconstruct the chessboard. The reconstructed chessboard is as shown in Fig. 10.1.

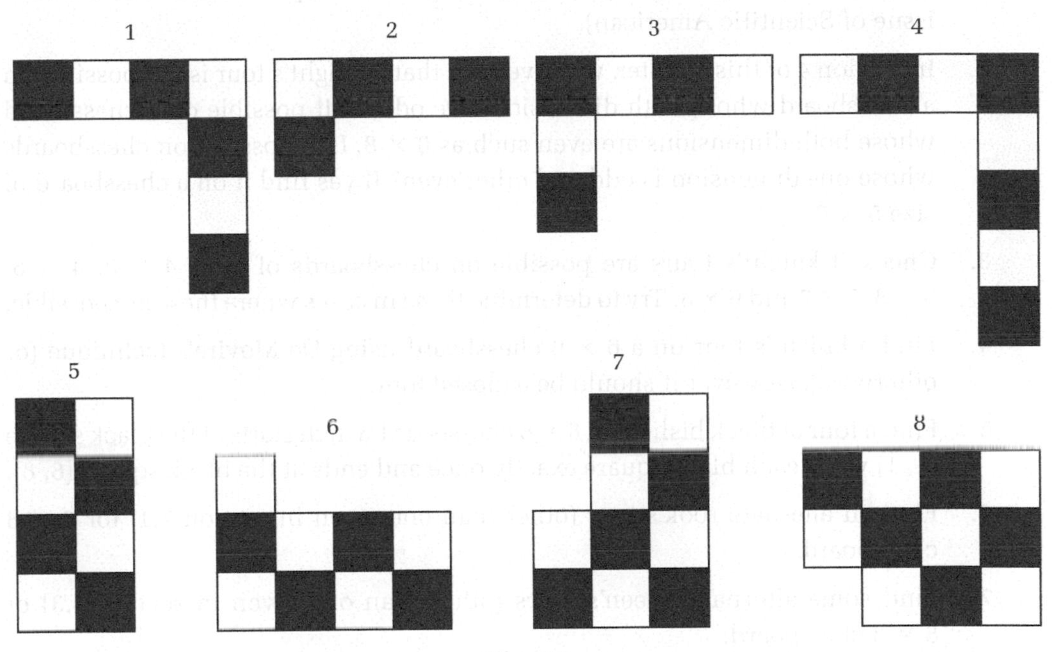

Fig. 10

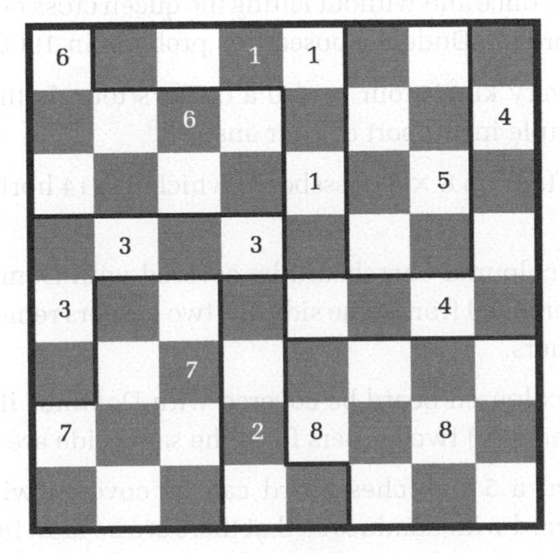

Fig. 10.1

Mind Teasers

1. In a little variation of Gurani's problem of section 2, suppose there is a 4×3 chessboard and 6 knights, three white and three black, placed on opposite side of the board. Is it possible to interchange the position of white and black knights using only knight's moves? (This problem appeared in December 1979 issue of Scientific American).
2. In section 4 of this chapter, we have seen that a knight's tour is not possible on a chessboard whose both dimensions are odd. Is it possible on a chessboard whose both dimensions are even such as 8×8. Is it possible on chessboards whose one dimension is odd and other even? If yes find it on a chessboard of size 5×6.
3. Check if knight's tours are possible on chessboards of sizes 4×8, 4×5, 3×8, 3×7 and 6×8. Try to determine these in cases where these are possible.
4. Find a knight's tour on a 6×6 chessboard using De Movire's technique (or otherwise). However it should be a closed tour.
5. Find a tour of black bishop on 8×8 chessboard which starts at the black square (1, 1) visits each black square exactly once and ends at the black square (8, 8).
6. Find an alternate rook's tour (other than one given in section 7.1) for 8×8 chessboard.
7. Find some alternate queen's tours (other than one given in section 7.3) of 8×8 chessboard.
8. Move a queen from square (3, 3) to square (6, 6) in 15 moves, visiting every square exactly once and without letting the queen cross her path (famous British puzzle master H.E. Dudeney posed this problem in 1906).
9. Show that every king's tour is also a queen's tour. Is the converse also true? Give an example in support of your answer.
10. Find a king's tour on 8×8 chessboard which has 14 horizontal and 14 vertical moves.
11. Can a 6×6 coloured chessboard be covered with Domino's? What if (i) two corners are removed from same side, (ii) two corners removed are diametrically opposite corners.
12. Can a 5×6 coloured board be covered with Dominos if (i) two diametrically opposite corners (ii) two corners from the same side are removed?
13. We know that a 5×6 chessboard can be covered with dominos. Cover a 5×6 chessboard with dominos so that there are no fault lines (i.e., no separating straight lines that go all the way across length or breadth of the board).

14. Can a 6×6 chessboard be covered with dominos so that there are no fault lines? What about 4×4 and 8×8 chessboards?

15. Which types of rectangular boards in general can be covered with dominos? What if there is a further restriction that in covering there be no fault lines?

16. Verify that it is possible to cover a 4×5 chessboard using each of the five types of tetraminoes shown in Fig. 8.2. What about a 2×10 board? (You can place a tetraminoes with either side facing up).

17. In chess knights and bishops are considered more powerful than rooks. Will the number of (i) knights, (ii) bishops needed to cover the board be more or less than rooks needed for this purpose?

18. Place 8 bishops on an 8×8 chessboard so that they dominate the entire chessboard.

19. Place 14 bishops on an 8×8 chessboard so that these are independent.

20. There is a square tract of land which has four oak trees growing in it. The trees are at equal distances from each other in a straight line starting from the middle of an edge moving parallel to the other edge going straight upto the centre of the field. This piece of land is to be equally divided amongst four siblings so that each gets one fourth of the land as well as one tree. Is this division possible? If yes how?

21. Can we construct a $6 \times 6 \times 6$ cube using only bricks of dimensions $1 \times 2 \times 3$? (This is known as bricklayer's problem).

Chapter 16
Magic Squares and Latin Squares

1. Introduction

In this chapter we discuss problems related to Magic Squares, Latin Squares and Eulerian squares. Such square based games and puzzles have been source of interest and curiosity since times immemorial. Besides providing entarement and excitement, these have also yielded results which have wider applications. We start with magic squares.

2. Magic Squares

Magic squares have been known to humanity since distant past. In fact magic squares are supposed to have been invented soon after natural numbers were invented and simple laws of basic arithmetic operations of addition, subtraction, multiplication and division framed and appreciated. Because of their peculiar special properties ordinary people started believing that these squares have magical properties and that is perhaps why the name 'Magic Squares'. Later on when mathematicians could understand the reasons behind these properties, the magic disappeared. However the name still contains the word magic. In fact they provide interesting and exciting mathematics based recreation.

A magic square consists of a set of natural numbers arranged in a specific order in the form of a square such that these numbers exhibit some specific properties. A 3×3 magic square formed by first nine numbers 1, 2, 3, 4, 5, 6, 7, 8 and 9 is

4	9	2
3	5	7
8	1	6

Fig. 1: A 3×3 magic square

This magic square has the property that sum of entries in each column, each row and both the diagonals is same. It is 15. This 15 is called its magic number. This magic square is very famous and also very old. There is even a legend associated with it. The legend is that over 4000 years ago a turtle emerged from Yellow river in China and it had this magic square inscribed on its shell. So this magic square is also called Lo-Shu magic square.

For all practical purposes (excluding rotation and/or reflection) this is the only third order magic square which can be formed with first nine integers (try for yourself). One can of course form from it other magic squares of order 3 by adding same positive integer to all its entries. For instance if we add 3 to each entry, we get the magic square.

7	12	5
6	8	10
11	4	9

Fig. 2: Another 3×3 magic squares

Its magic number is 15 + 9 = 24. In general if a 3 × 3 magic square is formed with integers $n + 1, n + 2, ..., n + 9$, then its magic number will be $15 + 3n$.

A famous fourth order magic square with magic number 34 was created by German artist Abrecht Dürer in 1514 using first sixteen integers. This magic square is

16	3	2	13
5	10	11	8
9	6	7	12
4	15	14	1

Fig. 3: Dürer's 4×4 magic squares

A notable thing about this magic square is that the artist could incorporate in it the year of its print '1514'.

Whereas only a unique magic square of order 3 can be created with natural numbers 1 to 9, alternative magic squares of order 4 × 4 are possible with numbers 1 to 16 (Try).

Perhaps not many are aware that Benjamin Franklin besides flying kites was also a proficient amateur mathematician. He is credited with the creation of one of the finest magic squares which is of order 16 (We shall discuss it in some detail later).

Definition: In general an nth order magic square is an $n \times n$ square array of integers from 1 to n^2 having the property that the sum of each of its rows, columns and diagonals is the same.

It can be shown that this sum is $\frac{1}{2}n(n^2 + 1)$ which is called the 'magic number' of the square.

3. Different Types of Magic Squares

We have more than one type of magic squares, each type having some specific properties. We discuss these briefly.

3.1 Magic Squares of Type I

Square of order 3 shown below may look normal (is it a magic square). However, it has some curious properties which can mystify the audience

26	27	28
29	30	31
32	33	34

Fig. 4.1

It consists of integers 26 to 34 written in sequence row wise (It is not a magic square). You may ask someone to suggest a number from the square. Say he suggests 26. Add 4 to it and multiply the same with 3 to get $(26 + 4) \times 3 = 90$. Now circle this number 90.

Now form a magic square of order 3 starting with this number 26 to obtain

29	34	27
28	30	32
33	26	31

Fig. 4.2

We observe that this number 90 is the magic number of this square.

Now ask another person in the audience to select any number in the original square (Fig. 4.1). Say he selects 31. Now mark it bold and rub the other numbers appearing in its row and column to obtain

26	27	
		31
32	33	

Fig. 4.3

Now ask another person in the audience to select any of the other surviving numbers in Fig. 4.3 (excluding 31). Suppose he/she selects 33. Wipe other numbers appearing in its row and column to obtain

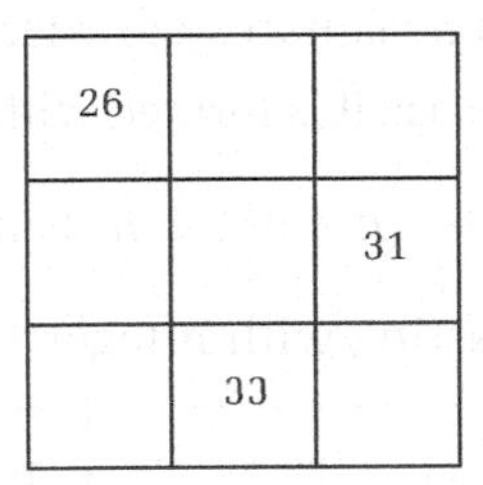

Fig. 4.4

Now add these numbers. We obtain 26 + 33 + 31 = 90.

Instead of 26, it can be repeated by starting with any other two digit number in Fig. 4.1. Magic squares possessing such a property are called magic squares of type I.

3.2 Magic Squares of Type II

A magic square of order $n(= 3, 4, ...)$ is said to be of Type II if it has the following properties:

(i) Cells of the square are filled with n^2 consecutive integers starting with any positive integer m.

(ii) Each integer between m and $m + n^2$ must appear once and only once (no repetition permissible).

(iii) Integers from m to $m + n^2$ need not appear in a sequence.

3 × 3 magic squares of order 3 discussed in section 2 as well as Dürer's magic square of order 4 discussed there have these properties and are thus of type II.

In fact Dürer's magic square of order 4 has some more interesting properties. These are:

(i) Four corner entries (16, 13, 4 and 1) also add up to magic number 34.

(ii) Central rectangle entries (10, 11, 6, 7) also add up to magic number 34.

(iii) Middle two entries of the first row (3, 2) and the last row (15, 14) also add up to magic number 34.

(iv) Middle two entries of the first column (5, 9) and the last column (8, 12) also add upto magic number 34.

(v) Each set of 4 numbers in the four quarters (16 + 5 + 10 + 3), (2 + 11 + 8 + 13), (9 + 4 + 15 + 6), (7 + 14 + 1 + 12) also add up to the magic number 34.

In fact some other symmetrical entries of four elements may also add up to 34 (search).

Magic squares of Type II of order higher than four also exist and can be framed (Try to form some). For a magic square of order 5 formed with integers from 1 to 25, the magic number is 65 (i.e., sum of entries in each row, each column and each of the two diagonals is 65). The magic number for magic squares of order 6 is 111. Magic number of magic square of order n is $\frac{1}{2}(n + n^3)$. In case it is formed with consecutive integers starting with $m + 1$, then its magic number is $\frac{1}{2}(n + n^3) + m \times n$.

A magic square of order 6 starting with integer 1 is:

4	36	29	13	18	11
30	5	34	12	14	16
8	28	33	17	10	15
31	9	2	22	27	20
3	32	7	21	23	25
35	1	6	26	19	24

Fig. 5

As can be checked that its magic number is 111.

3.3 Magic Squares of Type III

Magic squares of order 4 which possess some additional properties besides properties required of them as magic squares of Type II are called Magic Squares of Type III. Dürer's magic square of order 4 discussed earlier is a magic square of type III as it has several additional properties as discussed in section 3.2. Another magic square of type III is shown in Fig. 6. Its magic number is 139.

Besides sums of entries in each row, each column and each diagonal being 139, we also have

(i) Entries of the four corners also add up to 139.

(ii) Sum of middle four entries is also 139.

(iii) Sum of entries in corners of middle two columns (12 + 18 + 44 + 65) as well as middle two rows (28 + 12 + 19 + 80) is also 139.

A remarkable feature of this magic square is that numbers in the first row are the date of birth of legendary Indian mathematician Ramanujan (22.12.1887). In

22	12	18	87
28	59	40	12
80	3	37	19
9	65	44	21

Fig. 6

fact this magic square is regarded as a tribute to legendry Indian mathematician Ramanujan*.

4. Magic Squares of Higher Orders

From time to time people have been trying to construct magic squares of higher order and some of them even succeeded. Benjamin Franklin could construct a sixteenth order magic square. This is as under

200	217	232	249	8	25	40	57	72	89	104	121	136	153	168	185
58	39	26	7	250	231	218	199	186	167	154	135	122	103	90	71
198	219	230	251	6	27	38	59	70	91	102	123	134	155	166	187
60	37	28	5	252	229	220	197	188	165	156	133	124	101	92	69
201	216	233	248	9	24	41	56	73	88	105	120	137	152	169	184
55	42	23	10	247	234	215	202	183	170	151	138	119	106	87	74
203	214	235	246	11	22	43	54	75	86	107	118	139	150	171	182
53	44	21	12	245	236	213	204	181	172	149	140	117	108	85	76
205	212	237	244	13	20	45	52	77	84	109	116	141	148	173	180
51	46	19	14	243	238	211	206	179	174	147	142	115	110	83	78
207	210	239	242	15	18	47	50	79	82	111	114	143	146	175	178
49	48	17	16	241	240	209	208	177	176	145	144	113	112	81	80
196	221	228	253	4	29	36	61	68	93	100	125	132	157	164	189
62	35	30	3	254	227	222	195	190	163	158	131	126	99	94	67
194	223	226	255	2	31	34	63	66	95	98	127	130	159	162	191
64	33	32	1	256	225	224	193	192	161	160	129	128	97	96	65

Fig. 7: Benjamin Franklin's magic square of order 16

* Ramanujan was a legendry Indian mathematician of 19th century. Surprisingly, he was not a teacher or professor of mathematics. He worked as a clerk in *s* shipping company. Mathematics was his hobby and passion. Prof. Huxley, a well known mathematician of his times, on going through his work, called him the greatest mathematician of the century! Prof. Bell described him as 'a gift from heavens'. Prof. G.H. Hardy, a well known British mathematician invited him to work with him in U.K. He remarked, 'I have never met his equal and can compare him with Euler and Jacobi.' Unfortunately he died young. An institute exclusively devoted to studies in mathematics is named after him in Chennai. He was also fascinated by magic squares.

The magic number of this square is 2056 which is the sum of all entries in each row as well as each column. Some other properties possessed by this magic square are:

(i) Sum of the entries of 4×4 arrays at four corners is also magic number 2056.

(ii) For every 4×4 array of contiguous rows and columns, the sum is the magic number of the square 2056.

(iii) Starting with box from any border and counting eight boxes (half the size of magic square) in any direction: horizontal, vertical or even diagonal, the sum is 1028 half the magic number).

It has several other similar properties (search). However the sum of leading diagonal terms is not equal to the magic number! In that sense it is not a magic square but a semi-magic square.

5. Constructing a Magic Square

It is not easy to construct a new magic square by trial and error even though variations of an existing magic square can be obtained by interchanging rows and columns etc. For example as shown earlier seven variations of basic 3×3 magic square can be framed by rotation and/or reflection. For a 4×4 magic square the number of such variations is 880 and for a 5×5 magic square the number of such possible variations exceeds even a million! However we do not consider these as new magic squares. All variations have the same magic number. Similarly if we have a magic square constructed with first integers one can easily construct a magic square with integers from m to $n + m$. We do not regard even such much squares as new.

Methods have been proposed in literature from time to time to construct different types of magic squares of different orders, particularly magic squares of odd order. We discuss in this section some of these methods.

5.1 Muhammad Ibn Muhamad's Method for Constructing Magic Squares of Odd Order

At least a generation before Euler was doing work on magic squares in Switzerland, there was a Fulani mathematician and astronomer Muhammad Ibn Muhammad by name who lived in the city of Katsina in the region of west Africa (which is now in Nigeria). He wrote a manuscript proposing a method for constructing magic squares of odd order. His method is as under:

For constructing a magic square of an odd order using this method proceed as under. Write 1 in the cell immediately below the central cell of the square. Next move one cell to the right and one cell down to write 2 and so on. At the stage when this process gets stalled because the current entry is in the last row or last column of the square, write next entry in the first row and next column (if the earlier entry was in the last row) and first column and next row (if previous entry was in last column) (This is called swapping). Moreover if at any stage next entry gets stalled because

of an early entry being already there, then shift two rows (instead of one row) down (using wrapping when necessary).

We illustrate the working of this rule by creating magic squares of order 3 × 3, 5 × 5 and 7 × 7.

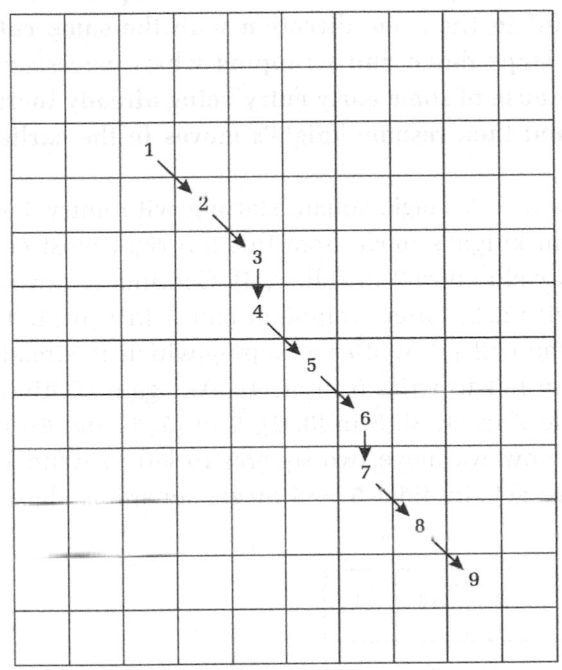

Fig. 9.1(a): Construction before wrapping

4	9	2
3	5	7
8	1	6

Fig. 9.1(b): Magic square of order 3 × 3 (after wrapping) (magic number 15)

11	24	7	20	3
4	12	25	8	16
17	5	13	21	9
10	18	1	14	22
23	6	19	2	15

Fig. 9.2: A 5 × 5 magic square (magic number 65)

22	47	16	41	10	35	4
5	23	48	17	42	11	29
30	6	24	49	18	36	12
13	31	7	25	43	19	37
38	14	32	1	26	44	20
21	39	8	33	2	27	45
46	15	40	9	34	3	28

Fig. 9.3: A 7 × 7 magic square (magic number 175)

5.2 Constructing a Magic Square Using Knight's Tour of Chess Board

Muhammad also proposed a method for construction of a magic square of odd order using knight's moves on a chessboard.

In this method one begins with writing 1 in a corner at the top and then making successive knight's moves, all in the same direction with the same rule (one step to left (or right) and two steps down and wrapping where necessary. When the next move gets blocked because of some early entry being already there, move two squares to left (or right) and then resume knight's moves in the earlier direction).

We illustrate this by generating a 5 × 5 magic square starting with entry 1 at the top rightmost cell (1, 5). We begin knight's move from this top right most cell (one step left and two steps down) to yield entry 2 in cell (3, 4). Continuing further we get entry 3 in cell (5, 3), 4 in cell (2, 2) (after wrapping) and 5 in cell (4, 1). At this stage next entry is to be in the cell (1, 5)(after wrapping)which is already occupied. So we move two squares to left to write 6 in (4, 4). We again continue with knight's moves as earlier to write 7 in (1, 3), 8 in (3, 2), 9 in (5, 1) and 10 in (2, 5). Next move is again stalled so now we move two squares to left to write 11 on (2, 3) cell. Continuing like this we get the final 5 × 5 magic square as shown in Fig. 10.

13	25	7	19	1
17	4	11	23	10
21	8	20	2	14
5	12	24	6	18
9	16	3	15	22

Fig. 10: A 5 × 5 magic square created by using knight's moves

An added bonus of this method is that not only rows and columns and two leading diagonal entries add up to its magic number 65, but entries of all other diagonals also add up to 65. For example, if we start with 19 in (1, 4), we have 19 + 10 + 21 + 12 + 3 = 65 (last three after wrapping).

5.3 A Variation in Knight's Tour Method

Muhammad Ibn Muhammad also provided a variation in the above knight's tour method. Instead of moving two cells to left when the next move gets stalled, move three cells straight up.

We illustrate this by forming a 7 × 7 magic square again starting with 1 in the top right most cell.

It may be noted that after writing 7 in cell (6, 1) knights move gets stalled as 1 is already there in (1, 7). So we move three cells up to write 8 in (3, 1) and then continue with knight's moves as usual. We again get blocked at 14 in (1, 2), so we move three steps up to write 15 in (5, 2) (on account of wrapping) and so on. In this magic square the magic number is 175. It may be noted that not only all the rows and columns but even all the fourteen diagonals sum up to this magic number. For example if we start with 14 in (1, 2) and move diagonal wise we have 14 + 5 + 45 + 36 + 34 + 25 + 16 = 175 (last entry 16 after wrapping).

32	14	38	20	44	26	1
48	23	5	29	11	42	17
8	39	21	45	27	2	33
24	6	30	12	36	18	49
40	15	46	28	3	34	9
7	31	13	37	19	43	25
16	47	22	4	35	10	41

Fig. 11: A 7 × 7 magic square generated using Muhammad's variation in knight tour method

5.4 Balof and Watkin's Method

In 1996 Watkin and one of his students Barry Balof discovered a method which is more or less identical with Muhammad Ibn Muhammad's method. However they were unaware of Muhammad's method when they discovered it. **The only difference in their method and that of Muhammad's method is that instead of moving left or right or up or down when a knight's move is blocked, they use a knight's move itself to get unblocked.** This is illustrated in Fig. 12, where a 7 × 7 magic square has been developed using their method.

They start at top left cell and make knight's moves to obtain cells for 2, 3, 4, 5, 6 and 7 where next move is blocked as 1 is already there in cell (1, 1). At this stage, they make knight's move 2 steps up and one to right (instead of 2 steps right and one down) to write 8 in cell (5, 7). From here they continue with knight's moves (2 steps right and one

1	24	47	21	37	11	34
12	35	2	25	48	15	38
16	39	13	29	3	26	49
27	43	17	40	14	30	4
31	5	28	44	18	41	8
42	9	32	6	22	45	19
46	20	36	10	33	7	23

Fig. 12: A 7 × 7 magic square by Balof and Watkin's method

down with wrapping where necessary) to reach 14. Here next move again gets stalled. So they move two steps up and one right to write 15 in (2, 6) and so on to obtain a 7×7 magic square presented in Fig. 12. This method works well to produce an $n \times n$ magic square as long as n is not divisible by 2, 3 or 5. If n is not divisible by 2 and 3 but is divisible by 5, even then every thing works well except that sums of two leading diagonal entries may not be the magic number.

5.5 Loubere's Method

Loubere a French mathematician proposed the following method for generating a magic square of odd order.

Consider a $n \times n$ magic square of odd order so that $n = 2m + 1$. Let e^{ij} denote entry in i^{th} row and j^{th} cell. Then set e^{ij} in the middle cell of the last row, i.e., for $i = 2m + 1$ and $j = m$ and write here 1. Now increase both the subscripts i and j by one each and put entry as 2 in this cell. Next put 3 in the cell obtained by increasing subscripts of cell where 2 is written by one each. Continue this (wrapping when necessary). In case this procedure leads to a cell which is already occupied, then reduce the first subscript by 2 and second by 1 and continue with the earlier procedure subsequently.

We illustrate this method by forming a magic square of order 5×5.

11	18	25	2	9
10	12	19	21	3
4	6	13	20	21
23	5	7	14	16
17	24	1	8	5

Fig. 13

Here $2m + 1 = 5$. So $m = 2$. So we write 1 in the cell which is in the middle of 5th row, i.e., in cell (5, 3).

Now on increasing each subscript by one we get cell (6, 4) or (1, 4) on wrapping and write 2 in cell (1, 4). Continuing we write 3 in cell (2, 5), 4 in cell (3, 1), 5 in cell (4, 2). Now 6 is to be in cell (5, 3) which is already occupied. So we reduce its first subscript by 2 and second by 1 to obtain cell (3, 2). So 6 is to be in cell (3, 2), 7 is to be in cell (4, 3), 8 in cell (5, 4), 9 in cell (1, 5), 10 in cell (2, 1) and 11 in cell (3, 2) which is occupied. Hence 11 is to be written in cell (1, 1). Next 12 is to be written in cell (2, 2), 13 in cell (3, 3), 14 in cell (4, 4), 15 in cell (5, 5), 16 in cell (6, 6), i.e., (1, 1) which is occupied. So 16 is written in cell (6–2, 6–1), i.e., (4, 5), 17 in (5, 1), 18 in(1, 2), 19 in (2, 3), 20 in (3, 4), 21 in (4, 5) which being occupied 21 is written in (2, 4), 22 in (3, 5), 23 in (4, 1), 24 in (5, 2) and finally 25 in (1, 3).

It can be checked that it satisfies all the requirements of a magic square.

6. Magic Squares of Even Order and Semi-magic Squares

So far we have discussed methods for generating magic squares of odd order. Magic squares of even order also exist. For example in section 2 we presented Dürer's

4×4 magic square. However, whereas specific rules are available for generating magic squares of odd order (most of these based on knight's move), no such general rules are available for generating magic squares of even order. Several persons have tried to develop magic square of order 8 using knight's tour (and some came very close to it) but failed. In figure 14 we present an 8×8 square found by Euler in his attempt to develop an 8×8 magic square. This is very close to magic square but not exactly the magic square. The sum of entries in each row and each column is same (260). However sum of entries of the two diagonals do not yield 260. Such squares are now called semi-magic squares.

1	48	31	50	33	16	63	18
30	51	46	3	62	19	14	35
47	2	49	32	15	34	17	64
52	29	4	45	20	61	36	13
5	44	25	56	9	40	21	60
28	53	8	41	24	57	12	37
43	6	55	26	39	10	59	22
54	27	42	7	58	23	38	11

Fig. 14: 8×8 semi-magic square of Euler

Definition: An array of numbers such that sum of entries in each row and each column is same is called semi-magic square.

A semi-magic square becomes a magic square when the sum of entries in each of its two diagonals is also equal to the sum of entries in its rows and columns.

It can be seen that this semi-magic square is primarily knight's open tour starting with entry 1 in (1, 1) but using mixture of appropriate knight's moves in different directions.

This semi-magic square however has some interesting properties. Some of these are:

(a) Four quadrants of this 8×8 semi-magic square are themselves semi-magic squares. Moreover in turn each 2×2 quadrant within each 4×4 quadrant contains numbers whose sum is 130.

In the following two figures (Fig. 14(a) and Fig. 14(b)), we present two more semi-magic squares. The first of these shown in Fig. 14(a) was developed by Jaenisch in 1862 and the second shown in Fig. 14(b) was developed by Wenzelides. Both are of order 8 × 8 and generated using knight's moves. As mentioned earlier Benjamin Franklin's magic square of order 16 × 16 presented in section 4 is also in reality a semi magic square.

46	55	44	19	58	9	22	7
43	18	47	56	21	6	59	10
54	45	20	41	12	57	8	23
17	42	53	48	5	24	11	60
52	3	32	13	40	61	34	25
31	16	49	4	33	28	37	62
2	51	14	29	64	39	26	35
13	30	1	50	27	36	63	38

Fig. 14(a): Jaenisch's 8 × 8 semi-magic square

50	11	24	63	14	37	26	35
23	62	51	12	25	34	15	38
10	49	64	21	40	13	36	27
61	22	9	52	33	28	39	16
48	7	60	1	20	41	54	29
59	4	45	8	53	32	17	42
6	47	2	57	44	19	30	55
3	58	5	46	31	56	43	18

Fig. 14(b): Wenzelide's 8 × 8 semi-magic square

6.1 Generating a Magic Square of Even Order

As stated earlier, it is not that magic squares of even order do not exist. However magic squares of even order (more particularly of higher order) are not easy to develop. In this very section we have seen earlier as to how various knight's moves in developing an 8 x 8 magic square resulted in semi-magic squares.

However Jaenisch who constructed 8 × 8 semi-magic square using knight's moves was also able to construct an 8 × 8 magic square using knight's tour. It is shown in Fig. 15. He first places numbers from 1 to 32 in a single cycle of knight's tour. Next he places numbers 33 to 64 in another completely separate cycle of knight's moves to fill the previously unfilled 32 cells.

15	20	17	36	13	64	61	34
18	37	14	21	60	35	12	63
25	16	19	44	5	62	33	56
38	45	26	59	22	55	4	11
27	24	39	6	43	10	57	54
40	49	46	23	58	3	32	9
47	28	51	42	7	30	53	2
50	41	48	29	52	1	8	31

Fig. 15: An 8 × 8 magic square generated by Jaenisch using two cycles of knight's moves.

However it is still an open question whether an 8 × 8 magic square can be formed using a single knight's tour. Same is the case with 12 × 12 magic square. However surprisingly knight's tour magic squares have been constructed in sizes 16 × 16, 20 × 20, 24 × 24, 32 × 32, 48 × 48 and even 64 × 64!. A 16 × 16 knight's tour magic square is shown in Fig. 16.

184	217	170	75	188	219	172	77	228	37	86	21	230	39	88	25
169	74	185	218	171	76	189	220	85	20	229	38	87	24	231	40
216	183	68	167	222	187	78	173	36	227	22	83	42	237	26	89
73	168	215	186	67	174	221	190	19	84	35	238	23	90	41	232
182	213	166	69	178	223	176	79	226	33	82	31	236	43	92	27
165	72	179	214	175	66	191	224	81	18	239	34	91	30	233	44
212	181	70	163	210	177	80	161	48	225	32	95	46	235	28	93
71	164	211	180	65	162	209	192	17	96	47	240	29	94	45	234
202	13	126	61	208	15	128	49	160	241	130	97	148	243	132	103
125	60	203	14	127	64	193	16	129	112	145	242	131	102	149	244
12	201	62	123	2	207	50	113	256	159	98	143	246	147	104	133
59	124	11	204	63	114	1	194	111	144	255	146	101	134	245	150
200	9	122	55	206	3	116	51	158	253	142	99	154	247	136	105
121	58	205	10	115	54	195	4	141	110	155	254	135	100	151	248
8	199	56	119	6	197	52	117	252	157	108	139	250	153	106	137
57	120	7	198	53	118	5	196	109	140	251	156	107	138	249	152

Fig. 16: A 16 × 16 knight's tour magic square

6.2 Generating Magic Squares of even order using Tours of Chess Board Pieces other than Knight

King's tour and Rook's tours have also been tried to generate even order magic squares and these have succeeded in generating 8 × 8 magic squares. In 1921, Ghersi constructed an 8 × 8 magic square using king's tour of chess board. This is shown in Fig. 17(a). (Note the pattern of number 1 to 8 in his tour). He then exactly reverses this pattern for generate numbers 9 to 16 in a symmetric fashion in the second quadrant. He then places numbers 17 to 32 in the lower half of the square as a mirror image of numbers 1 to 16, again reversing the order. He finally places numbers 33 to 64 as a mirror image (left to right) of 1 to 32 also in the reverse order.

61	62	63	64	1	2	3	4
60	11	58	57	8	7	54	5
12	59	10	9	56	55	6	53
13	14	15	16	49	50	51	52
20	19	18	17	48	47	46	45
21	38	23	24	41	42	27	44
37	22	39	40	25	26	43	28
36	35	34	33	32	31	30	29

Fig. 17(a): Ghersi's 8 × 8 magic square created using king's tour

61	62	63	64	1	2	3	4
12	11	10	9	56	55	54	53
20	19	18	48	17	47	46	45
60	59	58	8	57	7	6	5
37	38	39	25	40	26	27	28
13	14	15	49	16	50	51	52
21	22	23	24	41	42	43	44
36	35	34	33	32	31	30	29

Fig. 17(b): Rabinowitz's 8 × 8 rook's tour magic square

In 1985 Stanley Rabinowitz designed an 8 × 8 magic square shown in Fig. 17(b) using rook's tour. In this case it may be noted that the pattern of numbers 1 to 8 is repeated from numbers 9 to 16 in almost a mirror image form from left to right but shifted down. Rest of construction more or less is like Ghersi's method.

It may be of interest to know that in 2003 Gaunter Stertonbrink announced that a team of computer enthusiasts after exhaustively searching all possibilities has concluded that no single knight tour based magic square of order 8 × 8 is possible.

Magic squares is an interesting and fascinating topic. It has attracted attention of many. In fact entire books have been written about magic squares (For example 'New Recreations with Magic Squares', W.H. Benson and O. Jacoby, Dover New York, 1976).

7. Latin Squares

A Latin square is an $n \times n$ array of integers 0, 1, 2, 3, ..., $n - 1$ such that each integer appears once and only once in each row and each column. For example a 4×4 Latin square is

3	2	1	0
0	1	2	3
2	3	0	1
1	0	3	2

Fig. 18: A 4×4 Latin square

In a sense a Latin square is similar to magic square except that now we do not insist upon the sum of entries in each row and each column and each diagonal being exactly the same. There is nothing special about using integers 0, 1, 2, 3, ..., $n - 1$ in a square of order $n \times n$ except that occasionally their mathematical properties become handy. Other symbols such as letters of alphabet, can also be used and have been used. In fact the name 'Latin square' comes from the fact that Euler used Latin alphabet letters to label its different cells.

7.1 Constructing a Latin Square

It is not difficult to construct a Latin square of any size. W.W. Rouse Ball proposed the following simple method to construct an $n \times n$ Latin square. Choose any two integers, say a and b between 0 and $n - 1$ which are relatively prime to n i.e., are not factors of n. (These can be equal also). Then starting with the cell (1, 1) at the lower left most corner, fill the bottom row with numbers 0, b, $2b$, $3b$ (reducing modulo n where necessary). Next write left hand column starting from bottom and moving towards top using numbers 0, a, $2a$, $3a$ (again reducing modulo n where necessary). Remaining entries are to be formed as an addition table using entries of the left most column and entry in the bottom row of the column to be generated and doing addition modulo n.

For example in order to form a 5×5 Latin square, $n = 5$. So we may choose $a = 1$ and $b = 2$ none of these being divisor of 5 (one is not to be considered divisor of a number).

So the bottom row will be 0, 0 + 2, 2 + 2 = 4, 4 + 2 = 6 (modulo 5) = 1, and 1 + 2 = 3. The left most column will be 0, 0 + 1 = 1, 1 + 1 = 2, 2 + 1 = 3, 3 + 1 = 4. Now to form column 2 of the table (from top to bottom), 4 + 2 = 6 (mod 5 = 1), 3 + 2 = 5 (mod 5 = 0), 2 + 2 = 4, 1 + 2 = 3. Similarly we can form columns 3, 4 and 5 to obtain Latin square shown in Fig. 19(a).

One could frame another Latin square of same size by using Ball's method as shown in Fig. 19(b).

4	1	3	0	2
3	0	2	4	1
2	4	1	3	0
1	3	0	2	4
0	2	4	1	3

Fig. 19(a): 5 × 5 Latin square generated with $a = 1, b = 2$.

4	0	1	2	3
3	4	0	1	2
2	3	4	0	1
1	2	3	4	0
0	1	2	3	4

Fig. 19(b): 5 × 5 Latin square generated with $a = b = 1$.

8. Graeco-Latin (Eulerian) Squares

If two Latin squares (say A and B) of same size (as I and II of the previous section) are superimposed and the entries of each cell written as pairs of corresponding entries of the corresponding two cells of the two squares, (entry of cell (i, j) being $c_{ij} = (a_{ij}, b_{ij})$), then if in the resulting Latin square n^2 ordered pairs of elements occur once and only once, then it is called a **Graeco-Latin square** or **Eulerian square**. (It is called Graeco-Latin square because usually one entry in each cell is from Greek alphabet and the other from Latin alphanet).

For Example if

A =

3	2	1	0
0	1	2	3
2	3	0	1
1	0	3	2

Fig. 20(a)

B =

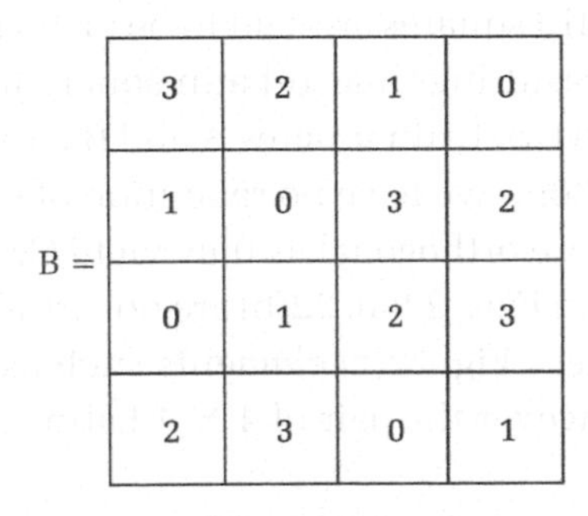

3	2	1	0
1	0	3	2
0	1	2	3
2	3	0	1

Fig. 20(b)

(3,3)	(2,2)	(1,1)	(0,0)
(0,1)	(1,0)	(2,3)	(3,2)
(2,0)	(3,1)	(0,2)	(1,3)
(1,2)	(0,3)	(3,0)	(2,1)

Fig. 20(c): Graeco-Latin square

Graeco-Latin square of Fig. 20(c) is in fact solution to the well-known Oznam's problem of cards in which one is asked to place four aces, four kings, four queens and four jacks from a deck of cards into a 4 × 4 array so that each row and each column contains a card of each rank as well as each suit. Another 5 × 5 Greaco-Latin square is shown in Fig. 21.

(4,4)	(1,0)	(3,1)	(0,2)	(2,3)
(3,3)	(0,4)	(2,0)	(4,1)	(1,2)
(2,2)	(4,3)	(1,4)	(3,0)	(0,1)
(1,1)	(3,2)	(0,3)	(2,4)	(4,0)
(0,0)	(2,1)	(4,2)	(1,3)	(3,4)

Fig. 21: 5 × 5 Graeco-Latin square formed by I and II

9. Orthogonal Latin Squares

Two Latin squares are said to be orthogonal if each of the ordered pairs occur only once in the resulting Graeco-Latin square formed with these Latin squares.

The two Latin squares A and B written in Fig. 20(a) and 20(b) are orthogonal because in Fig. 20(c) we have no repetition of elements. Similarly 5 × 5 Latin squares I and II of Fig. 19 are orthogonal as they yield Graeco-Latin square. However pair of Latin squares shown in Figs. 22(a), 22(b) are not orthogonal because in the Greaco-Latin square formed with these Fig. 22(c) elements such as (0,1) and (3,2) repeat.

However the pair of 4 × 4 Latin squares

A =

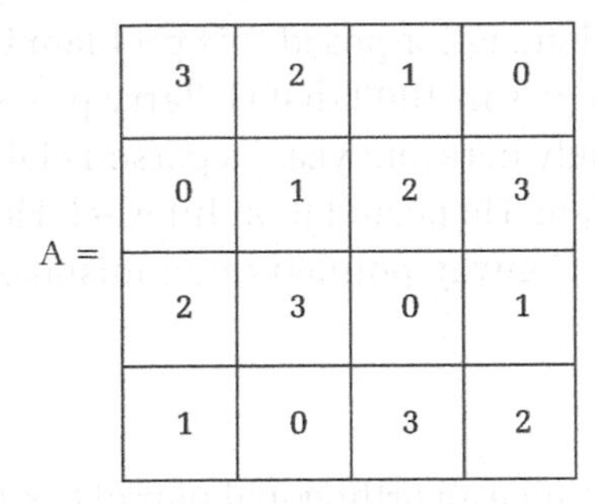

3	2	1	0
0	1	2	3
2	3	0	1
1	0	3	2

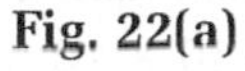

Fig. 22(a)

and B =

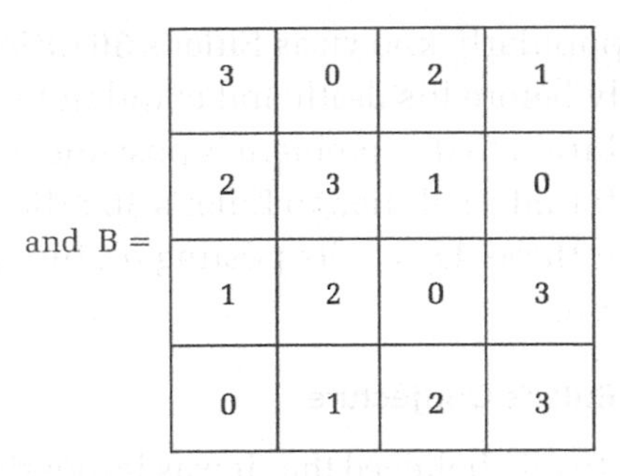

3	0	2	1
2	3	1	0
1	2	0	3
0	1	2	3

Fig. 22(b)

which yield Eulerian square

C =

(3,3)	(2,0)	(1,2)	(0,1)
(0,2)	(1,3)	(3,1)	(3,0)
(2,1)	(3,2)	(0,0)	(1,3)
(1,0)	(0,1)	(3,2)	(2,3)

Fig. 22(c): Graeco-Latin square

9.1 Constructing Orthogonal Latin Squares

Euler observed that it was not difficult to construct an orthogonal pair of Latin squares when n is odd. Such a Latin square could be constructed using **Ball's method**. According to this method, (say in the case of 5×5 squares) generate a Latin square by first setting $a = 1$ and $b = 1$ and then by setting $a = 1$ and $b = 2$. Latin squares thus generated are always orthogonal if n is odd.

For $n = 2$ only two Latin squares are possible and these are not orthogonal (check). However if k is some power of 2 (2^k), say $k = 3$, 4, i.e., magic squares of order $2^3 = 8$ or $2^4 = 16$, even then we can still use Ball's method to generate two orthogonal Latin squares merely by replacing 0, 1, 2, 3, ... ($n - 1$)by what are now known as Galio's Field GF (n).

Again according to Ball, if we have one $j \times j$ and another $k \times k$ Eulerian square, it is easy to construct a $jk \times$ jk Eulerian square. Therefore the only values of n where one might not be able to construct a pair of orthogonal Latin squares would be $n = 2$, 6, 10, 14, 18.... In fact Euler was unable to find a pair of 6×6 orthogonal Latin squares. Euler looked at this problem in the following way:

Suppose there are 6 regiments and each regiment has 6 officers and each officer has one of 6 different ranks, then can these 36 officers be placed in a 6×6 square so that each row and each file contains one officer of each rank and each regiment. This is now popularly known as Euler's 36 officers problem. Euler posed this problem in 1782 shortly before his death and could not solve it. It was in 1900 that G. Tarry proved that no solution to this problem is possible. Interestingly in recent years a person claimed to have found a solution to Euler's 36 officers problem. He posted it on Internet. However within three days of its posting on the net, John Conway pointed out a mistake in the solution.

9.1.1 Euler's Conjecture

Euler firmly believed that it was impossible to produce an orthogonal pair of 6×6 Latin squares and it was similarly impossible to produce pairs of 10×10, 14×14 or 18×18 such squares. This bold prediction based on the simple fact that it was not true for 2×2 size Latin squares and he was also unable to produce a pair of 6×6 orthogonal Latin squares is now known as **Euler's conjecture**. According to it **orthogonal Latin square of order $n = 2, 6, 10, 14, 18$.... (in general for $n = 2 + 4k$, $k = 1, 2, 3, 4, ...$) is not possible.**

Tarry's proof in 1900 that 6×6 orthogonal Latin squares are not possible added support to Euler's conjecture. For the next half century nothing further happened. However in 1958, Parker, Bose and Shrikhande showed that Euler was wrong beyond $n = 6$ and obtained orthogonal Latin squares for $n = 10, 14, 18$ and 22. Interestingly, Bose and Shrikhande first found orthogonal Latin squares of order $n = 22$!. Later Parker found it for $n = 10$. The 10×10 Eulerian square found by Parker is shown in Fig. 23.

(0,0)	(4,7)	(1,8)	(7,6)	(2,9)	(9,3)	(8,5)	(3,4)	(6,1)	(5,2)
(8,6)	(1,1)	(5,7)	(2,8)	(7,0)	(3,9)	(9,4)	(4,5)	(0,2)	(6,3)
(9,5)	(8,0)	(2,2)	(6,7)	(3,8)	(7,1)	(4,9)	(5,6)	(1,3)	(0,4)
(5,9)	(9,6)	(8,1)	(3,3)	(0,7)	(4,8)	(7,2)	(6,0)	(2,4)	(1,5)
(7,3)	(6,9)	(9,0)	(8,2)	(4,4)	(1,7)	(5,8)	(0,1)	(3,5)	(2,6)
(6,8)	(7,4)	(0,9)	(9,1)	(8,3)	(5,5)	(2,7)	(1,2)	(4,6)	(3,0)
(3,7)	(0,8)	(7,5)	(1,9)	(9,2)	(8,4)	(6,6)	(2,3)	(5,0)	(4,1)
(1,4)	(2,5)	(3,6)	(4,0)	(5,1)	(6,2)	(0,3)	(7,7)	(8,8)	(9,9)
(2,1)	(3,2)	(4,3)	(5,4)	(6,5)	(0,6)	(1,0)	(8,9)	(9,7)	(7,8)
(4,2)	(5,3)	(6,4)	(0,5)	(1,6)	(2,0)	(3,1)	(9,8)	(7,9)	(8,7)

Fig. 23: Parker's 10×10 Eulerian square

It appeared on the cover page of November 1959 issue of Scientific American. A colour version of it is available on Dormouth College Mathematics department website: www.math.dormouth.edusphere.

In fact it has now been even proved that for a value of n for which orthogonal Latin squares exist will never have more than $n - 1$ mutually orthogonal Latin squares. But to have as many as n–1 is sometimes possible. For example for $n = 5$, it is easy to construct 4 mutually orthogonal Latin squares using Ball's method by setting $a = 1$ and successively setting $b = 1, 2, 3$ and 4. However as Tarry has shown for $n = 6$, even two mutually orthogonal Latin squares are not possible. Even for $n = 10$, not even three mutually orthogonal Latin squares have been found even though theory predicts that up to 9 are possible.

10. Practical Applications of Latin Square

Euler was fascinated by Latin squares and played with these regarding these as just mathematical diversions having no practical use. However in 1920's, it was realized that Latin squares could be helpful in design of experiments, particularly in the field of agriculture. A Latin square is a perfect design for an experiment that seeks to test different treatments of a particular crop by equalizing any possible local effects due to reasons such as varying soil fertility, amount of moisture received, and other similar uncontrollable factors.

Latin squares are now widely being used in several applied areas ranging from design of experiments in medicine to product testing in business. Graeco-Latin squares (i.e., Eulerian squares) are especially useful form of Latin squares as these automatically allow one to include an additional variable in the experiment.

Puzzles based on Latin squares have also been designed. Eric Anderson produced such puzzles and posted these on website www.latinsquare.com. We present one of his puzzles in exercises.

Mind Teasers

1. Build 3×3 and 5×5 magic squares such that each of these has entry one in the middle cell of the first row. What are their magic numbers?
2. Build 3×3 and 5×5 magic squares starting with number 11 in place of one in the middle cell of the first row. What are their magic numbers now?
3. Can 3×3 and 5×5 magic squares be built starting with 1 in the last cell of the last row. If yes then build these.
4. Form a 3×3 magic square using a sequence of 9 consecutive integers. What is its magic number?

5. Build a 3×3 magic square using first nine odd numbers in place of first 9 integers. What is its magic number?

6. Can a magic square of size 3×3 be formed using any nine consecutive even integers. If yes, form it with integers $2n$, $2n + 2$, $2n + 4$, ..., $2n + 16$. What is its magic number?

7. Build a 4×4 magic square using first 16 integers such that 1 appears at the last cell of the last row. What is its magic number?

8. Construct 5×5, 7×7 magic squares using 25 and 49 consecutive integers respectively starting with 11. What are their magic numbers?

9. Construct a 13×13 magic square using knight's tour.

10. Construct an 8×8 semi-magic square using king's tour starting with numbers 1 to 8 placed in 2×4 rectangular block at the upper hand right corner.

11. Construct a pair each of 3×3, 4×4 and 5×5 Latin square which are not orthogonal.

12. Construct a pair each of 3×3, 4×4 and 5×5 Latin squares which are orthogonal.

13. Construct all possible orthogonal Latin squares of order 4×4.

14. Use Ball's method to construct an 8×8 Latin square taking $a = 1$ and $b = 3$.

15. Obtain another Latin square from Latin square obtained in question. 14 by interchanging third and fifth rows and check if the two Latin squares are orthogonal or not.

16. Show that 4×4 Eulerian square shown in the following figure yields a magic square if we think of ordered pair of numbers in each cell as two digit numbers written in base 4 (For example (3, 2) in base 4, is $3 \times 4 + 2 = 14$ etc.).

(3,3)	(2,2)	(1,1)	(0,0)
(0,1)	(1,0)	(2,3)	(3,2)
(2,0)	(3,1)	(0,2)	(1,3)
(1,2)	(0,3)	(3,0)	(2,1)

17. A 6×6 square has been divided into 6 parts drawing zig-zag lines as shown in Fig. 24. Each part has 6 cells. Six letters A to F have been placed in different cells as shown there. Complete this Latin square by filling remaining cells with these very letters so that each of these 6 parts contains all these six letters once and only once.

18. Rework the puzzle of question 17 by replacing zig-zag lines by straight lines at 2nd and 4th rows and 3rd column as shown in Fig. 24(a).

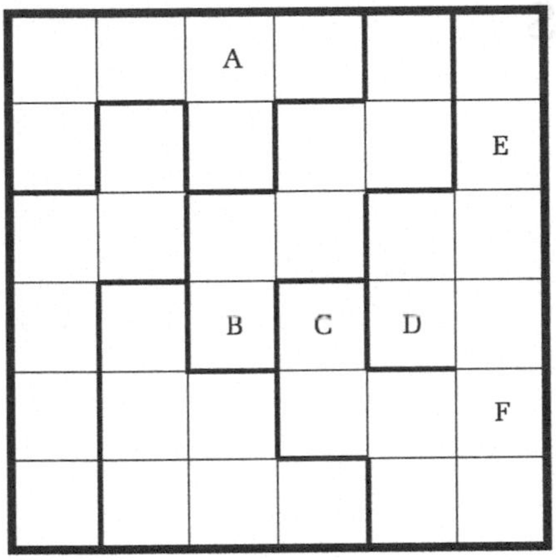

		A			
					E
		B	C	D	
					F

Fig. 24

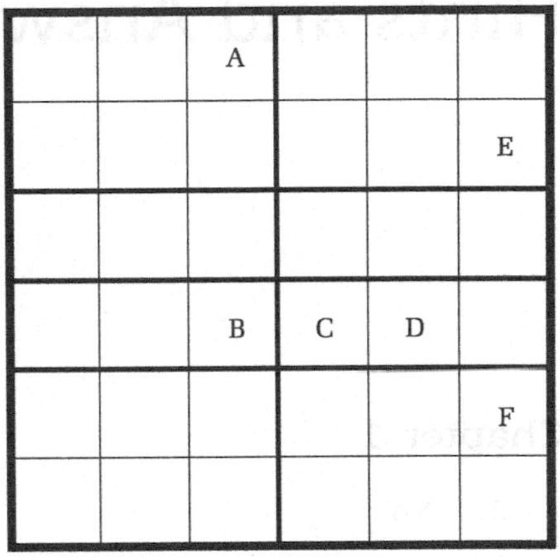

		A			
					E
		B	C	D	
					F

Fig. 24(a)

Hints and Answers

Chapter 1

5. No
6. No, (example $n = 2$)
7. 59
8. Let the numbers be $n - 1, n, n + 1$ then their sum is $3n$ which is divisible by 3.
9. Let the numbers be n, $100 - n$. Then their product is $n(100 - n)$. If it is to be 3000 then

   $n(100 - n) = 3000$ or $n^2 - 100n + 3000 = 0 \Rightarrow n = 50 \pm \sqrt{2i}$. Which is not real?
10. It may not be possible to guess. Let it be N. Suppose it has n digits then when we shift 2 from last to first place it becomes

$$2 \times 10^{n-1} + \frac{N - 2}{10} \text{ which has to be } 2N.$$

So we have to solve $2 \times 10^{n-1} + \dfrac{N - 2}{10} = 2N.$

or $\qquad 2(10^n - 1) = 19N$

Solution is possible for $n = 19$. For this value of n, is

$$N = \frac{2(10^{19} - 1)}{19} = 105263157894736842\text{, a very big number!}$$

12. For printing each of first 9 pages one digit is needed, for next 90 pages (from 10 to 99) 2 digits per page are needed and for pages beyond 99, 3 digits per page will be needed.

    $9 + 180 + 3n = 1890$ or $3n = 1890 - 189 = 1701$ or $n = 567$ so pages are $567 + 99 = 666$.
13. No. Let numbers printed on torn pages be

$$n, n + 1, n + 2 \ldots\ldots\ldots\ldots n + 49.$$

Then $\qquad n + n + 1 + n + 2 + \ldots\ldots\ldots\ldots + n + 49 = 1990$

or $\qquad 50n + (1 + 2 + \ldots\ldots\ldots\ldots + 49) = 1990$

or $$50n + \frac{49}{2}[2 + (49 - 1) \times 1] = 1990.$$

This yields $n = \frac{765}{50}$ a non integer. So a solution is not possible.

14. $\sqrt{3}$ Draw a cube of unit side each. Its diagonal OB represents length

$$OC = \sqrt{OA^2 + AB^2 + BC^2} = \sqrt{1+1+1} = \sqrt{3}$$

Similarly $\sqrt{5} = \sqrt{4+1}$. So $\sqrt{5} = \sqrt{4+1}$ is the hypoteneous of a right angled triangle with sides 1 and 2

17. With base 7,

$$\begin{array}{r} 2\,3\,4\,5\,1 \\ +\,1\,5\,6\,4\,2 \\ \hline 4\,2\,4\,2\,3 \end{array}$$

18. (i) 7 (ii) 6 (iii) 12

19. $(1 - i)(1 + i) = 1 - i^2 = 1 + 1 = 2$

21. (i) ∞ (ii) ∞ (iii) not defined (iv) ∞ (v) not defined (vi) not defined

Chapter 2

1. No. Suppose N is the sum of integers 0, 1, 2, 3, ..., up to 9 which have been just used once.

   Then N – (1 + 2 + 3 + ... + 9) = N – 45 should be divisible by 9.

   But for N = 100, N – 45 = 55 is not divisible by 9.

2. $4^{4^{4^{4}}}$

6. No. Let three consecutive numbers be $a - 1$, a, $a + 1$, then we have to show that $(a - 1)^2 + a^2 + (a + 1)^2$ is a perfect square i.e. $3a^2 + 2$ is a perfect square for some integer value of a. Which is not so.

7. For a number to be perfect square sum of its digits must be divisible by 9 if this sum is divisible by 3. In the present case the sum of digits is 300 which are divisible by 3 but not 9.

8. 9

9. Integers m and n should satisfy $m + n = m \times n$ or $m = n(m - 1)$.

   So $n = \frac{m}{m-1}$. Now $\frac{m}{m-1}$ is integer only for $m = 2$ and in that case n is also 2.

10. $100! = 1 \times 2 \times 3 \times 4 \times 5 \times .... \times 98 \times 99 \times 100$. Zero occurs when 1 is multiplied with 10 or 5 is multiplied with 2, 4, 6 or 8. So break it into groups of 10's and count .It will have 24 zeros.

13. $3^1 = 3, 3^2 = 9, 3^3 = 27, 3^4 = 81, 3^5 = 243$ etc. It is noticed that 1, 3, 7, 9 are only possible last digits. Now $34789 = (3^4)^{11}.3^2$. So last digit is $1 \times 9 = 9$.

16. When divided by 3 remainder can only be 0, 1 or 2. If n has remainder 1, n^3 will also have remainder 1 and $2n$ remainder 2. So $n^3 + 2n$ has remainder 3 i.e. $n^3 + 2n$ is divisible by 3. If n has remainder 2, $2n$ will have remainder 1 and n^3 again remainder 2. So $n^3 + 2n$ will have remainder $2 + 2 \times 2 = 6$ which is again divisible by 3. If n has remainder zero, $n^3 + 2n$ will also have remainder zero.
17. Factors of 45045 are 3, 3, 5, 7, 11, 13. Use these to form a, b, and $a - b$.
18. No. For this we must have integer n such that $(n - 1) + n + (n + 1) = a$, $(n + 2) + (n + 3)+(n + 4) = b$. So $3n \times (3n + 9) = ab = 11, 11, 11, 11$ so 11, 11, 11, 11 should be divisible by 9 which is not.
19. 2^{32} mm or $2^{32} \times 10^{-6}$ km. Indeed very very thick. Its size will be 2^{-32} of the original size (Very very small). It may be of interest and a bit surprise to know that it is infact impossible to fold a sheet of nonzero thickness (howsoever large it be) more than eight times (each time to half of its original size) without breaking or cutting it.
20. 6(1 + .02) 16 billion in 2015
In n years where $6 \times 10^{12} (1 + .02)^n = 150 \times 10^6 \times (10^3)^2 = 150 \times 10^{12}$ or $(1.02)^n = 25$ or $n \approx 16$.

Chapter 4

1. Paul, John, Eric, William.
3. Shyam.
4. Vikram clerk, Faruq teacher, John dentist.
5. Total length is 9 meters.

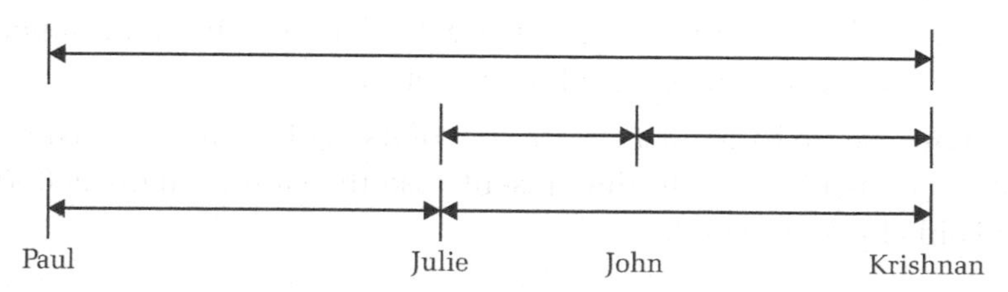

7.

Upward +13		**Downward –8**	
Switch 2	$13 - 8 = 5$	Switch 1	$5 + 13 = 18$
Switch 2	$18 - 8 = 10$	Switch 2	$10 - 8 = 2$
Switch 1	$2 + 13 = 15$	Switch 2	$15 - 8 = 7$
Switch 1	$7 + 13 = 20$	Switch 2	$20 - 8 = 12$
Switch 2	$12 - 8 = 4$	Switch 1	$4 + 13 = 17$
Switch 2	$17 - 8 = 11$	Switch 2	$11 - 8 = 3$
Switch 1	$3 + 13 = 16$	Switch 2	$16 - 8 = 8$

8. Peaches.

9. Tarun.

11. Suppose the person at the end has seen both persons sitting in front of him wearing white hats, then the middle person will have also seen the front person wearing white hat. Now the person at the back will argue, as both people in his front have white hats, he must be having either a white or a black hat. If he guesses the color of his hat to be white, then person sitting in the middle can again argue that as the person sitting in his front and back have both white hats, he must be having either a white or black hat. As both persons sitting at the back of the person sitting in front are undecided about the color of their hats. He will also be undecided. Same is the case if one person in front of the person sitting at the back has white and the other black has. However if both the persons sitting in front of the last person have black hat then he will be sure that he has white hat as there are only two black hats. However persons sitting in the middle and front will still be undecided.

12. (i) No.
    (ii) Does this road lead to village of truth tellers? A liar will reply no but truth teller yes.
    (iii) If the statement of A is true then B is a liar.
    (iv) A liar, B outsider, C tells truth.

13. Pointing to one of the roads that he has not come along, he may ask: 'If I ask your sister whether this road leads to the city Y or not', will she say 'yes' or 'no'? In case the sister whom he has asked this question is a truth teller and road indeed leads to city Y, she will say that her sister who is a liar will say No. However in case the sister whom he has asked this question is a liar and other speaks truth again she will say No. So in both the cases the tourist can take No as the correct answer.

14. Deputy Secretary.

17. White shirt.

20. Place Rs. 1, 2, 4, 8,16,32 and 64 in the seven wallets. Any amount upto Rs. 127 can be paid by just choosing appropriate wallets (binary representation for digits upto 127)

Chapter 5

1. (i) 14+8–4=18, (ii) 18+5=23.

2. Let age of Mohan be x, age of father y, then $x + y = 58$, and $y - (x + 5) = 23$ or $y - x = 28$. Therefore $2y=58 + 28 = 86$. Hence $y = 43$ and $x = 58 - 38 = 20$. So father is 43 yrs and Mohan is 15 yrs old.

3. No.
4. Prince should name x, y, z as 1, 100, 10000 (i.e. 1, 10^2,10^4) in this case number $ax + by + c$ will be $a + b \times 100 + c \times 10000$ i.e. cba. So he can easily tell a, b, c.
5. It will be so if $x + (x + 1) + (x + 2) + ... + (x + 24) = 1990$ for some positive integer x. Now L.H.S. is an A.P.
So $\frac{25}{2}[2x + 24 \times 1] = 1990$. So $25x + 300 = 1990$ or $25x = 1690$. Hence x is not integer. So it is not possible.
6. Sum of series $1^2 + 2^2 + ... + 99^2 = \frac{99\,(99 + 1)\,(2 \times 99 + 1)}{6} = 33 \times 50 \times 199$ $= .... 7$. So clearly the last digit is 7.
7. One second before one minute. As bacteria double themselves every second.
8. Let r be 55th radius of the container then its volume is $\frac{4}{3}\pi r^3$. This should be same as volume of 55th cubic container of sides 940 km each

So $$\frac{4}{3}\pi r^3 = (940)^3$$

or $$r = 940 \times \left(\frac{3}{4\pi}\right)^{1/3} = 940 \times \left(\frac{3}{4} \times \frac{7}{22}\right)^{1/3} \approx \frac{940}{(4)^{1/3}}$$

9. (i) $4! = 24$
(ii) $(4 - 1)! = 6$. When of the same colour, one possible way in each case.
10. (i) $k\,(k - 1)$ (ii) $k!$ (iii) $\frac{k!}{(k - m)!}$ (iv) ${}^kC_m = \frac{k!}{m!\,(k - m)!}$
11. $9!\ {}^{10}P_4$ (seating 10 girls in 9! ways and then seating 4 students in between 10 girls).
12. ${}^{26}P_7 - 2 \times {}^{24}P_5 \times 6$. If no restriction then ${}^{26}P_7$. (Two chosen can be placed with remaining 5 in 6 ways which can be placed in ${}^{24}P_5$ ways. 2 chosen can be interchanged.)
13. She can use 4, 3 or 2 colors. $\left(\frac{1}{4}\right){}^8P_4 + \left(\frac{1}{2}\right){}^8P_3 + \left(\frac{1}{2}\right){}^8P_2 = 420 + 168 + 28 = 616$
14. 5, 10, 10, 5.
15. 8, 12; 10, 20; 12, 18; 30, 36.
16. Show that $a^3 + b^3 + c^3$ can be expressed in a form which has $a + b + c$ as a factor.
17. a, b, c are Pythagorean triplets. So they will be of the form $a = m^2 - n^2$, $b = 2$ mn, $c = m^2 + n^2$, where m, n are relative primes. It can be verified that for different choices of m and n as relative primes, one of $m^2 + n^2$, $m^2 - n^2$ and 2 mn will have 3 as a factor.

Chapter 6

7. 7 in general, 6 if all pass through the same point.
9. For a connected graph E ≥ 2F. This does not hold in the present case.
10. Apply Euler's Theorem.
11. Try analytic reasoning used by Euler.
12. Yes.
13. Yes. Area is ½ base×altitude. Make triangle such that altitude is less than 1/50 cm.
14. Sum of altitudes drawn from an arbitrary point on sides is constant in an equilateral triangle.
15. As sides are of integral lengths these will be Pythagoreans triplets.
16. No.
17. Square.
18. Square. Let perimeter be $4p$ and lengths and breadth be $(p + x)$ $(p - x)$.

    Area of such rectangle is $(p + x)$ $(p - x)= p^2 - x^2$. For this area to be maximum clearly x must be zero.
19. Use triangular inequalities AB > BC + CA etc.

    For the second part use the fact that the external angle of a triangle is equal to the sum of opposite angles, to express all angles in terms of internal angles of the triangle and use the fact that the sum of the internal angles of a triangle is 180°.
20. Use the fact that ABC and ABD being angles in semicircle are 90° each.
22. Yes the length of each edge is 10 cm. Number of cuts 11.
23. Six.
24. Four. $V + F - E = 2$. Since three edges meet at each vertex, so $E = \frac{3}{2}V$. Similarly $F = \frac{3}{2} \times 3V = V$. Thus $V + V - \frac{3}{2}V = 2$ or $V = 4$.
27. No. Why?
28. Yes. South Pole as well as points 1 km south of a circle of circumference 1 km near North Pole.
29. (i) Yes, points between 23½° north and 23½° south latitude.

    (ii) Yes, points on the Equator.
30. Yes, the pole star.

Chapter 7

1. $\sqrt{a^2+b^2}$ and $-\sqrt{a^2+b^2}$ (max when $\tan 2\theta = \frac{b}{a}$)
2. $\sqrt{7}$ km (use $\cos A = \frac{b^2+c^2-a^2}{2bc}$)
3. It will be equilateral triangle. So $a = b = c = 2$ and $s = \frac{b+c+a}{2} = 3$

   Hence Area $= \sqrt{3(s-2)(s-2)(s-2)} = \sqrt{3}$
4. Third angle is obtuse 105°. Using sine formula $b = \sqrt{2}$, $c = 2$.

   Therefore $s = \frac{\sqrt{3}+1+\sqrt{2}+2}{2} = \frac{3+\sqrt{2}+\sqrt{3}}{2}$

   So the Area is $\sqrt{s(s-a)(s-b)(s-c)} = \frac{1}{\sqrt{3}-1}$
6. $10\sqrt{3}$ meters
7. $5\frac{\sqrt{3}}{2}$ meters
8. $10(1+\sqrt{2})$ meters.
9. $100\sqrt{3}$ meters.
10. 20 meters.

Chapter 8

1. $\left(\frac{1}{2}\right)^3 = \frac{1}{8}$
2. Total numbers of ways in which 4 slips can be drawn out of 8 is ${}^8C_4 = \frac{8!}{4!4!} = 70$.

   Of these only one namely 1, 2, 3, 4 is favorable. So desired probability is $\frac{1}{70}$
3. Possible outcomes are: (H, H), (H, T), (T, H), (T, T). Of these head appears in first trial in only the first two. Hence probability of second being also head given that first is head is $\frac{1}{2}$
4. It is intuitively clear that if $a = b$ then probabilities of exit at A or B would have been equal. However if $a > b$ then there is greater chance of his exit at B and vice versa.

   Let $P(A) = kb$ and $P(B) = ka$. Since $P(A) + P(B) = 1$. Therefore $k = \frac{1}{a+b}$

   Thus $P(A) = \frac{b}{a+b}$ and $P(B) = \frac{a}{a+b}$

5. $P(A) = \dfrac{a}{a+b}$ and $P(B) = \dfrac{b}{a+b}$. (reverse of answer of question 4).

6. Whereas A can win in 1st, 3rd, 5th, 7th, ... (odd throws), B can win in 2nd, 4th, 6th, 8th, ... (even throws). So

$$P(A) = \frac{1}{2} + \frac{1}{2} \times \frac{1}{2} \times \frac{1}{2} + \frac{1}{2} \times \frac{1}{2} + \frac{1}{2} \times \frac{1}{2} \times \frac{1}{2} + ...$$

$$= \frac{1}{2}\left(1 + \frac{1}{2^2} + \frac{1}{2^4} + ...\right)$$

$$= \frac{1}{2} \times \left(\frac{1}{1 - \frac{1}{2^2}}\right) = \frac{1}{2} \times \frac{4}{3} = \frac{2}{3}$$

Hence $\quad P(B) = 1 - P(A) = 1 - \dfrac{2}{3} = \dfrac{1}{3}$

7. Possible outcomes are: (1, 1), (1, 2), (1, 3), (1, 4), (1, 5), (1, 6); (2, 1), (2, 2), (2, 3), (2, 4), (2, 5), (2, 6); (3, 1), (3, 2), (3, 3), (3, 4), (3, 5), (3, 6)....... with sums (2, 3, 4, 5, 6, 7), (3, 4, 5, 6, 7, 8) (4, 5, 6, 7, 8, 9)...... Total six such groups. Clearly only 7 appears in all the six groups. So 7 has greatest prob. of appearing as sum.

8. (i) $P(A \cup B) = .4 + .2 = .6$ (ii) $P(A \cup B) = P(B) = .4$
(iii) $P(A \cup B) = P(A) + P(B) - P(A \cap B) = .2 + .4 - .15 = .45$

9. $\dfrac{1}{{}^8C_4} = \dfrac{1}{70} = .014$

10. (i) $\dfrac{1}{8} \times \dfrac{1}{8} \times \dfrac{1}{8} = \dfrac{1}{8^3}$ (ii) $\dfrac{1}{{}^8C_3} = \dfrac{2 \times 3}{6 \times 7 \times 8} = \dfrac{1}{56} \approx (2.018)$

11. (i) $\dfrac{1}{{}^{20}C_5}$ (ii) $\dfrac{{}^{15}C_5}{{}^{20}C_5}$

12. (i) This is one minus none has the same birthday. (A single person can not have same birthday).

$$= 1 - \frac{365 \times 364 \times 363 \times .... \times 356}{(365)^{10}} = .117$$

(ii) None has birthday on Jan 1 $= 1 - \dfrac{(364)^{10}}{(365)^{10}} = .0271$

So at least one has birthday on Jan 1 $= 1 - .0271 = .9729$

13. Assuming that each of 10^5 possible ordered quintlets has same probability, the probability is

$$\frac{10^3}{10^5} = \frac{1}{100} = .01$$

14. (i) $P_{70} = P_{65}\,(1 - .160) = .746 \times .840 = .627$

(ii) $\dfrac{P_{50}}{P_{25}} = \dfrac{.85}{.95} = \dfrac{17}{19}$ (conditional Probability)

15. $.42 \times .42 + .33 \times .33 + .18 \times .18 + .07 \times .07 = .1724 + .1098 + .0364 + .0049 = .2235$

16. He should put one white ball in one jar and the rest 199 (100 black, 99 white) in the other jar.

17. Probability that it will not rain this weekend is $\dfrac{1}{2} \times \dfrac{1}{2} = \dfrac{1}{4}$

So the probability that it will rain this weekend is $1 - \dfrac{1}{4} = \dfrac{3}{4}$ i.e. 75%

19. Since in reality no tyre was punctured probability of their naming any of the four tyres was equal for both of them. Hence the probability that Shyam named same tyre as Ram $= \dfrac{1}{4} \times \dfrac{1}{4} = \dfrac{1}{16}$

20. Not likely to succeed. If such a policy is followed, the family structures will be of type M, FM, FFM, FFFM. If family size is not to be large.

21. $\dfrac{1}{3} \times \dfrac{1}{3} = \dfrac{1}{9}$

22. Both arguments make sense. If the woman has randomly chosen a child to accompany her then answer is ½. However if she prefers taking male child with her in her outings, then second answer seems more justified.

23. Probability of head not appearing in 15 flips is $\dfrac{1}{2^{15}}$

Probability of head appearing in 15 flips is $\dfrac{1}{2} + \dfrac{1}{2^2} + \dfrac{1}{2^3} + \ldots\ldots + \dfrac{1}{2^{15}}$

Clearly prob. of head appearing in 15 flips is more than its not appearing in 15 flips. So one is advised not to accept the bet.

$$\text{Expected pay off} = 100 \times \frac{1}{2^{15}} + \left(\frac{1}{2} + \frac{1}{2^2} + \ldots\ldots + \frac{1}{2^{15}}\right) \times 100$$

$$= 100 \times \left(\frac{1}{2} + \frac{1}{2^2} + \ldots\ldots + \frac{1}{2^{15}}\right) + 100 \times \frac{1}{2^{15}}$$

$$= 100 \times \frac{1}{2}\left(1 - \frac{1}{2^{16}}\right) + 100 \times \frac{1}{2^{15}}$$

$$= 50 \times \left(1 - \frac{1}{2^{16}} + \frac{1}{2^{15}}\right)$$

24. $\dfrac{15 \times 65 + 69 \times 1}{16} = \dfrac{1044}{16} = \dfrac{261}{4} = 65.25''$ or (5 feet 5.25 inches)

25. Every student need not have identical marks in both the tests. What can at most be said is that if a student x is scoring m marks in test 1, there is some student (he need not be x) who is also scoring same m marks in test 2. In case tests are given to two different classes it implies that their average performance in the subject is identical.

26. In the first case $\overline{x} = \dfrac{\Sigma x_i f_i}{\Sigma f_i}$

$$= \frac{0 \times 3 + 1 \times 4 + 2 \times 6 + 3 \times 6 + 4 \times 4 + 5 \times 3 + 6 \times 2 + 7 \times 2}{30}$$

$$= \frac{91}{30} = .3$$

In the second case it is $= \dfrac{\Sigma x_i f_i}{\Sigma x_i} = \dfrac{91}{28} = 3.25$

27. After 5 seconds bug can reach (5, 0), (4, 1), (3, 2), (2, 3), (1, 4) and (0, 5). However probabilities are not the same. These are $\frac{1}{32}, \frac{5}{32}, \frac{5}{16}, \frac{5}{16}, \frac{5}{32}$ and $\frac{1}{32}$ respectively.

(terms of binomial expansion of $\left(\frac{1}{2} + \frac{1}{2}\right)^5$)

28. Average of 1, 2, and 3 is 2 lakhs kilometers. If the tyres can last only 2 lakh km each and the cost of one tyre is c, then his profit on sale of one tyre of each variety is $4500+6000 - 3c = 10500 - 3c$. It is profitable as long as $3c < 10500$ or $c < 3500$. So it will be profitable per sale of one tyre of each variety as long as $c < 3500$. If the tyre being produced is of the lowest quality then for sale of one tyre of each type two will be returned and there would be loss of $4500 - 3c$. In case the tyre being manufactured is of the best quality, then none would be returned and there would be profit of $10500 - 3c$ per sale of one tyre of each variety.

29. Let Y_1 be the payment received for the ith draw x_i. Then

$$Y_i = \begin{cases} 0 & \text{If } x_i \neq i \ (i = 1, 2, 3, 4) \\ 100 & \text{If } x_i = i \end{cases}$$

Now $P(Y_i = 1, 2, 3, 4) = \frac{1}{4}$. Hence $E[Y_i] = 100 \times \frac{1}{4} = 25, i = 1, 2, 3, 4.$

Hence expected pay off $= E[Y_1] + E[Y_2] + E[Y_3] + E[Y_4] = 4 \times 25 = 100.$

30. Not necessarily. In fact one cannot conclude anything about ρ_{xy} from this information. By definition

$$\rho = \frac{E[(X-\overline{X})(Y-\overline{Y})]}{\sigma_x \times \sigma_y}$$

Though $\overline{X} = \overline{Y}$, $\sigma_x = \sigma_y$ still x, y need not be identical.

31. Expected return on shifting is $\frac{1}{2} \times 50 + \frac{1}{2} \times 450 = 25 + 225 = 250$. So risk is worth taking as the amount in hand is just Rs. 150. In the second case expected return is $\frac{1}{2} \times 75 + \frac{1}{2} \times 300 = 187.50$. So even now the risk is still worth taking.

Chapter 9

1. Such an year will start on Tuesday. Thus Jan1 will be Tuesday and 20th will be Sunday.
2. Earliest on 2nd November if first of November is Monday. Latest 8th Nov. if Nov. one is Tuesday.
3. First day after completion of 100 years = 99 + 24 + 1 = 124 mod7= 5 Friday. First day after completion of 200 years = 199 + 48 +1 = 248 mod7 = 3 Wednesday.
   First day after completion of 300 years = 299 + 72 + 1 = 372 mod7 = 1 Monday.
   First day after completion of 400 years = 399 + 97 + 1 = 497 mod 7 = 0 Sunday.
   So Sunday, Monday, Wednesday or Friday.
4. There is a cycle of 400 years after which days of the week start repeating. Every year new year day of the week shifts one day (365 mod7 = 1). However in a leap year it shifts two days. So in 400 years it shifts 400 + 97 = 497 mod7 = 0 days. In each cycle of 28 years, new year day shifts 28 + 7 = 35 mod7 = 0 days. There are 14 complete cycles of 28 years in 400 year period with 8 extra years. So in 392 years new year falls on each day of the weak exactly 4 × 14 = 56 times. In 400th year it falls on Sunday. It also falls on Sunday in 395th year and Saturday on 394th year. Thus in this 8 year period it falls on Sunday twice and Saturday only once. Hence in a cycle of 400 years it falls once more on Sunday compared to Saturday.

Chapter 10

1. 1/3 days.
2. 30 cattle 5 attendant boys.
3. 3 men can cut in 2 days so one men can cut the same field in 6 days. 2 women can cut the field in 3 days so 1 woman is able to cut the field in 6 days. So the

rate of working of both men and women is same. Each can cut 1/6 of field in a day. So two of them will be able to cut in 3 days.

4. No.
5. Suppose there were x draws. Then $4x + 5 \times (10 - x) = 40$. Hence $x = 10$ There is no other solution.
6. Total sum of numbers on given cards is: $2 + 4 + 6 + 8 + 10 + 12 + 14 = 56$. Sum of numbers on Dolly's Paul's and Amit's cards is $20 + 10 + 14 = 44$.

   So the left out card is $56 - 44 = 12$. Dolly's cards are 6, 14 Paul's cards are 2, 8 and Amit's cards are 4 and 10.
7. Let there be x boys and y girls. If a chocolate costs c than sandwich costs $c + 1$. So

   $$cx + (1 + c)\,y + 10 = (1 + c)\,x + cy.$$
   $$c\,(x + y) + y + 10 = c\,(x + y) + x. \text{ So } x = 10 + y$$

   Hence there are 10 more boys then girls in the class.
8. First fold 1 makes thickness $\frac{2}{10}$ mm, 2nd makes thickness $\frac{4}{10} = \frac{2^2}{10}$ mm, 3rd makes thickness $\frac{8}{10} = \frac{2^3}{10}$ mm and so on. So the 32nd fold will make thickness $\frac{2^{32}}{10}$ mm or $\frac{2^{32}}{10^4}$ meters. (quite thick).
9. If all were to get equal wages. Then their monthly wages will be Rs. 30,000 each and none will be able to buy the bike. However if the wages of all are not equal but their average wages are Rs. 30000 pm obviously not all the workers can have wages Rs. 32000 or more per month. The persons whose wages are less than Rs. 32000 pm. will have to wait and these can not be less than two.
10. Average age of each person is 322/7= 46 years. Clearly age of some will be more than 46 years age and age of some less than 46 years. It will be possible to choose three people the sum of whose ages exceeds 140 years if the age of each of these three exceeds 46 years.
11. Keeping aside one coin put 50 of the rest in one pan and 50 in the other. If both pans balance, the left out coin is counterfeit. Weigh it against one of genuine coins to decide if it is lighter or heavier (2 weighings). However if two pans do not balance the counterfeit coin is among the remaining 100. Now put 25 coin from one pan on one side of the balance and the remaining 25 on the other pan (3rd weighing). If they balance counterfeit is in the other 50 else it is in these 50. After deciding which 50 contain counterfeit, keep out one coin and put 25 in one pan and the rest 24 along with one more from the other 50 non counterfeit coins in the other pan. If they balance left out coin is fake else decide in which

lot of 25 contains it. Again take out one coin out of the lot of 25 which contains fake coin and weigh the rest 12 against 12 and decide whether left out coin is fake or are decide the lot of 12 which has counterfeit. Continue like this. At each stage either the fake coin is detected or the lot that contains fake coin gets reduced in size to half. Ultimately we are left with a lot of 3 which has one fake coin. Now take out one coin and weigh the remaining two against each other to decide if the left out coin is counterfeit or one of these two. If the taken out is genuine weigh these against the genuine to decide the counterfeit coin and whether it is lighter or heavier.

12. Let the distance be s then $\frac{s}{45} + \frac{s}{90} = 4$ or $2s + s = 360$. So s is 120 km.

13. Out of 25, 8 can be of one variety, 8 of another and 9 of the third. If crates of any variety are less than 8, then deficit number will get added to one or both of the remaining two varieties. So in each case there are at least 9 crates of one variety.

14. One more than the number of beads of colors whose number is less (pigeon whole principle)

15. $8 + 6 + 2 = 16$.

16. We can divide the number into 10 disjoint groups in which one number divides the others. These are (1 divides all), (2 divides 4, 6, 8, 10, 12, 14, 16, 18, 20), (3 divides 6, 9, 12, 15, 18), (4 divides 8, 12, 16, 20), (5 divides 10, 15, 20), (6 divides 12, 18), (7 divides 14), (8 divides 16), (9 divides 18), (10 divides 20). Numbers beyond 10 i.e. 11 onwards divide none.

    Now since we are given 12 natural number from 1 to 20 all different, these will contain at least two numbers from the same group of which one number divides the other.

17. Each person can have no friend, one friend, 2 friends, 3 friends or 4. If A is a friend of B then B is a friend of A.

20. The way the books are stacked in library almirahs, the first page of volume 1 is separated from the last page of volume 2 by the two cover pages of these volumes. Similarly last page of volume 3 is separated from first page of volume 2 by cover page of these two volumes. So total distance to be traveled by the worm is just 5 cm. For this it will need only 5 weeks.

22. It is clear that he is making a net profit of Rs. 10 each day on each item purchased for Rs. 100. So he is making a profit of 10% each day. Again money invested by him on the first day is Rs. 100. He sells the purchased item same evening for Rs. 110 making a profit of Rs. 10. He invests the same 100 rupees next day to make profit of Rs. 10 on them. He continues like this for the whole year. Thus making a profit of 3650% for the whole year on just Rs. 100 invested in the beginning.

23. If the rain is falling vertically the best strategy for him is to walk steadily. However if the rain is falling at an angle to the vertical, the best strategy would be to move in the direction of rain at the same speed at which the rain is falling. This will result in essentially the rain falling on him vertically.

24. No. The sum of $1 + 2 + 3 + 4 \ldots\ldots 1985 = \frac{1985}{2}(2 \times 1 + (1985 - 1) \times 1)$ $= 1985 \times \frac{1986}{2} = 1985 \times 993$

which is an odd number. Now if the sign of one or more of these terms is changed to negative, the sum will decrease by an even number (twice of the number whose sign is changed) and this can never lead to zero answer.

25. At each inner crossing the person has two choices either driving east or north However at junctions on the outer roads, he has only one choice. However on reaching the north most street, he will have no choice. At starting point he has two choices. So $2 \times (m - 1)(n - 1) = 2 \times (8 - 1) \times (6 - 1) = 2 \times 7 \times 5 = 70$

27. Yes 241. ($241 \times \frac{1}{3} = 80 + 1$, so Ram takes 80, monkey 1 and 160 are left, next $160 \times \frac{1}{3} = 53 + 1$, so Shyam takes 53 monkey 1 and 106 are left, next $106 \times \frac{1}{3} = 35 + 1$ so Mohan takes 35, monkey 1 and 70 left, finally $70 \times \frac{1}{3} = 23 + 1$. So each gets 23 in the morning, and monkey 1).

28. Using coordinate geometry after the room has been opened (figure 2), we notice that if 0 is chosen origin horizontal direction as x-axis, vertically downward direction as Y-axis then the coordinates of spider S are:

$$\left(10 + 4, \frac{1}{2}\right) \text{ i.e.} \left(14, \frac{1}{2}\right) \text{ and that of F} \left(-\frac{1}{2}, 9 + 4\right) \text{i.e.} \left(-\frac{1}{2}, 13\right)$$

Hence shortest distance between S and F is

$$\sqrt{\left(14 + \frac{1}{2}\right)^2 + \left(-\frac{1}{2} + 13\right)^2} = \sqrt{196 + \frac{1}{4} + 14 + \frac{1}{4} + 169 - 13}$$

$$= \sqrt{359.2} \approx 19.88$$

29. No since sum or difference of an even and odd number cannot be even.

30. Minutes hand will coincide with the hour hand after making one circle and traveling further to meet hour hand which has also traveled some distance further mean while. Since minutes hand travels 12 times faster then the hour hand, they will meet x hours later where $1 + x = 12x$ or $x = 1/11$ hours or 60/11 minutes after one hour i.e. at one hour and 60/11 minutes after 12. They will again coincide at 2 hours 120/11 minutes after 12 and so on.

Chapter 12

12.2 1. $4 + 4 + 4 + 4 + 4 + 4 + 44 + 44 + 444 + 444$

2. Make following consecutive moves

   (i) 1 2 3 7 8 9 4 5 6

   (ii) 1 2 3 4 5 6 7 8 9

3. Squares I, II of dimensions (3×3), square III dimension (2×2) Squares IV to VIII each of dimension (1×1).

4. 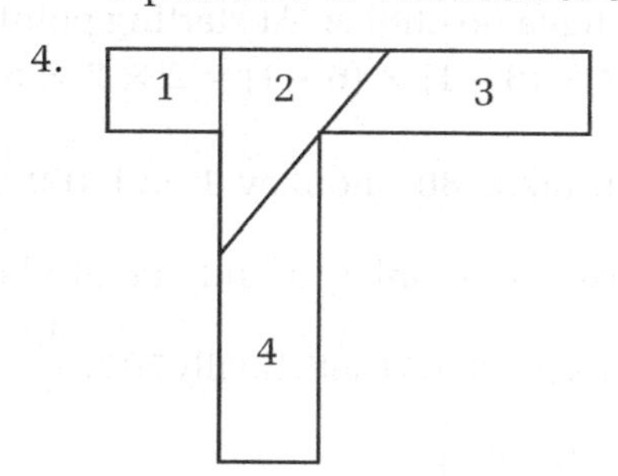

5.

6. 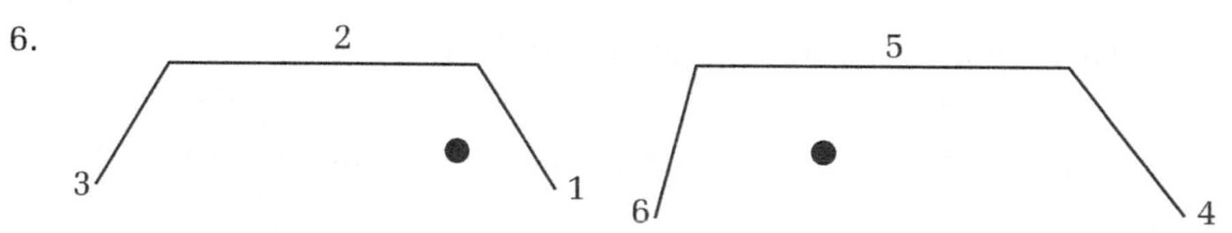

7. 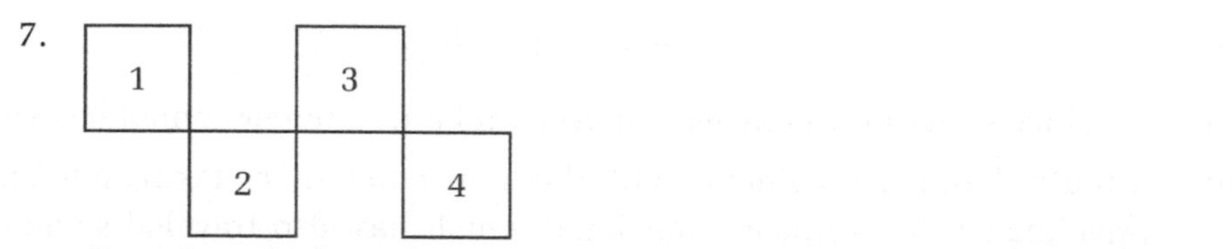

   (Two dotted edges shifted to form four squares).

9. Latin squares have been considered in chapter 16 and this problem is discussed there.

10. Step (i) change

0	0	0
0	0	0
0	0	0

to

4	4	0
4	4	0
0	0	0

by applying step 4 times on the top left most 2×2 square.

Step (ii) next change

4	4	0
4	4	0
0	0	0

to

4	9	5
4	9	5
0	0	0

by applying step 5 times on the top right most 2×2 square.

Step (iii) next change

4	9	5
4	9	5
0	0	0

to

4	9	5
10	15	5
6	6	0

by applying step 6 times on the bottom leftmost 2×2 square.

Step (iv) finally change

4	9	5
10	15	5
6	6	0

to obtain desired table

4	9	5
10	22	12
6	13	7

by applying step 7 times on the right most bottom 2×2 square.

12.3 1. Pick 3 and guess 1. If both make correct guesses or both make wrong guesses, then no payment is to be made to anyone. If you guess wrong and other guesses right you pay Rs. 1. He gets this plus what he picks (maximum 4). On the contrary if you guess right and other guesses wrong then you get Rs. 3 plus what he guesses (minimum 4).

2. Optimal strategy will be for both A and B to choose 2. If that happens, then A will have to pay B rupee one only.

   The argument in favour of this strategy is the following. A wants to maximize his gain and B wants to minimize his loss (or vice versa). Consider player A. He will argue that if he chooses 1, then he can win 4 if player B also chooses 1. However he can loose Rs. 4 if player B chooses 3. Similarly if A chooses 2, he will gain Rs. 3 if B chooses 1 but will lose Rs. 2 if B chooses 4. Similarly if A chooses 3, he will gain Rs. 3 if B chooses 2 and lose Rs. 2 if B chooses 4. In case of A choosing 4, he gains 1 if B chooses 4 but will lose Rs. 3 if B chooses 2 or 3. Clearly he would like to make a move in which he is sure of minimum loss and that move is for him to choose 2. Similar arguments apply to choice of B. Arguing in a similar way he will also prefer choosing 2.

3. Initially any number from 1 to 6 can be guessed. However as the total is about to cross 40 (or has crossed 40) only such a number be added so that sum is less than 44. In this example if A was to win he should not have added 6 when total was 39. Instead he should have added 4.

4. Central square is important as it controls 4 out of 8 possible winning combinations (one column, one row and two diagonals). Therefore the player should try to start from this position as far as possible. It increases his winning possibility. However the other player always has a chance of making it a draw.
5. Clearly the person whose turn comes when there is only one marble in the box will loose, as he cannot remove more than half of marbles in the box. For this to happen in the previous turn of other player there must have been two marbles. Had there been three, then since as he could not remove more than half, he would have removed one leaving two. In order to win the player must ensure that when he removes marbles in the last but one trial he must leave 3 marbles in the box.
6. The player should ensure that in his last but one move he leaves back exactly 7 chips. In that case when the other player removes any number of chips from 1 to 6, he will be leaving chips (from 1 to 6) which the first player can remove in one go.
7. When plus sign is put between two consecutive numbers, the sum results in an odd number. However if minus sign is put between two consecutive numbers, they yield result as –1. So one plus and one minus make sum even.
8. Effort should be made at each trial to ensure that there remain three or more not joined points on one side of an earlier drawn chord.
9. Suppose there are n_1 (say 101), n_2 (say 60) and n_3 (say 47)stones in these three heaps. Writing this in binary notation we have:

$n_1 = 101$	1100101
$n_2 = 60$	111100
$n_3 = 47$	101111
Final	0000000

When a player removes stones from a certain heap, he changes some of its zeros into ones and some ones into zeros. Hence number of ones in those columns change by one each. Thus parity of those columns changes from odd to even and vice versa. A winning strategy would be to ensure that number of ones in each column is even.

12.4 1. 1, 2, 4, 8, 16, 32.

2. Percentage is same since the quantity in each tumbler does not change.
3. Looking backwards if glass is full after 60 seconds, it must have been half full after 59 seconds.

4. Since he arrived 10 minutes early, so driving time from place where he met his wife to home must be 5 minutes. So he must have been walking 55 minutes before his wife met.

5. Let number of sons be x and number of daughters y, then M + 2My + 6Mx = 21,09,000 with M = 11,1000. Simplifying $2y + 6x = 18$ or $y + 3x = 9$. Hence $y = 3$ and $x = 2$. So he has 2 sons and 3 daughters.

6. At –40°. We know that 0°C is same as 32°F both being freezing points of water. Also 100°C equals 212°F (boiling point). Thus $1°C = \frac{9}{5}$. If temperature at which both are same is x, then $\frac{9}{5}x + 32 = x$ or $\frac{4}{5}x = -32$ i.e. $x = -40°$

   So –40°C = –40°F (Strange!)

7. 3 and 2. Let the cost be x 679.y0. Then it has to be divisible by 72. This requires $x = 3$, $y = 2$. Cost per item Rs. 51.20.

8. Start burning both ropes at the same time. However light both ends of one rope and only one end of the other rope. First rope will be fully burnt in half hour while the second rope will be only half burnt. Now ignite second end of the half burnt rope. It will now get fully burnt in next 15 minutes.

9. When car travels 1 km each of its four tyres travel 1 km. Hence when car travelled 20,000 km, four of its working tyres travelled 80,000 km. Thus each of the 5 tyres travelled only 16,000 km.

10. 4 boys and 3 girls.

11. 46; $(30 \times 34 - 26 \times 18)/(30 - 18) = 552/12 = 46$.

12. This is a bit tough question based on prime numbers. 999919 has factors 991 and 1009 which are both prime. So the number of cats is 991 and number of mice each cat killed is 1009.

13. 5 stamps of Rs. 2, 30 stamps of Rs. 1 each and 7 stamps of Rs. 5 each.

14. 84.

15. (i) 9, 81, 324, 576.
    (ii) Smallest: 139854276 (square root 11826), Largest : 923187456 (square root 30384)
    (iii) 386, 2114, 3970 and 10430.

16 Let there be $n + m$ houses in the street and my house number be n. Then we have to find n and m such that

$$1 + 2 + 3 + \dots\dots + (n - 1) = (n + 1) + (n + 2) + \dots\dots + (n + m)$$

i.e. $\frac{n-1}{2}[2+(n-2)] = \frac{m}{2}[2(n+1)+(m-1)]$

or $(n-1)n = m(2n+m+1)$

Now for $n = 204$, $n(n-1) = 204 \times 203 = 41412$

so $m(409+m) = 41412$

or $m^2 + 409m - 41412 = 0$

Hence $m = \dfrac{-409 \pm \sqrt{(409)^2 + 16568}}{2} = 84$

So the total number of houses in the street is 204 + 84 = 288.

17. 10968 when turned upside down and reversed it reads 89601.

18. 7 and 5.

19. Let it be x then $100(30-x) = 150x$

or $250x = 3000$ or $x = 12$

So he worked for 18 days and was absent for the remaining 12 days.

20. From each corner of a polygon of n sides we can draw $n-3$ diagonals.

So their number is $\frac{1}{2} \times n(n-3)$ as diagonal from corner i to corner j is same as diagonal from corner j to i corner. For hexagon $n = 6$. So their number is $\frac{1}{2} \times 6(6-3) = 9$

21. Second is preferable.

22. 31″, 41″ and 49″.

23. 20

24.

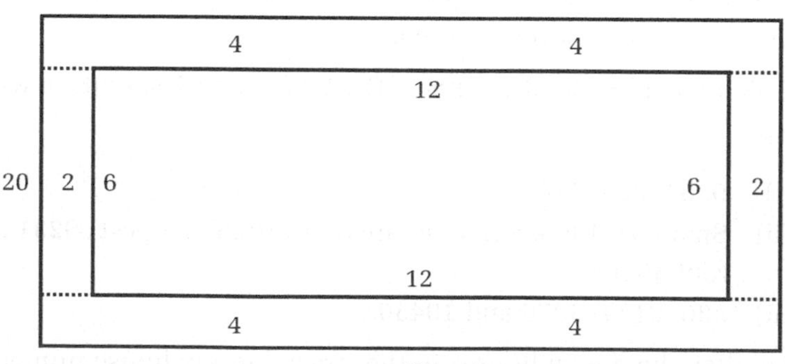

length = 12 + 2 + 2 = 14″, Height 6 + 4 + 4 = 14″

25.

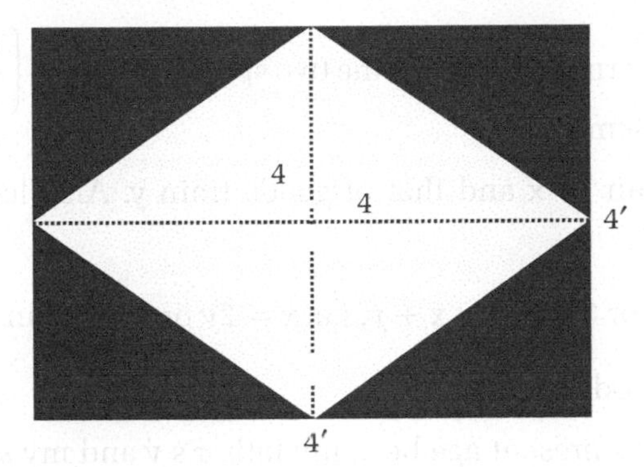

26. Area 194.4

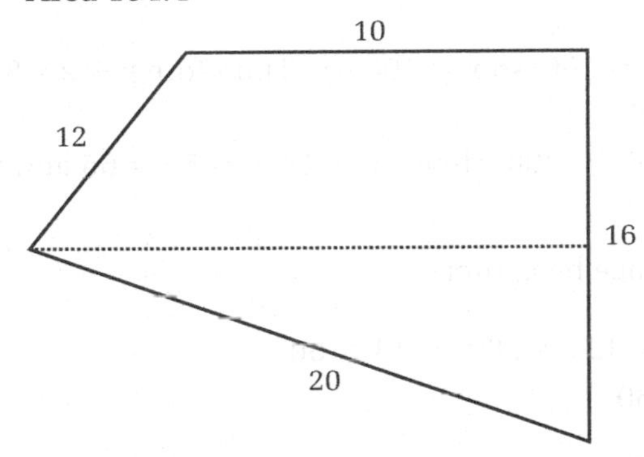

27. 60, 62, 64 and 70 inches.

28. (i) C.P. of first is 10000 and C.P. of second is 15000.
Total C.P. = 10000 + 15000 = 25000.
Total S.P. = 12000 × 2 = 24000. Loss is Rs. 1000.

(ii) First time loss Rs. 100. Second time again loss Rs. 100. So total loss Rs. 200.

(iii) He loses as long as number of times he wins equals the number of times he loses irrespective of the order of winning and losing.

(iv) Two for a rupee means one for $\frac{1}{2}$ rupees. Three for a rupee means one for $\frac{1}{3}$ rupees.

Five for two rupees means one for $\frac{2}{5}$ rupees.

So he first loses $\frac{1}{2} - \frac{2}{5} = \frac{5-4}{10} = \frac{1}{10}$ or Rs. 1 on the sale of his 10 marbles.

Second gains $\frac{2}{5} - \frac{1}{3} = \frac{6-5}{15} = \frac{1}{15}$ or Rs. 1 on the sale of his 15 marbles.

29. This is the inverse of harmonic mean of the two speeds. Since $\frac{1}{2}\left(\frac{1}{40}+\frac{1}{60}\right)$ $= \frac{1}{48}$. So speed is 48km/hr.

30. Let speed of mail train be x and that of goods train y. Also let length of goods train be l.

    Then $\frac{3l}{x+y} = \frac{l}{x-y}$ or $3(x-y) = x + y$, i.e. $x = 2y$ or mail train is running twice as fast as the goods train.

31. (i) 65, 35, 13. Let my present age be x, my father's y and my son's z. Then $x + (y - x) = 5z$, so $y = 5z$. Also $(y - x) + z = 8 + x$ or $y + z = 8 + 2x$.

    Moreover $x + y = 100$, so $x = 100 - y$. Therefore $y + z = 8 + 200 - 2y$ i.e. $3y + z = 208$.

    Using $y = 5z$, $16z = 208$. Hence $z = 13$, $y = 5z = 65$ and $x = 100 - y = 35$.

    (ii) Let his present age be x, then $x - \left(\frac{x}{4}+\frac{x}{5}+\frac{x}{3}\right) = 13$.

    i.e. $60x - (15x + 12x + 20x) = 13 \times 60$

    or $13x = 13 \times 60$

    i.e. $x = 60$

(iii) Let the age of lady be xy. So her age is $10x + y$ and the age of her husband is $10y + x$.

Now $(10y + x) - (10x + y) = \frac{1}{11}(10y + x + 10x + y)$. i.e. $9y - 9x = x + y$ or $8y = 10x$.

Therefore $\frac{x}{4} = \frac{y}{5}$. So the age of lady 45 and that of her husband 54.

32. As it rises 4 feet while completing one circuit so in ascending 4 feet it has travelled 5 feet.(Had it been moving straight on a horizontal surface it would have travelled only 3 feet in completing a circuit. As it has travelled 4 feet vertically it has travelled one feet extra because of spiral path). Since pole is 16 feet high it will have to travel 20 feet to reach its top.

33. After first round 1, 2, 4, 5, 7, 8, 10, 11, 13, 14, 16, 17, 19, 20, 22, 23, 25, 26, 28, 29, 31, 32, 34, 35, 37, 38, 40, 41.

    After second round 2, 4, 7, 8, 11, 13, 16, 17, 20, 22, 25, 26, 29, 31, 34, 35, 38, 40.

    After third round 2, 4, 8, 11, 16, 17, 22, 25, 29, 31, 35, 38.

    After fourth round 2, 4, 11, 16, 22, 25, 31, 35.

    After fifth round 2, 4, 16, 22, 31, 35.

After sixth round 4, 16, 31, 35.
After seventh round 16, 31.

So answer be (i) is No.
(ii)Yes, if they position themselves at 16, 31

In case they sit in a straight line, then

After first round 1, 2, 4, 5, 7, 8, 10, 11, 13, 14, 16, 17, 19, 20, 22, 23, 25, 26, 28, 29, 31, 32, 34, 35, 37, 38, 40, 41.

After second round 1, 2, 5, 7, 10, 11, 14, 16, 19, 20, 23, 25, 28, 29, 32, 34, 35, 37, 41.

After third round 1, 2, 7, 10, 14, 16, 20, 23, 28, 29, 34, 35, 41.

After fourth round 1, 2, 10, 14, 20, 23, 29, 34, 41.

After fifth round 1, 2, 14, 20, 29, 34.

After sixth round 1, 2, 20, 29.

After seventh round 1, 2, 29.

After eighth round 1, 2.

In this case persons at positions one and two will escape and the two should sit at positions 1, 2.

34. Sounds very complicated but in reality it is not so. Just reverse the process. Answer is 28.

Now $(10y + x) - (10x + y) = \frac{1}{11}(10y + x + 10x + y)$. i.e. $9y - 9x = x + y$ or $8y = 10x$.

$2 \times 10 = 20 - 8 = 12$, $122 = 144$, $144 + 52 = 196$, $196 = 14$, Hence $\frac{x}{7} - \frac{1}{3} \cdot \frac{x}{7} = 14$

So $x = 21 \times 7 = 147$

Now $147 = y + \frac{3}{4}y = \frac{7}{4}y$.

Thus $y = 21 \times 4 = 84$ and again $\frac{84}{3} = 28$.

35. Answer 72 bees. Let bees be x in number, then

$x = \sqrt{\frac{1}{2}x} + \frac{8}{9}x + 2$ writing $\sqrt{\frac{x}{2}} = t$, we have $x = 2t^2$

so, $2t^2 = t + \frac{8}{9} \times 2t^2 + 2$ i.e. $\frac{2t^2}{9} - t - 2 = 0$ or $2t^2 - 9t - 18 = 0$

Hence $t = \frac{9 \pm \sqrt{81 + 144}}{4} - \frac{9 \pm 15}{4} = 6$ (negative answer not being admissible).

Therefore $x = t^2 = 2 \times 36 = 72$

Mind Teasers

1. No
4. 14, 8.
7. 72 km p.h. (it covers 6 km in 5 minutes)
8. Let them meet when A has travelled the distance with speed s_1. Let the speed of B be s_2. Then when they meet

$$\frac{x}{s_1} = \frac{(24 - x)}{s_2} + 2 \text{ or}$$

This yields $x = 14$, $s_1 = 3\frac{1}{7}$, $s_2 = 4\frac{4}{5}$

So point of meeting is 14 km from A. A walks at the speed of $3\frac{1}{7}$ km/hr and B at a speed of $4\frac{4}{5}$

9. Let the numbers be $x - 1, x, x + 1$, Then $(x - 1)\,x + x\,(x + 1) + (x - 1)(x + 1) = 74$

or $3x^2 = 75$ or $x = 5$,

so the numbers are (4, 5, 6)

12. Second.
13. CDAB
14. Three
15. First trip: Farmer and his son (time 2 minutes)

Return: Son stays father returns (time 1 minute)

Second trip: Farmer's mother and wife (time 10 minutes)

Return: Both stay and son returns (time 2 minutes)

Third trip: Farmer and his son (time 2 minutes)

Total time: 17 minutes.

(NOTE: *Alternative solutions are possible. Is there any solution with total time less than 17 minutes?*).

Chapter 13

(1)

3	9	6	1	4	8	5	7	2
7	5	8	3	6	2	9	1	4
1	2	4	7	5	9	8	3	6
4	3	2	8	1	6	7	5	9
5	7	9	2	3	4	1	6	8
6	8	1	5	9	7	4	2	3
8	4	5	6	2	1	3	9	7
2	1	7	9	8	3	6	4	5
9	6	3	4	7	5	2	8	1

(2)

7	4	2	3	1	9	6	5	8
9	5	8	6	7	4	2	1	3
6	3	1	2	5	8	4	9	7
8	7	4	5	6	2	9	3	1
2	1	5	9	3	7	8	4	6
3	9	6	8	4	1	5	7	2
4	2	7	1	9	6	3	8	5
5	8	9	7	2	3	1	6	4
1	6	3	4	8	5	7	2	9

(3)

8	9	1	7	3	5	2	4	6
2	3	6	8	4	1	5	7	9
7	5	4	2	6	9	8	3	1
1	7	3	9	5	6	4	2	8
9	6	8	3	2	4	1	5	7
4	2	5	1	8	7	9	6	3
5	4	9	6	1	3	7	8	2
6	1	2	4	7	8	3	9	5
3	8	7	5	9	2	6	1	4

(4)

3	4	9	2	7	8	6	1	5
6	1	2	3	4	5	7	8	9
7	5	8	9	1	6	4	2	3
1	3	4	5	6	2	8	9	7
8	6	7	4	3	9	1	5	2
9	2	5	1	8	7	3	6	4
5	7	6	8	2	3	9	4	1
2	8	1	7	9	4	5	3	6
4	9	3	6	5	1	2	7	8

(5)

3	4	8	7	2	6	1	9	5
5	1	7	8	9	4	3	6	2
2	6	9	3	5	1	8	4	7
1	9	4	2	3	8	7	5	6
7	5	3	6	4	9	2	1	8
6	8	2	5	1	7	9	3	4
4	3	6	1	8	2	5	7	9
8	7	1	9	6	5	4	2	3
9	2	5	4	7	3	6	8	1

(6)

5	9	2	1	7	6	4	8	3
3	1	8	2	5	4	6	7	9
4	7	6	3	9	8	1	5	2
2	8	1	4	3	5	9	6	7
9	4	7	6	8	2	3	1	5
6	3	5	9	1	7	8	2	4
8	5	3	7	4	1	2	9	6
7	6	9	8	2	3	5	4	1
1	2	4	5	6	9	7	3	8

(7)

7	9	4	8	1	3	6	5	2
6	5	3	7	2	9	4	1	8
2	8	1	6	5	4	9	3	7
5	3	8	4	9	1	2	7	6
9	1	7	2	6	8	5	4	3
4	2	6	5	3	7	8	9	1
8	4	2	3	7	5	1	6	9
1	7	5	9	8	6	3	2	4
3	6	9	1	4	2	7	8	5

(8)

4	7	2	5	3	9	8	6	1
3	8	6	2	4	1	5	9	7
9	1	5	6	7	8	2	3	4
6	4	8	7	9	3	1	2	5
5	3	1	8	2	4	9	7	6
2	9	7	1	6	5	3	4	8
1	6	4	3	8	2	7	5	9
7	5	3	9	1	6	4	8	2
8	2	9	4	5	7	6	1	3

(9)

2	9	1	4	5	8	6	7	3
3	8	6	7	2	9	4	1	5
7	5	4	3	6	1	9	8	2
5	1	7	6	8	2	3	9	4
8	6	3	9	4	7	2	5	1
9	4	2	1	3	5	7	6	8
6	2	9	8	1	4	5	3	7
1	3	5	2	7	6	8	4	9
4	7	8	5	9	3	1	2	6

(10)

3	5	1	4	8	6	7	2	9
7	9	6	2	5	1	3	4	8
2	4	8	3	7	9	6	1	5
6	1	2	9	4	8	5	3	7
5	8	4	1	3	7	9	6	2
9	3	7	5	6	2	1	8	4
8	2	5	7	1	3	4	9	6
1	7	9	6	2	4	8	5	3
4	6	3	8	9	5	2	7	1

(11)

4	6	1	8	9	5	2	7	3
8	9	3	6	7	2	4	5	1
7	2	5	3	4	1	8	6	9
3	5	9	1	6	8	7	4	2
1	4	7	2	5	9	6	3	8
2	8	6	4	3	7	9	1	5
5	3	4	9	8	6	1	2	7
6	1	8	7	2	3	5	9	4
9	7	2	5	1	4	3	8	6

(12)

4	6	9	1	8	7	3	5	2
1	2	7	6	3	5	8	9	4
8	3	5	9	4	2	7	1	6
2	8	3	4	6	1	9	7	5
5	4	1	7	9	8	2	6	3
9	7	6	2	5	3	4	8	1
6	1	2	3	7	9	5	4	8
7	5	4	8	2	6	1	3	9
3	9	8	5	1	4	6	2	7

(13)

9	7	5	8	1	6	4	2	3
2	8	3	5	7	4	6	9	1
1	4	6	9	3	2	7	8	5
8	1	4	2	6	9	5	3	7
6	3	9	7	5	8	2	1	4
5	2	7	1	4	3	9	6	8
4	6	2	3	8	5	1	7	9
7	9	8	4	2	1	3	5	6
3	5	1	6	9	7	8	4	2

(14)

3	6	8	9	5	7	1	2	4
4	9	5	8	1	2	3	7	6
7	2	1	4	3	6	9	5	8
2	4	9	7	8	3	6	1	5
5	8	3	1	6	4	2	9	7
1	7	6	2	9	5	8	4	3
6	5	7	3	2	1	4	8	9
9	1	4	6	7	8	5	3	2
8	3	2	5	4	9	7	6	1

(15)

9	5	4	7	2	8	6	1	3
8	1	7	3	6	4	5	2	9
3	2	6	5	9	1	8	4	7
6	4	1	8	7	2	3	9	5
7	9	8	4	5	3	1	6	2
2	3	5	9	1	6	4	7	8
4	7	9	1	3	5	2	8	6
1	6	3	2	8	7	9	5	4
5	8	2	6	4	9	7	3	1

(16)

4	2	8	9	3	6	1	7	5
5	3	7	1	8	2	4	6	9
6	1	9	5	7	4	3	8	2
2	7	4	6	1	8	5	9	3
8	9	1	4	5	3	6	2	7
3	6	5	2	9	7	8	4	1
9	4	3	8	2	1	7	5	6
1	5	6	7	4	9	2	3	8
7	8	2	3	6	5	9	1	4

(17)

2	5	7	6	8	9	4	3	1
1	3	8	2	4	7	5	9	6
4	6	9	5	1	3	2	7	8
7	9	2	8	3	1	6	4	5
6	8	1	7	5	4	3	2	9
3	4	5	9	6	2	1	8	7
8	1	4	3	9	6	7	5	2
5	7	3	1	2	8	9	6	4
9	2	6	4	7	5	8	1	3

(18)

5	2	7	9	4	6	8	3	1
8	9	3	5	1	2	4	6	7
6	4	1	3	7	8	9	2	5
4	8	2	7	5	1	6	9	3
1	6	9	2	8	3	5	7	4
7	3	5	4	6	9	2	1	8
3	7	6	8	9	4	1	5	2
9	5	4	1	2	7	3	8	6
2	1	8	6	3	5	7	4	9

(19)

4	5	1	8	9	3	2	7	6
2	8	3	6	4	7	9	5	1
6	9	7	5	1	2	4	8	3
8	2	4	9	3	1	7	6	5
7	1	5	2	6	8	3	4	9
9	3	6	4	7	5	1	2	8
3	7	8	1	5	4	6	9	2
5	4	9	3	2	6	8	1	7
1	6	2	7	8	9	5	3	4

(20)

1	7	4	5	2	6	3	9	8
8	5	9	3	7	4	2	6	1
2	3	6	9	8	1	4	5	7
7	8	2	4	6	5	9	1	3
5	4	1	7	9	3	6	8	2
6	9	3	2	1	8	7	4	5
9	1	5	6	3	7	8	2	4
3	2	8	1	4	9	5	7	6
4	6	7	8	5	2	1	3	9

4. (1)

3	6	5	7	1	2	4	9	8				9	1	2	3	6	4	5	7	8
2	4	1	3	9	8	5	6	7				7	8	3	1	2	5	4	9	6
8	9	7	6	5	4	1	3	2				4	5	6	8	9	7	2	3	1
1	3	2	8	7	6	9	4	5				6	7	8	9	1	2	3	4	5
9	7	4	5	3	1	2	8	6				5	9	4	7	3	8	6	1	2
6	5	8	2	4	9	3	7	1				3	2	1	5	4	6	7	8	9
5	1	6	9	8	3	7	2	4	9	3	8	1	6	5	4	8	3	9	2	7
4	2	9	1	6	7	8	5	3	6	1	7	2	4	9	6	7	1	8	5	3
2	8	3	4	2	5	6	1	9	5	4	2	8	3	7	2	5	9	1	6	4
						3	4	5	2	7	6	9	8	1						
						1	6	8	3	9	5	4	7	2						
						9	7	2	8	1	4	3	5	6						
8	7	5	1	9	4	2	3	6	8	5	1	7	9	4	5	3	6	1	8	2
3	9	6	5	2	7	4	8	1	7	6	9	5	2	3	1	8	7	6	4	9
2	4	1	6	3	8	5	9	7	4	2	3	6	1	8	4	9	2	5	7	3
4	5	3	8	6	2	7	1	9				1	8	9	6	2	4	3	5	7
1	6	7	9	4	5	8	2	3				4	3	5	8	7	1	2	9	6
9	2	8	3	7	1	6	4	5				2	6	7	9	5	3	4	1	8
7	3	9	4	8	6	1	5	2				3	4	1	7	6	8	9	2	5
5	8	2	7	1	9	3	6	4				9	7	2	3	1	5	8	6	4
6	1	4	2	5	3	9	7	8				8	5	6	2	4	9	7	3	1

(2)

1	7	5	9	3	8	2	4	6				2	1	8	4	7	9	5	3	6
4	8	9	1	2	6	3	7	5				9	4	5	2	3	6	7	8	1
6	3	2	5	4	7	1	8	9				3	7	6	8	5	1	9	4	2
8	5	1	3	9	4	7	6	2				6	9	4	3	2	5	1	7	8
2	6	3	8	7	1	5	9	4				1	3	7	9	6	8	2	5	4
7	9	4	2	6	5	8	3	1				5	8	2	1	4	7	3	6	9
9	4	8	7	5	2	6	1	3	2	7	4	8	5	9	6	1	3	4	2	7
3	2	7	6	1	9	4	5	8	3	9	6	7	2	1	5	8	4	6	9	3
5	1	6	4	8	3	9	2	7	5	1	8	4	6	3	7	9	2	8	1	5
						3	4	6	8	5	7	9	1	2						
						5	9	1	6	4	2	3	7	8						
						8	7	2	9	3	1	6	4	5						
5	4	3	2	8	1	7	6	9	1	8	5	2	3	4	7	1	5	6	8	9
6	8	2	7	5	9	1	3	4	7	2	9	5	8	6	2	4	9	7	1	3
7	1	9	4	3	8	2	8	5	4	6	3	1	9	7	6	8	3	4	5	2
3	2	4	5	7	6	9	1	6				8	6	9	3	5	1	2	4	7
8	5	6	9	1	3	4	2	7				7	5	2	9	2	4	1	6	8
1	9	7	6	4	2	3	5	8				4	1	3	8	6	7	3	9	5
2	6	1	8	9	7	5	4	3				9	7	5	4	3	6	8	2	1
4	7	8	3	2	5	6	9	1				6	2	1	5	7	8	9	3	4
9	3	5	1	6	4	8	7	2				3	4	8	1	9	2	5	7	6

Chapter 14

1. (i)
$$\begin{array}{r} 825 \\ +476 \\ \hline 1301 \end{array}$$

(ii)
$$\begin{array}{r} 427 \\ -356 \\ \hline 71 \end{array}$$

(iii)
$$\begin{array}{r} 340 \\ \times 12 \\ \hline 680 \\ 340 \\ \hline 4080 \end{array}$$

(iv)
$$\begin{array}{r} 132 \\ \times 74 \\ \hline 528 \\ 924 \\ \hline 9768 \end{array}$$

(v)
$$\begin{array}{r} 853 \\ 749\overline{)638897} \\ 5992 \\ \hline 3969 \\ 3745 \\ \hline 2247 \\ 2247 \\ \hline \end{array}$$

2. (i)

2× 1	2	7+ 3	4	75× 5
40× 4	3÷ 1	8× 2	5	3
5	3	4	2÷ 1	2
2	60× 4	15× 5	2÷ 3	4÷ 1
3	5	2× 1	2	4

(ii)

2÷ 1	24× 2	3	4	5÷ 5
2	60× 3	14+ 4	5	1
1– 3	4	5	3+ 1	2
4	5	2× 1	5+ 2	1– 3
6+ 5	1	2	3	4

(iii)

12× 3	8× 1	4	3– 2	40× 5
1	8+ 3	2	5	4
4	5	2– 3	4+ 1	2
1– 2	4	5	3	1– 1
7+ 5	2	4÷ 1	4	3

(iv)

12× 3	4	3– 5	2	7+ 1
3+ 2	8+ 3	60× 4	1	5
1	2	3	5	2÷ 4
3– 4	4– 5	1	36× 3	2
5	2÷ 1	2	4	3

(v)

8+ 2	3	80× 4	4 5	5× 1
3	3+ 1	2	4	5
10+ 4	10× 2	5	4+ 1	3
1	1– 5	5+ 3	2	6+ 4
5	4	3÷ 1	3	2

(vi)

1– 3	4	3+ 2	1	25× 5
11+ 4	2	24× 3	5	1
5	3× 1	4	2	5+ 2
8+ 1	3	4– 5	48× 4	3
2	5	1	3	4

(vii)

7+ 3	20× 4	5	15× 1	8× 2
4	5+ 2	3	5	1
5÷ 1	5	6+ 2	3	4
3− 5	3× 1	4	8× 2	2− 3
2	3	1	4	5

(viii)

6+ 2	8+ 3	3÷ 1	1− 4	5
4	5	3	3+ 1	9+ 2
3× 1	1− 4	5	2	3
3	1	3− 2	5	4
11+ 5	2	4	3÷ 3	1

(ix)

12+ 4	2− 5	3	11+ 2	2× 1
3÷ 1	3	4	5	2
2− 2	4	5× 5	1	2− 3
2− 3	3+ 2	1	48× 4	5
5	2× 1	2	3	4

(x)

4+ 1	2÷ 4	2	45× 3	20× 5
2	1	3	5	4
12× 4	3	5× 5	1	2÷ 2
2− 3	9+ 5	4	24× 2	1
5	2× 2	1	4	3

3. (i) $\begin{array}{r} 122 \\ +322 \\ \hline 444 \end{array}$ (ii) $\begin{array}{r} 765 \\ +567 \\ \hline 1332 \end{array}$ (iii) $\begin{array}{r} 345 \\ +814 \\ \hline 1159 \end{array}$ (iv) $\begin{array}{r} 41 \\ -14 \\ \hline 27 \end{array}$

(v) $\begin{array}{r} 3249 \\ -2394 \\ \hline 855 \end{array}$ (vi) $\begin{array}{r} 92 \\ \times 9 \\ \hline 828 \end{array}$ (vii) $\begin{array}{r} 21 \\ \times 12 \\ \hline 252 \end{array}$ (viii) $\begin{array}{r} 1029 \\ \times\ 9 \\ \hline 9261 \end{array}$

(ix) $\begin{array}{r} 2 \\ 23 \\ 234 \\ +2341 \\ \hline 2600 \end{array}$ (x) $\begin{array}{r} 321 \\ \times 231 \\ \hline 321 \\ 963 \\ 642 \\ \hline 74151 \end{array}$ (xi) $\begin{array}{r|l} & 54 \\ \hline 321 & 17334 \\ & 1605 \\ \hline & 1284 \\ & 1284 \\ \hline \end{array}$

4. (i)
$$\begin{array}{r} 251 \\ \times 521 \\ \hline 251 \\ 502 \\ 1255 \\ \hline 130771 \end{array}$$

(ii)
$$\begin{array}{r} 129 \\ \times 129 \\ \hline 1161 \\ 258 \\ 129 \\ \hline 16641 \end{array}$$

(iii)
$$\begin{array}{r} 462 \\ \times 462 \\ \hline 924 \\ 2772 \\ 1848 \\ \hline 213444 \end{array}$$

(iv)
$$\begin{array}{r} 3201 \\ \times 3201 \\ \hline 3201 \\ 0000 \\ 6402 \\ 9603 \\ \hline 10246401 \end{array}$$

(v)
$$\begin{array}{r} 64434 \\ \times 34 \\ \hline 257736 \\ 193302 \\ \hline 2190756 \end{array}$$

(vi)
$$\begin{array}{r|l} & 132 \\ \cline{2-2} 432 & 57024 \\ & 432 \\ \cline{2-2} & 1382 \\ & 1296 \\ \cline{2-2} & 864 \\ & 864 \\ \cline{2-2} \end{array}$$

(vii)
$$\begin{array}{r|l} & 951 \\ \cline{2-2} 951 & 904401 \\ & 8559 \\ \cline{2-2} & 4850 \\ & 4755 \\ \cline{2-2} & 951 \\ & 951 \\ \cline{2-2} \end{array}$$

5. (i)
$$\begin{array}{r} 231 \\ +231 \\ \hline 462 \end{array}$$

(ii)
$$\begin{array}{r} 431 \\ +6420 \\ \hline 6851 \end{array}$$

(iii)
$$\begin{array}{r} 85254 \\ +50671 \\ \hline 135925 \end{array}$$

(iv)
$$\begin{array}{r} 24153 \\ +24153 \\ \hline 48306 \end{array}$$

(v)
$$\begin{array}{r} 1567 \\ +9085 \\ \hline 10652 \end{array}$$

(vi)
$$\begin{array}{r} 45881795 \\ +302758 \\ \hline 46184553 \end{array}$$

(vii)
$$\begin{array}{r} 582472 \\ 272429 \\ 31650 \\ +32429 \\ \hline 919250 \end{array}$$

(viii)
$$\begin{array}{r} 3238550 \\ 3238550 \\ +37984 \\ \hline 6515084 \end{array}$$

(ix)
$$\begin{array}{r} 19564882 \\ 78082 \\ 482 \\ +482 \\ \hline 19643928 \end{array}$$

(x)
$$\begin{array}{r} 14963 \\ +15823 \\ \hline 30786 \end{array}$$

(xi)
$$\begin{array}{r} 613 \\ 804 \\ +79508 \\ \hline 80925 \end{array}$$

6. (i) $4 \times AYE = 3 \times YES \rightarrow 400\ A + 40\ Y + 4\ E = 300\ Y + 30\ E + 3\ S$
or $400\ A = 260\ Y + 26\ E + 3\ S$ which is satisfied for $A = 5$, $Y = 7$, $E = 6$, and $S = 8$.
So $4 \times 576 = 3 \times 768$

(ii) $6268449 + 6268449 + 14349 + 14349 + 14349 + 849 = 12580749$

(iii) $15 + 67 = 82$, $23 + 40 = 63$, $67 + 40 = 107$

Chapter 15

1. Yes, it is possible in total of 16 moves. Seven for each color and two back moves to clear the way for the opposite knight.
2. In case of chess board where one dimension is odd and other even the number of squares of one color is same as that of the other. For example in case of the

chess board of the size 5×6 we have 15 white and 15 blacksquares. So a knight tour should be possible.

B	W	B	W	B	W
W	B	W	B	W	B
B	W	B	W	B	W
W	B	W	B	W	B
B	W	B	W	B	W

Fig. 1

3. These are possible in case of chess boards of sizes 4×8, 3×8 and 6×6.
4. A possible tour is shown in Fig. 2

34	7	24	15	32	1
23	14	33	36	25	16
6	35	8	17	2	31
13	22	29	26	9	18
28	5	20	11	30	3
21	12	27	4	19	10

Fig. 2

5. A possible tour is shown in Fig. 3.

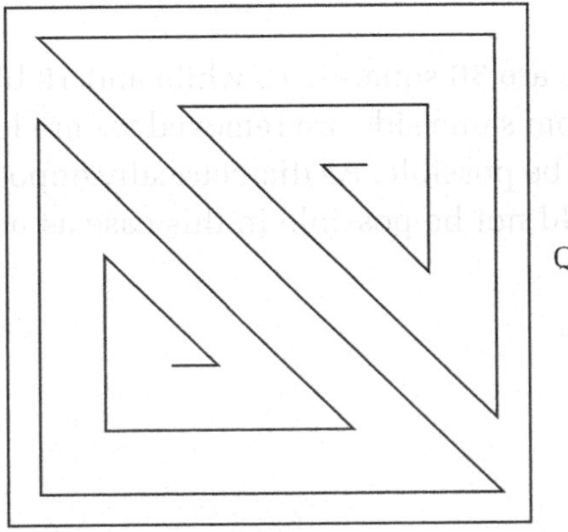

Fig. 3

8. A possible tour is shown in Fig. 4

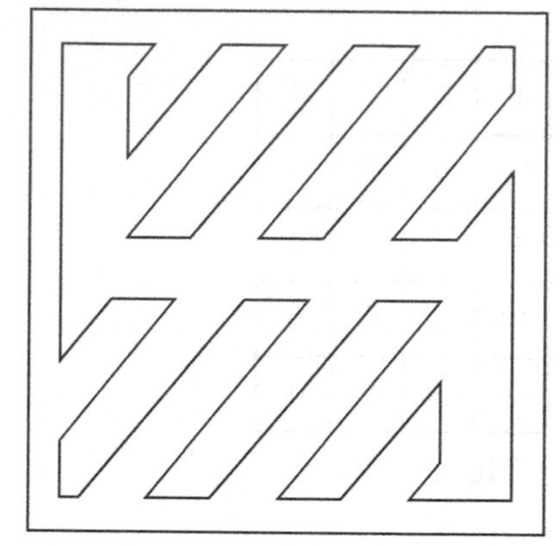

A King's tour with fourteen horizontal moves and fourteen vertical moves

Fig. 4

10. A possible tour is shown in Fig. 5

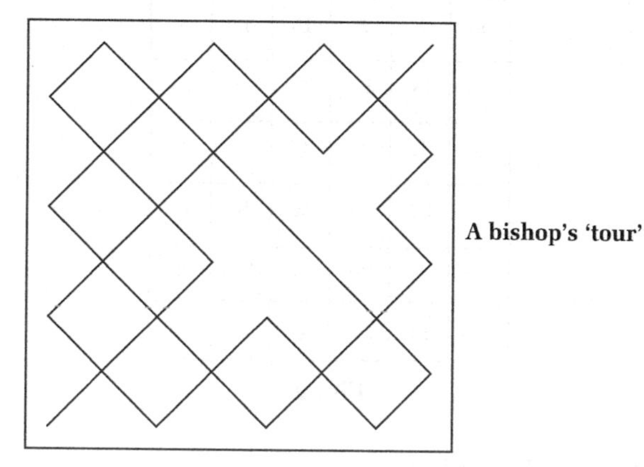

A bishop's 'tour'

Fig. 5

11. In a 6×6 board there are 36 squares, 18 white and 18 black. So, it should be possible. If corners from same side are removed we are left with 17 white and 17 black. So it should be possible. As diametrically opposite corners are of the same color so it should not be possible in this case as now we have 18 black and 16 white squares.
12. Possible in both cases.

13. Two possible fault free domino coverings are:

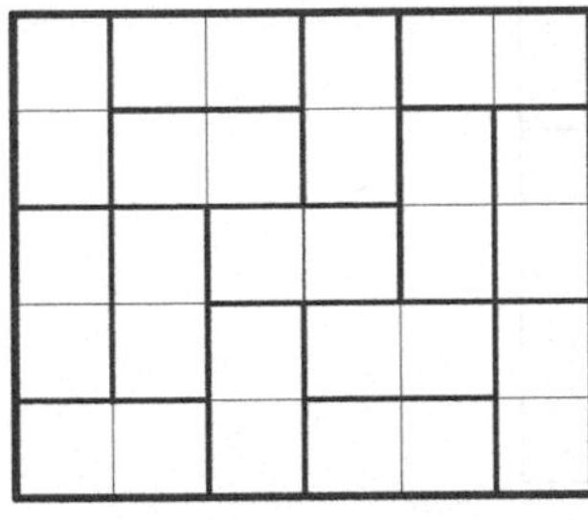
Fig. 6(a)

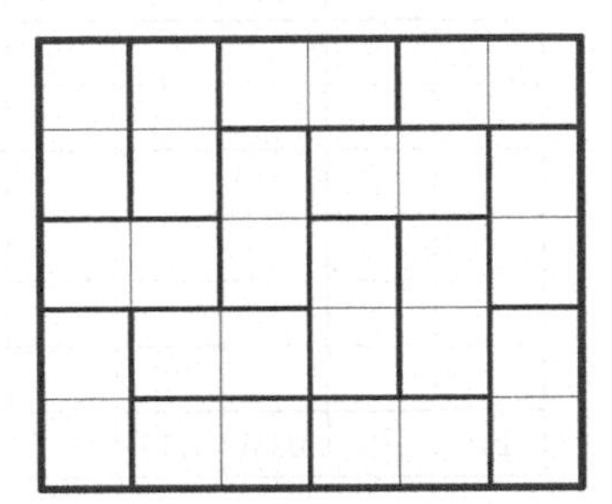
Fig. 6(b)

14. Any domino covering of 6×6 board will have fault lines. So will be the case with 4×4 and 8×8 boards.

15. Boards in which one dimension is odd and other even or both dimensions are even. Only boards in which one dimension is odd and other even can be covered with dominoes such that there are no fault lines.

17. A knight controls 8 squares (besides itself) surrounding it but of opposite colors to where it is located leaving 8 squares around it. So many more will be needed. A bishop can move diagonally any distance. Eight of these can be placed to dominate the board

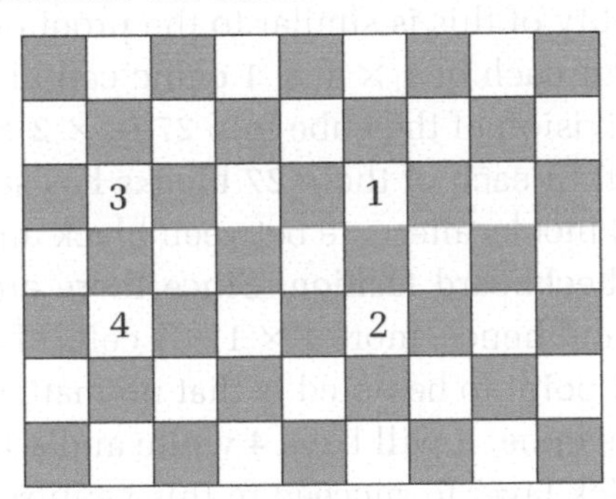

Fig. 8

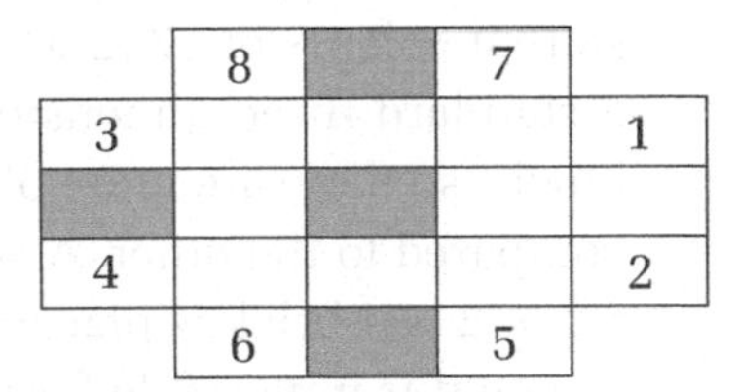

Fig. 8(a)

18.

			B1				
			B2				
			B3				
			B4				
			B5				
			B6				
			B7				
			B8				

19.

B1	B2	B3	B4	B5	B6	B7	
B8	B9	B10	B11	B12	B13	B14	

20. Yes it is possible as shown in the figure.

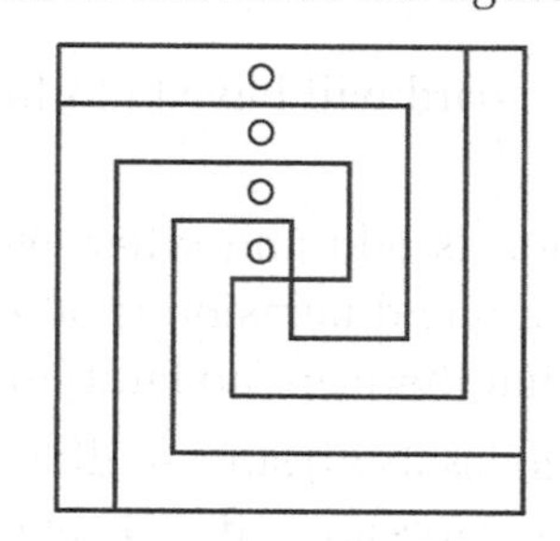

Fair distribution of the property

21. The answer surprisingly is no. Even though 27 such bricks can cover the volume of the desired cube. ($6 \times 6 \times 6 = 216$ also $(1 \times 2 \times 4) \times 27 = 8 \times 27 = 216$).

Justification of the impossibility of this is similar to the proof of Domino's puzzle, through colouring. Colour each of $1 \times 1 \times 1$ cubic cell of $6 \times 6 \times 6$ cube so that in the natural subdivision of the cube into 27 $(2 \times 2 \times 2)$ blocks each of the eight $(1 \times 1 \times 1)$ cell in each of these 27 blocks has same colour so that colours of 27 $(2 \times 2 \times 2)$ blocks alternate between black and white in a standard three dimensional checkboard fashion. Since there are in all 27 blocks so there are more blocks and hence, more $1 \times 1 \times 1$ cells of one colour compared to the other. A second point to be noted is that no matter wherever a $1 \times 2 \times 4$ brick is placed in the cube, it will have 4 white and 4 black cells. Hence it is impossible for the brick layer to succeed in this venture.

Chapter 16

1. (i) Magic number 15.

8	1	6
3	5	7
4	9	2

(ii) Magic number 65.

18	10	1	22	14
5	17	13	9	21
12	4	25	16	8
24	11	7	3	20
6	23	19	15	2

2. (i) Magic number $15 + 10 \times 3 = 45$ (ii) Magic number $65 + 10 \times 5 = 115$

18	11	16
13	15	17
14	19	12

28	20	1	32	24
15	27	23	19	31
22	14	35	26	18
34	21	17	13	20
16	33	29	25	12

4. Let the sequence of 9 consecutive integers be $k, k + 1, k + 2, \ldots k + 8$. Then a magic square built with these is

$k + 7$	k	$k + 5$
$k + 2$	$k + 4$	$k + 6$
$k + 3$	$k + 8$	$k + 1$

Its magic number is $3k + 12$

5. Form any 3×3 magic square with numbers 1 to 9. Now multiply each entry with 2 and subtract 1 from each of these.

15	1	11
5	9	13
7	17	3

Magic number $15 \times 2 - 3 = 27$

6. In a magic square formed with first 9 integers, replace integer i with $2i + 2k$, $i = 1, 2, \ldots.. 9$ where $2k + 2$ is the first even integer to appear in the magic square. Its magic number will be $(16 + 2k) + (2 + 2k) + (12 + 2k) = 30 + 6k$.

Example: suppose we want to form a magic square with 24, 26, 28, 30, 32, 34, 36, 38, 40, then $2 + 2k = 24$ yields $k = 11$

38	24	34
28	32	36
30	40	26

Its magic number is 96.

7.

16	2	3	13
5	11	10	8
9	7	6	12
4	14	15	1

Magic Number is 34

8. Magic numbers for 5×5 is $65 + 50 = 115$, for 7×7 it is $175 + 70 = 245$

9.

1	116	49	151	84	17	119	65	167	100	33	135	68
136	69	2	117	50	152	85	18	120	53	168	101	34
102	35	137	70	3	105	51	153	86	19	121	54	169
55	157	103	36	138	71	4	106	52	154	87	20	122
21	123	56	158	104	37	139	72	5	107	40	155	88
156	89	22	124	57	159	92	38	140	73	6	108	41
109	42	144	90	23	125	58	160	93	39	141	74	7
75	8	110	43	145	91	24	126	59	161	94	27	142
28	143	76	9	111	44	146	79	25	127	60	162	95
163	96	29	131	77	10	112	45	147	80	26	128	61
129	62	164	97	30	132	78	11	113	46	148	81	14
82	15	130	63	165	98	31	133	66	12	114	47	149
48	150	83	16	118	64	166	99	32	134	67	13	115

A 13 × 13 knight's tour magic square

10.

61	62	63	64	1	2	3	4
60	59	58	57	8	7	6	5
12	11	10	9	56	55	54	53
13	14	15	16	49	50	51	52
20	19	18	17	48	47	46	45
21	22	23	24	41	42	43	44
37	38	39	40	25	26	27	28
36	35	34	33	32	31	30	29

An 8 × 8 king's tour semi-magic square

14.

7	2	5	0	3	6	1	4
6	1	4	7	2	5	0	3
5	0	3	6	1	4	7	2
4	7	2	5	0	3	6	1
3	6	1	4	7	2	5	0
2	5	0	3	6	1	4	7
1	4	7	2	5	0	3	6
0	3	6	1	4	7	2	5

16.

3, 3	2, 2	1, 1	0, 0
0, 1	1, 0	2, 3	3, 2
2, 0	3, 1	0, 2	1, 3
1, 2	0, 3	3, 0	2, 1

→

$3 \times 4 + 3$	$2 \times 4 + 2$	$1 \times 4 + 1$	$0 \times 4 + 0$
$0 \times 4 + 1$	$1 \times 4 + 0$	$2 \times 4 + 3$	$3 \times 4 + 2$
$2 \times 4 + 0$	$3 \times 4 + 1$	$0 \times 4 + 2$	$1 \times 4 + 3$
$1 \times 4 + 2$	$0 \times 4 + 3$	$3 \times 4 + 0$	$2 \times 4 + 1$

→

15	10	5	0
1	4	11	14
8	13	2	7
6	3	12	9

17.

D	F	A	E	C	B
B	D	C	F	A	E
A	B	E	D	F	C
F	E	B	C	D	A
E	C	D	A	B	F
C	A	F	B	E	D

Suggested Collateral Readings

1. **Mathematics in Nature:** Modelling Patterns in Natural World, John. Aldam, Princeton Univ. Press, 2003,(Indian print) Univ. Press, Hyderabad, 2005.
2. **Have Some Sums to Solve:** The Complete Alphametics book, Steven Kahan, Baywood Pub. Co. New York, 1978 (Indian print) Univ. Press, Hyderabad. 1978.
3. **Understanding Mathematics,** K.B. Sinha, R.L. Karandinkar, C Musili, S Pattanayak, D. Sing and A. Dey, Univ. Press, Hyderabad.
4. **Techniques of Problem Solving,** Steven G. Krantz, American Mathematical Society, 1996 (Indian print) Univ. Press, Hyderabad.
5. **Vedic Mathematics,** Jagadguru Bharti Tirathji Maharaj, (General Editor V.B Agarwala) Motilal & Banarasidass Publishers Pvt. Ltd; Delhi 2nd revised edition 2009.
6. **The Math Explorer, (A journey through Beauty of Mathematics)** Jefferson Hane Weaver, Amazon.com (Indian print) Univ. Press Hyderabad.
7. **Mathematmatical Circles** (Russian Experience) Dmitri Fomin Sergey Genkin and Ilia Itenberg. (Translated from Russian by Mark Saul, for American Mathematical Society, 1996, Indian print) Univ. Press Hyderabad, 1998.
8. **Across the Board (Mathematics of Chess Board Problems,** John J. Watkins, (Indian Edition) Univ. Press, Hyderabad
9. **Math Charmers: Tantalizing Titbits for the mind,** Alfred, S. Posamentier, Univ. Press Hyderabad 2007 under license from Prometheus Books 2006.
10. **Fun and Fundamentals of Mathematics,** Jayant V. Narlikar, Mangala Narlikar (Univ Press). Hyderabad 2001.
11. **Mathematical Puzzles and Diversions,** Penguin Books, England, 1965.
12. **Mathematics and Imaginations,** Simon and Schvestre, New York, 1952.
13. **The Mathematical Century,** Piergiorgio Odifreddi, (English translation) 2004, Princeton Univ.Press, (Indian print) Univ. Press Hyderabad 2005.

14. **Towing Icebergs, Falling Dominoes and other Adventures in Applied Mathematics,** Robert B. Bank. (Indian print) Univ. Press Hyderabad 2001.

15. **An Imaginary Tale:** The story of $\sqrt{-1}$, Paul J. Nahin. (Indian edition) Univ. Press Hyderabad 2002.

16. **Understanding Probability and Statistics,** Ruma Falk, Publisher, A.K. Peters 1993 (Indian Edition), Univ. Press. Hyderabad.

17. **Explorations in Mathematics,** A.A. Hattangadi, Univ. Press. Hyderabad, 2002.

18. **Mathematical Recreations and Essays,** Ball, W.W.R, and H.S.M. Caxeter 1987, Dover, New York

19. **Amusements in Mathematics,** Dudney H.E., 1958, Dover, New York.

20. **Mathematical Puzzles,** Sam Loyad, GardnerM, 1959, Dover, New York.

21. **Mathematical Recreations,** Karitchik M, 1953, Dover, New York.

22. **Journal of Recreational Mathematics,** USA.

23. **Challenging Mathematical Problems with Elementary Solutions,** Holdar-Day 1964, San Francisco.

24. **One Two Three Infinity,** George Gamow, Dover Publication, New York, 1947, 1988.

25. **Websites such as:**

    http://ia600504us.archive.org/23/items/amusementsinmaths16713gut/16713-h/16713-h.htm

    http://03indiatimes.com/mindsport/archive/2005/12/26/390329aspx